Introduction to Physical and Biological Oceanography

Introduction to Physical and Biological Oceanography

Cuchlaine A. M. King

Professor of Physical Geography, University of Nottingham

Edward Arnold

First published 1975 by
Edward Arnold (Publishers) Ltd
25 Hill Street, London W1X 8LL

This book is a fully revised edition of the second part of
Oceanography for Geographers (*Introduction to Oceanography*),
first published 1962 by Edward Arnold (Publishers) Ltd

Reprinted 1964, 1965, 1968, 1969 (twice)

Cased Edition ISBN 0 7131 5735 6
Paper Edition ISBN 0 7131 5736 4

Printed in Great Britain by
Butler & Tanner Ltd, Frome and London

Contents

Preface

The second edition of *Oceanography for Geographers* is being divided into two volumes, a course that has been made necessary by the great volume of new material that has been published since the first edition was published over a decade ago. The first volume concentrates on the geomorphology of the oceans in relation to the exciting new ideas of global tectonics and sea floor spreading, an advance that owes a great deal to Oceanographical work. This second volume deals with some aspects of Physical and Biological Oceanography. The physical aspects that are considered include an introduction to the character of ocean water, its circulation in the form of surface currents and deep water movements. The tides are considered in one chapter and the various waves that disturb the surface of the ocean in another. Two chapters are devoted to the biological aspects of oceanography, the first dealing with productivity and the second with exploitation. This chapter leads on to the final one, which is concerned with the uses and problems of the oceans. These problems stem to a large extent from the fact that the open ocean belongs to all mankind, and legal problems concerning the use of the biological and inorganic resources of the oceans are increasing in importance and complexity. For this reason an appendix by an expert in oceanic law, Edgar Gold, has been specially prepared, and this forms a valuable addition to the volume. We shall depend more and more in the future on the riches of the oceans, including such vital raw material as oil, which is very plentiful in certain areas in the offshore zone, and on the food resources of the sea. These organic resources are very liable to over-exploitation—including such interesting creatures as the Blue Whale, the largest animal to exist on earth—unless international control can be effectively agreed amongst the fishing nations of the world. The oceans are a major field of scientific endeavour at present, and their importance to the world at large can only increase in the future. It is, therefore, increasingly essential that the oceans become better known and that their intricacies are studied more deeply.

Acknowledgements

The author and publishers gratefully acknowledge permission given by the following to reprint or to modify copyright material:

The Editor, *Advances in Marine Biology*, for figure 7.3; the American Fisheries Society for figures 7.4 and 7.5; the Editor, *Deep Sea Research* and Pergamon Press Ltd., for figures 3.7, 3.8, 3.9, 3.11, 3.12 and 3.21; the French Embassy Information Service for plate 3; Gordon & Breach Science Publications, New York, for figures 3.13, 6.1, 6.2, 6.3, 6.4 and 6.11; Her Majesty's Stationery Office for figure 3.10; Icelandic Photo & Press Service for plate 11; The Institute of Navigation for figure 5.1; The Institute of Oceanographic Sciences for plate 1 and plate 2; Interscience Publications Inc., John Wiley for figure 4.7; Japanese Information Office for plate 18; the Natural History Museum for plates 8, 9, 11, 12, 13 and 19; the Editor, *New Scientist*, for figure 6.18; Oxford University Press, London and New York, for figures 3.18 and 3.22; Pergamon Press Ltd. for figure 2.8; G. W. Potts, Marine Biological Association of the United Kingdom, for plates 7, 14, 15 and 16; Prentice-Hall, Inc., for figure 5.13; the Royal Society of Arts for figures 4.6 and 5.6; The Editor, *Science*, for figures 6.6 and 6.7; the United States Army Corps of Engineers, Coastal Engineering Research Center, for plate 5, and the United States Naval Oceanographic Office for plate 4.

1   Introduction

1   The scope of this volume

The subject of oceanography covers a wide range of topics. One of the most clear-cut of these is marine geology and geomorphology, which deals with the form of the crust and mantle, and with processes operating within them including those responsible for the distribution of sediment in the ocean basins. These aspects of oceanography were con- considered in volume 1, *Introduction to Marine Geology and Geomorphology*, together with factors concerning the accumulation of and variation in the amount of water in the ocean basin receptacles.

Other major fields of oceanography include physical oceanography, chemical oceano-graphy, and biological oceanography. All these major branches of the subject are inter-related with each other and with geological and geomorphological oceanography. They are only treated separately for convenience, and it is always necessary to bear in mind the connections between them. In this volume the emphasis is placed on physical oceanography and to a lesser extent on biological oceanography, although some aspects of chemical oceanography are also relevant in both fields (for instance, in a study of the fertility of the oceans). Because of the present active interest in all aspects of man's environment, the rapidly growing field of applied oceanography gains importance as a field that draws upon knowledge from studies in pure oceanography and focusses on man's direct involvement as the main user of oceanic resources and wealth. In order to use this wealth wisely, man must learn more and more about the oceanic environment in all the diverse fields of oceanography, and appreciate the ways in which they are linked together. As technology increases in complexity and power, so man's impact on the oceans becomes more and more marked. It is vital that control of use should be agreed upon and implemented as soon as possible, and this control must be based on sound knowledge. Some aspects of the human-based problems (such as pollution and conservation) are considered briefly in this volume, as are some of the legal problems associated with them.

2 Interrelationships in oceanography

2.1 Character of ocean water

In this section some of the links between the different aspects of oceanography that are mentioned in this volume, the broader field of oceanography, and the other sciences with which it comes into contact, will be referred to. The character of ocean water is dependent both upon outside influences and internal processes. The major external influence is that of the atmosphere which affects the temperature and salinity of ocean water. Salinity, however, is a complex property and lies more properly within the field of chemical oceanography, which is not dealt with here in any detail. Complications in salinity variations are concerned with the dynamic character of the chemical components of sea water and the processes that are involved, which include biological activity as well as geological processes, such as volcanic eruptions and the entrance of chemicals into the oceans via the rivers. An important point in this connection is that each chemical element has a specific residence time in the chemical cycle. The result, however, is to provide a steady-state system approximately, as far as the character of the chemical constitution of sea water is concerned. Thus the proportions of the different elements in sea water are maintained in a constant ratio, which applies throughout the mass of ocean water. There are, however, variations in the total amount of dissolved matter in ocean water, leading to variations both in time and space in the total salinity of the water. It is these variations in total salinity that are closely related to the atmospheric processes that determine the zones of greater and lesser precipitation and evaporation over the ocean surface. Thus the zones of higher precipitation and lower evaporation are those of lower salinity and vice versa. These zones can be directly related to the major pattern of atmospheric circulation discussed in chapter 2. The temperature of the water is also directly affected by the atmosphere, both by direct insolation and through the weather, which is dynamic and hence variable. There are daily, seasonal, and longer-term variations, again both in time and space, of temperature conditions in the ocean waters.

Salinity and temperature together determine the density of the water and hence have an effect on its movement, because water that is dense will sink, and this induces further movement of deeper water. The vertical movement of water, related to density, are of considerable significance also in the biological field of oceanography, because the fertility of the water is often directly related to the degree of stirring in the vertical sense. This is because the phytoplankton and other marine plants, the seaweeds, can only make use of nutrients when they are in the upper layers of water to which sufficient light for photosynthesis can penetrate. Density variations are not, however, the only cause of vertical movement of ocean waters, and other interrelationships of vital importance to the fertility of the sea will be mentioned later.

2.2 The role of the oceans in the hydrological cycle

One of the most fundamental cycles that alone maintains life on earth is the hydrological cycle. It is this cycle that makes possible life on land because of the transference of water from the great reservoir of the oceans to the land via the atmosphere. The hydrological cycle depends fundamentally on the interrelationship between the circulation of the ocean and that of the atmosphere, which is one extremely complex, coupled system with numerous highly complicated feedback relationships between the different aspects of the system. Water is withdrawn from the oceans into the atmosphere by processes of evaporation, dependent on air and sea temperatures and on wind strength and humidity. Another important relationship concerning chemical oceanography and the hydrological cycle is the raising into the atmosphere of minute particles of salt and other minerals from the sea as spray. This process plays a significant part in the residence time of sodium chloride in the ocean; it also provides many of the hygroscopic nuclei that enable precipitation to take place more effectively. The salt then returns to the oceans via rainfall, rivers, and other phases of the land stage of the hydrological cycle.

The contrast between an oceanic and continental climate, characteristic of the west and east sides of the continents respectively, also demonstrates very effectively the importance of the oceans in the operation of the hydrological cycle: precipitation is higher on the west, and temperatures are much milder than on the east side.

Still larger climatic events (including major ice ages) have, according to the theory of Ewing and Donn, been brought about by changes in the characteristics of the oceans. They argue that the conditions of the north polar Arctic Ocean control the climate of the areas around its shores and initiate ice ages when conditions are suitable. The most important variable in this connection is the extent of ice cover in the Arctic Ocean. When ice is widespread, precipitation is reduced on surrounding lands, but when ice cover is reduced, precipitation in the form of snow allows ice-sheets to build up, which in turn reduce the temperature, thus causing more widespread Arctic freezing, the reduction of precipitation, and perhaps the onset of a cold interglacial period. This is a complex feedback situation in which positive elements are self-generating up to a point beyond which negative controls come into operation, leading to a reversal of the former change. These ideas are still controversial, but they do serve to illustrate the complex interrelationships between atmospheric and ocean circulation. The theory also provides a link with geological oceanography in the suggestion that the movement of the continents (on their crustal plates) to suitable geographical positions is responsible for the initial change in ice conditions in the Arctic, starting the chain reaction that could lead to an ice age.

The wider repercussions of this major climatic event are so widespread (and are felt in all branches of earth science and biological science) that it is very difficult to identify them all. One or two are mentioned in volume 1, for example the effect of climate on ocean sediments, and the information that can be obtained from a study of ocean cores concerning climatic change and ice conditions in the Arctic. The links between ocean sediment and the climate include important biological ones, because temperature and other characteristics of the surface water determine the species of creatures that can flourish under those conditions, and which form an important element in the sediment that eventually collects on the bottom.

The hydrological cycle is concerned with water in all its forms; as ice both in the form of glaciers and ice-sheets (as well as sea-ice), as water in the liquid form (fresh water on land and salt water in the sea), and as vapour in the atmosphere. The fact that water occurs naturally in all three states is of immense importance in the operation of the hydrological cycle, and also has repercussions in all aspects of oceanography as well as other branches of earth sciences. The interaction between ocean and atmosphere involves all three states of water—the solid, liquid, and gaseous.

2.3   The interaction of the oceans and atmosphere

The air is the medium by which water is transferred from the oceans to the land. The wind blowing over the water also generates waves, which eventually exert a very significant effect on the edge of the ocean, bringing into being all the diversity of coastal landforms. Wind also plays an important part in the movement of the surface waters and in setting up the circulation of the ocean currents, whereby water of different characteristics is carried away from its place of origin to modify the shore against which it may eventually flow. In this way the distribution of warmth received on the earth is spread more evenly over the surface. The wind also plays its part in inducing zones of upwelling, which are important in connection with the very rich fishing areas that result from the bringing of nutrient-rich waters to the surface by this process.

There are many features in common between the surface circulation of the oceanic waters and the atmosphere above them. There is also another most important interplay between the oceans and the air, which can be seen by studying the hydrological cycle and the part it plays in life on earth. Wind blowing over the ocean takes up water vapour from the surface into the air by evaporation. It can then be distributed over the land surface by the wind and the processes that lead to precipitation. The oceans, therefore, play a vital part in the character of the earth's climate. It is one of the fundamental facts of climatology that the oceanic and continental climates differ greatly, and through their influence on the climate, the oceans play an important indirect part in many aspects of geography. It is difficult, however, to separate cause and effect in the interrelationship between the atmosphere and the oceans. They interact in a very complex way with many feedback loops. The ocean currents and the general circulation of the oceans depend to a great extent on the wind and on heating, cooling, and evaporation, which in their turn depend on the energy of the sun received through the atmosphere. On the other hand, by their transfer of heat to higher latitudes, the currents supply energy which helps to keep the atmospheric circulation going. The relationship between the atmosphere and oceans is in a state of dynamic equilibrium, whereby a steady state of movement is more or less achieved.

The essential difference between the oceans and the land, climatically, is the ability of the oceans to store heat to a much greater degree than the land. The processes of mixing allow the heat to spread through a greater thickness of water, where it is stored to be given off when the air temperature falls; from this follows the effect of the oceans in reducing the extremes of the continental climates. Such effects will be most marked in the zones where the winds carry warmer air from the sea onto the land. This is on the western coasts in the temperate west wind belt, where the warmer water from lower latitudes can penetrate further poleward than normal. Thus the contrast arises between the east and west coast continental climates.

In considering the effect of the oceans on weather, the amount of energy given off by evaporation is strongly localized. It is greatest where the water is relatively warmer than the air, which is mainly off the east coasts of the continents in the temperate regions in winter (for example, where the warm waters of the Gulf Stream give abnormally high sea temperatures). These regions of excess energy loss from the sea are associated with zones of active frontal development. The ocean currents account for the position of these zones of maximum energy loss. The presence and character of water beneath an air-mass can also change its character fundamentally. Polar air moving south to lower latitudes, will gain water vapour over the oceans, while tropical air, moving in the opposite direction in the northern hemisphere, will be cooled and may well lose water vapour by condensation and precipitation. The water vapour that is carried from the oceans into the air is then partly transferred to land and this transfer is of even greater importance. It supplies the capital on which the hydrological cycle operates, bringing water to the land as rain, snow, and other forms of precipitation. It can then go through the cycle by a more or less direct route according to its subsequent movements. Some may be withheld from the ocean for a considerable time as snow, then ice, in which state it may be kept out of circulation for prolonged periods. Some water may return directly to the ocean via the rivers, while some may pass slowly through the earth as groundwater, before re-entering the cycle, via springs and wells. Much water is used by plants, animals, or man on its way through the cycle, but eventually it will return to the ocean.

One effect of the hydrological cycle has been indicated by Thornthwaite (Deacon et al., 1955). He relates Munk's estimate of a reduction of $5 \cdot 10^{19}$ grams of water in the ocean in March, compared with the October volume, to an increase of water on land in the form of groundwater, snow, and other forms after the northern winter period. This demonstrates the dominant part played by the greater amount of land in the northern hemisphere. Defant (1961) gives the following analysis of the water budget of the earth:

Table 1.1 Water budget of the earth

	Precipitation		Evaporation		Outflow — Inflow +	
	10³ km³/year	cm/year	10³ km³/year	cm/year	10³ km³/yr	cm/yr
Ocean	324	90	361	100	−37	−10
Continent	99	67	62	42	−37	+25
Whole earth	423	83	423	83	—	—

The effect of climate is such that the temperature and circulation of the oceans are affected by the external change of radiation as much as the temperature on land, but often these changes can be recorded in the sea more completely than they can on land, where all the deposits are liable to subsequent erosion. In favourable sites in the oceans, the changes of climate can be read with considerable accuracy from the character of the deep sea sediments (see volume 1). The oceans, therefore, also provide a means of studying the changes of climate which they help to bring about.

On a small scale variations in climatic elements are also associated with variations of sea level of relatively small dimensions. The amount of water vapour in the atmosphere, differences of temperature and salinity, and variations in evaporation and precipitation all cause changes in sea level. An increase of $1°F$ ($\frac{5}{9}°C$) of a water column 183 m deep will cause a rise of 2·54 cm, and if the salinity is decreased from 35·0 to 34·9‰, the level would rise 1·9 cm. The north Atlantic is about 25·4 cm higher in summer and autumn than in spring for this reason. Evaporation and precipitation cause the Pacific to be about 20 cm higher than the mean level of the Atlantic over the whole year, the difference being greatest in October, which is the wet season. This is due to the lower salinity and less dense water of the Gulf of Panama, compared with the Caribbean Sea, in the region where this difference can most readily be assessed. There is also a rise of sea level up the west coast of the USA from southern California to the state of Washington of about 30 cm (Deacon, 1960c).

Regular variations of sea level with about a four-day period have been recorded at Canton Island in the Pacific. These have an amplitude (half range) of 3 cm and can be related to atmospheric waves in the easterly wind zone. The variations of the sea level can be correlated with the north–south wind component related to these atmospheric waves. This example illustrates the capacity of the atmosphere to affect the level of the sea directly. Sea level is rising at present, but the precise cause is not known. It is partly the result of melting ice and partly due to the warming of the water, but other factors probably influence it as well. Sea level is unlikely to become stable as long as large volumes of water are locked up as ice. Sooner or later the ice will melt, but before this there may be another glacial period. When the ice does eventually melt, sea level will rise by about 100 m. Such a change would have profound effects on land, but it may be preceded by a fall if another ice age develops on the scale of the Pleistocene ones.

The relationship between atmospheric and oceanic climatic fluctuations on a small scale is discussed by Dickson and Lee (1969). Sea surface temperature during the period 1958–60 warmed by 1·1°C in the northeast Atlantic (compared with norms for 1900–1940), while the northwest Atlantic cooled by 0·8°C. At the same time precipitation was less in Norway. This was due to blocking of depressions over Europe and their concentration west of Iceland as a result. Anticyclonic conditions lead to horizontal convergence, hence higher sea surface temperatures, while cyclonic conditions lead to thermal doming and cooler water reaches the surface. The increased sea surface temperature gradient could be responsible for the changes in air circulation. In autumn 1958 an abnormally large amount of Atlantic water moved into the Norwegian Sea, spreading into the North Sea in the summer of 1959. The warming spread to the Greenland and Barents Seas in summer 1960. Salinity was also unusually high around Iceland. It is not clear how far this warming initiated the blocking action in the atmosphere. The amount of feedback in the interaction between ocean and atmosphere is difficult to calculate, but probably an important aspect of the coupling of the two systems is connected with feedback.

3   Water movement on different scales

There are many different scales on which the ocean waters move and these have different causes. Some of the movements are cyclic, and can be described as waves, while others are long-term systematic circulations, both in the horizontal and vertical senses. Some appear to be more random fluctuations. In many of these movements, there is a close relationship between the ocean and the atmosphere, but not all of them are due to such a relationship.

3.1   Ocean currents

Of the largest extent, forming a world-wide system, is the dynamic ocean current pattern. This is very closely tied to global atmospheric circulation, since major ocean currents are wind-driven. There is, therefore, a close similarity between the major elements of the two systems; each consists of circulations dependent on the pattern of energy coming from the sun and causing the resultant movements of air and water. Although the similarities between the two systems are striking, even to the reversal of current direction in the Indian Ocean with the reversal of the monsoon wind system, there are significant differences, such as the westward intensification of the ocean current system which has not got an exact atmospheric equivalent. This and several other important major current characteristics are related to variations of the Coriolis effect with latitude, itself due to the rotation of the earth around its axis. This effect influences both air and water, but with rather different results, owing to variations in the character of the two fluids. The relationship between the major ocean currents and the major atmospheric circulation, like so many oceanographical phenomena, is not only a one-way relationship. There are important feedback systems operating, whereby the two systems are intimately coupled together. The oceans provide the water, the evaporation of which supplies much energy to the atmospheric system, and without which precipitation could not occur. Heat is distributed around the earth both by air and water, the air accomplishing about nine-tenths of the work. Further details of the close interaction between the atmosphere and the oceans in setting up the global current system are considered in chapter 3.

Oceanic circulation also affects many other branches of oceanography, such as ocean fertility, and through this, chemical and biological oceanography and sedimentology. The pattern of land and sea also influences the details of ocean circulation, indicated in the contrast in pattern between the Atlantic and Pacific Oceans. The wide connection between the Atlantic and the Arctic Oceans contrasts strongly with the narrow and shallow Bering Straits that link the Pacific with the Arctic Ocean. The high, very cold southern Antarctic continent also plays an important part in the general pattern of oceanographic circulation, both on the surface and at depth. Thus oceanic circulation, horizontal and vertical, on the surface and at depth, is dependent both on atmospheric circulation and the pattern of world relief and land and sea distribution. Again the relationship is not just one way; the oceans play a vital part in the general air circulation and determine the pattern of land and sea.

3.2   Cyclic movement: waves

As more detailed and accurate direct measurements of ocean water movement accumulate it is evident that major ocean currents are only one of many causes of ocean water flow. The irregularity and variability of flow is to a large extent the result of the superimposition of many different types of movement on each other to form a very complex spectrum. Many movements are cyclic in character in that they are wave motions. Chapters 4 and 5 are devoted to a consideration of a number of different wave motions. Some of them are regular and predictable, but many of them are complex and variable and are caused by a wide range of forces. The most regular and ubiquitous waves are those that give rise to the tides. Their widespread occurrence is due to their extraterrestrial cause in the gravitational attraction of the moon and sun on the ocean water. The action of the gravitational attraction is well known and amenable to mathematical analysis, so that tide-producing forces can be accurately defined and predicted. The reaction of the irregular oceans to these forces is, however, less easy to define, and although the broad pattern of interaction between the forces and the oceans is understood, details still remain to be filled in. Useful computer models have been developed to study these interactions. The response of the ocean basins to the tide-producing forces depends on the size and shape of the basins in relation to the forces, leading to variation in response in the different oceans and seas.

One possible way in which waves can be classified is by their length. Tides are very long waves with periods of semi-diurnal and diurnal lunar and solar dimensions. Period is related to length in that long waves have a longer period ($L = CT$, where L is length, C is velocity, and T is period). Other types of waves of progressively shorter length include surges, seiches, tsunamis, internal waves of various types, surf beat, wind-generated gravity waves, and microseisms, the wind waves themselves covering a considerable range of period from about 30 seconds to about 3 seconds. At the shorter end of the scale there are ripples.

The longer of these waves, the surges, have periods comparable to that of the tide, but they are the result of abnormal meteorological conditions, and thus demonstrate another form of interaction between the atmosphere and the oceans. They are normally generated by intense depressions, such as hurricanes and similar storms. Hurricanes are normally formed over the ocean and depend on the evaporation of moisture from the ocean to maintain their energy; thus the interrelationship and interaction between the two systems is close.

Tsunami, on the other hand, are the result of submarine earthquakes, so that they demonstrate the interaction between ocean waves and the geophysical processes operating in the crust and mantle beneath the ocean floor. Tsunami are long waves that can travel over great distances of water. They move across the Pacific and have lengths of about 800 km. Being long and low they pass unnoticed in the open ocean, but they may become refracted on approaching a low coast and build up to heights that result in major coastal damage and flooding, at a very great distance from their origin.

Perhaps the most readily apparent movement of the ocean water is that resulting from the generation of gravity waves by the wind blowing over the ocean surface. This is another example of the very close interaction between the atmosphere and the ocean: the waves are the direct result of transfer of energy from the moving air to the water across which it is flowing. The precise mechanism by which this transfer of energy is achieved still eludes exact mathematical analysis; the general principles are becoming better known although the matter is highly complex and difficult to test by measurement in the ocean. The result is a wide range of waves of different length, moving in different directions. This situation is best analysed by spectral methods, whereby the complex pattern of waves can be split into the simpler components of which it consists. Each component is regular and sinusoidal in form in deep water, with properties that are clearly defined.

The study of internal waves, which can only be observed indirectly, is less advanced than that of the more obvious surface waves, but interesting data have become available and a wide range of internal wave types have been identified. Another type of interaction is demonstrated in the development of microseisms. These waves have exactly half the period of the wind waves that cause them. They are formed and travel through the earth's crust along the ocean bed when two sets of wind waves interfere with each other on the surface by meeting head on. In this way the atmospheric circulation has an effect on the ocean floor through the medium of wind-generated gravity waves on the sea surface.

4 Life in the ocean

The life of the oceans, as life on land, is intimately related to its environment. There is a very close connection between the environment and all the processes that determine its character and the life it can support. Marine life is extremely varied, and adaptations have developed that allow creatures to occupy almost every available niche in the environment—from the deepest ocean trenches to the intertidal zone, and from the polar to the equatorial regions, from the land border to the open ocean, and from rocky substrates to the finest clay floor. Life first started in the oceans so that time in which to adapt to this wide variety of environments has been long, and many varieties have evolved. As well as adapting their way of life to their environment, marine creatures must also learn to live with their neighbours to their mutual advantage as far as possible. Thus the biological aspects of oceanography must be considered in terms of ecosystems, the living complex of plants, animals, and their habitat or environment. Man has his place in the marine ecosystem, but he cannot hope to exploit it to his best advantage until he understands its complexities and has the ability to control it (at least to a certain extent) as he does with farming on land. There is still a long way to go before the sea is farmed as efficiently as the land. Such control requires a detailed knowledge both of the quality of the environment and of the life cycle of its living inhabitants, and their interdependence one upon the other.

As on land, the basis of life in the oceans is the plant community, which alone can synthesize energy and living tissue from the raw materials of mineral fertility in the presence of sunlight by photosynthesis. The circulation of the oceans determines those areas where nutrients can reach the upper levels in which alone there is sufficient light for photosynthesis to proceed. Thus those parts of the oceans where upwelling is active are the fertile parts. For this reason the processes by which upwelling takes place are very significant in the marine environment, as only by this means can nutrients be supplied continuously for use by marine plants. Upwelling is caused by the general pattern of circulation, by the effects of the earth's rotation (for example along the equator), and by a combination of rotational effects on wind action. In shallow water rather different processes are involved, and wind stirring as well as nutrient addition from land sources are locally important.

Most marine plants are minute, consisting of the smaller elements of the floating, drifting plankton. Many of the food chains in the ocean are rather long, in that tiny plants, diatoms, and others provide sustenance for small members of the zooplankton, which in turn feed invertebrates and fish, the former often being food for fish. There is a large loss of production of nourishment at each link in the food web; the system is not very efficient if only the higher parts of the web are exploited. Some important species have relatively short food chains, such as the Antarctic whales, who feed directly on the zooplankton which are prolific in the upwelling waters of the southern ocean.

Seaweeds are the other primary producers of organic matter. They have the advantage of being mostly fixed in habitat, attached to the bottom in shallow water. This characteristic makes them more susceptible to exploitation than the more mobile members of the oceanic community. It is, therefore, in this field that some progress in exploitation and commercial development (which could be called cultivation) has been achieved. But on the whole the ocean is still hunted, rather than farmed, for food.

Chapter 6 is a brief introduction to the topic of marine productivity and some of the more important types of marine creatures. These organisms have widely different life habits: some are sessile, others are mobile and move great distances, some change character at different phases of their life, and all vary in their food requirements and living conditions. Many species can, therefore, live together in a community, building up the complex web of the living community or ecosystem by preying on or being prey for others. Chapter 7 introduces ways in which man has exploited and is exploiting this living marine wealth. Further problems associated with the living environment are mentioned in chapter 8, which deals briefly with such matters as pollution of ocean waters and conservation of resources and their fair utilization by different communities of the world. These problems give rise to legal complications, which will only be solved with good-will on the part of all nations. Men have used the sea for a long time, but there is still much to learn concerning its proper use and exploitation.

5   The ocean in exploration and transport

The broad horizons of the sea have always lured the more curious and adventurous members of the human race to seek what lay beyond. From the very earliest days of history and prehistory, men have set out to seek the lands beyond the sea. The ancient Egyptians began marine exploration, but their lack of suitable ship-building material rather limited their vessels and voyages. Heyerdal has, however, recently shown that their papyrus boats were seaworthy and could have been capable of sailing across the Atlantic. The Minoans sailed right round Africa in 600 BC according to some accounts. Many of these early voyages were largely coastal exploration, and although the Phoenicians had sailed to Britain earlier, Pytheas first explored England from the sea in 310 BC. In this way explorers spread out from the more restricted waters of the Mediterranean into the stormier expanse of the open Atlantic Ocean, although they mostly kept fairly close to the shore.

Early boats were difficult to control in rough weather and navigational equipment was virtually lacking; familiar coast landmarks provided the best means of location. Such problems, however, did not prevent some rather later and more seaworthy boats from setting out boldly across the oceans. The earliest Atlantic crossings, if not by the papyrus boats of the Egyptians, were probably by the Irish sailors, in their light but seaworthy skin boats, using the various stepping stones provided by islands on their journeys. This took them to the north Atlantic via the Faroes, Iceland, and Greenland to North America, where they preceded the hardy Vikings in discovering America. Both of these early explorers reached there long before Columbus sailed westwards to reach the Bahama Islands in 1492. The Irish and Norse voyages took place before the eleventh century, and nearly all knowledge of them was forgotten by the fifteenth century.

In the early days of oceanic exploration, ships were often at the mercy of the winds and waves. The set of currents exerted a powerful influence on the ease and speed of a journey by sea, which could not be too long in those days because suitable provisions could not be taken in and kept fresh on board. Some of the earliest winds and currents to be recognized and used by the early navigators were those of the monsoon, which greatly facilitated the journey from north Africa to India, if the voyage was timed to make use of the changing monsoon winds and currents. It was also not long before the Spaniards and other sailors in the north Atlantic began to be aware of, and to take advantage of, the circulation of the north Atlantic Ocean. By 1519, not long after Columbus had reopened the route to America, navigators were going to America by a more southerly route, which enabled them to make use of the north equatorial current, and the favourable wind to carry them westwards. On the return they sailed through the Straits of Florida, where the Florida current would help them on their way. They then followed the fast-flowing Gulf Stream to the latitude of Cape Hatteras, before turning eastwards towards Europe.

The importance of the currents to ships in the days of sail is clear, and the many voyages across the Atlantic in the sixteenth and seventeenth centuries provided a fair impression of the nature of the ocean current in this part of the ocean. The counter-current landwards of the Gulf Stream was first recorded in 1590, while the sudden change of temperature across the Gulf Stream was noted in 1606. A chart showing the Gulf Stream was published in 1665, but at this time a knowledge of the ocean current pattern was far in advance of a reasonable explanation of them (Stommel, 1958a).

Some of the most remarkable ocean voyages of discovery were those of Cook in the southern ocean (1772–1775) when he discovered New Zealand, charted parts of the Australian coast, and set a limit to the conjectural southern continent, which made it much smaller than previous guesses had suggested. He made use of the west to east drift in these southern latitudes as did most of his successors. One of his major achievements was the conquest of scurvy, the disease that had for long been a menace to prolonged oceanic voyages of discovery. He put oceanic exploration on a much more scientific basis and produced very accurate charts of all his discoveries, assisted by the greatly improved navigational aids available by this time. These included chronometers to measure longitude for the first time.

Other early voyages (for which direct evidence is very limited, but for which circumstantial evidence is strong) are the journeys made by the Polynesian peoples, largely in the south and west Pacific. These people had the advantage of travelling in very stable balsa wood rafts, which were virtually unsinkable. Although some of these voyages may have been unintentional (the sailors being swept out to sea), they were nevertheless extensive. That such long passages were possible, even in a raft with limited steering capacity, has been shown by the voyage undertaken by Thor Heyerdal from Peru to the Polynesian Islands of the south Pacific. In journeys of this type it is essential to make use of the ocean current, and in this region the steadiness of the south equatorial current carried the raft, at a steady speed westwards to cover 6900 km in 101 days at sea, an average of 68 km/day. The Peru current carried the raft from the coast into the south equatorial current. The oceans, therefore, are no barrier even to primitive craft (provided that they are seaworthy) and the currents are sufficiently constant to carry them steadily across the ocean.

Now in the days of steam, diesel power, and stabilizers, ships are more independent of their oceanic environment. The oceans, nevertheless, still provide the most convenient highway between distant parts of the earth for bulky goods and people who are not in a hurry. Under modern conditions, however, the competition of air transport is reducing the essential part played by the oceans for many centuries in trade and communications between distant lands.

6   The strategic role of the oceans

The part played by the oceans in strategy and warfare throughout the years has varied considerably. In the early days of the history of Britain, most invaders came by sea. The Romans and Normans landed on the south coast, while the Danes and Norsemen came from the north, also by sea for their raids and later settlement. The original Celtic peoples (who originally came across the sea) were driven more and more to the western highlands by the later comers. The earliest inhabitants of Britain, however, could have reached the country by land, since during the Palaeolithic and Mesolithic periods the Straits of Dover were at least at times dry land because of the Pleistocene lowering of sea level in the last glaciation.

The sea was no barrier to the invaders who crossed it. Their boats were small and they could sail up the rivers while the defending population was sparsely scattered over the country and could not readily congregate together to repel the invader; they also had no navy to defend their shores. Another disadvantage of defence was the difficulty of communication, despite an elaborate system of signal towers set up as early as Roman times. Towers were often located on prominent headlands, such as at Scarborough. Organization of this type declined after the Romans departed and so left the country relatively defenceless against the later Norman and northern invaders. These were, however, the last invasions of the British Isles.

Although the great Spanish Armada set out to invade the country, the sea provided an effective barrier. This was partly due to the strength of the British Navy and the defences set up against the invader, and partly due to the action of the sea itself. Stormy conditions probably played almost as great a part in dispersing the invaders and wrecking most of their ships on the inhospitable shores of western Britain as did Drake and his ships.

The effectiveness of even a narrow sea barrier was clearly felt during the Napoleonic wars and the 1939–1945 Second World War. As Bryant (1944) has pointed out, the strategy of Britain is determined to a considerable degree by her geographical position as an island. Sea power can give freedom of movement over three-quarters of the surface of the earth, which is as effective in the days of sail as of the aeroplane. Water-borne supply lines are more economical than land lines, and a distant base may be more easily supplied by a long sea line than a shorter land line. Often at the beginning of a war, Britain faced a territorially expanded adversary, such as Napoleon or Hitler. In such circumstances Britain could choose the area to attack by sea, where the enemy's land communications are most strained, provided she had mastery of the sea. This made the best use of a small force, for example Wavell's campaign in the winter of 1940 and Moore at Corunna in the Napoleonic wars. Such an assault at a distant point is possible for the power controlling the sea; this may be the easiest way to attack an enemy rather than by land.

It might appear easier to cross 32 km of sea, but in some circumstances an attack across 3200 km may be more effective. It is much more difficult to maintain the use of a long land line effectively than a long sea line of communications. Air power extended and supported sea power, enabling Britain to retain such control of the sea in 1944 that the adversary, who was widely deployed, could not resist attack on all fronts. Sea power defeated the invasion threat of Napoleon, who was waiting to cross the channel in 1803 and 1804. Nelson's victory at Trafalgar in 1805 put an end to his plans because of the supremacy of Britain on the sea.

During the two world wars, however, being an island brought its problems, as well as its safeguards. It proved impossible to invade the British Isles, but attempts to prevent ships reaching the country to bring essential supplies from overseas nearly proved effective at some stages of both world wars. Only the efficient and powerful British Navy, built up to protect far-flung dominions, could prevent the starving of the British people.

The necessity to mount an elaborate invasion by sea during the Second World War in order to re-enter Europe provided a very good reason for examining the forces at work in the ocean, as these could make such an important contribution to the success or failure of such a complex undertaking. One of the problems brought to light as a result of the preliminary study of the invasion areas, was the lack of knowledge of the effects of waves on the character of the beaches on which landings were to be made. In the Mediterranean, for example, it was found that there were submarine bars beneath the sea on which the landing craft might run aground, leaving a trough of deeper water between them and the shore. In some of the landings in Italy these bars were successfully cleared on the first day, but changing wave conditions increased the height of the bars overnight, so that on the following day the landing craft grounded on them. A great deal has since been learnt about the influence of the waves on these features, but the basic research that was done in planning for these landings will be of permanent value. Similarly, the beaches selected for the landings in Normandy provided problems which have stimulated research into the character of such beaches and the effects of the waves on the ridges which are typically found on them.

Other basic research work, which was found to be essential for the successful prosecution of the landings, was a study of the behaviour of waves. Information on their generation, subsequent movement, and shallow water modification were all needed to produce reasonably reliable forecasts of the wave pattern. The complexity of the planning required for these invasions shows that even a narrow belt of sea is a very strong defence under the conditions of warfare of the last two great wars. It is much more difficult to anticipate the part the ocean might play in any future conflict, but it is sure to be important. The part played by the sea in strategy has changed markedly, from being an easy route for invaders, to being a strong line of defence against them. In the modern days of complex life the ocean is now a vital link between the island fortresses and their overseas sources of supply, such as it was not in the days of self-sufficient communities.

7 The development of oceanography

The seeds of oceanography were sown during the Renaissance period, but have only come to fruition in the later nineteenth and twentieth centuries, during which period increasingly accurate and numerous measurements of oceanographic data have been made. One of the features of the oceans that early attracted attention was the tide, and by the thirteenth and fourteenth centuries it was generally recognized that the moon supplied the force responsible for tides. A wide range of theories, however, sought to explain the relationship. Other matters that attracted interest at this time was the saltiness of the sea, and how equilibrium was maintained between the land and sea, in view of the currently held idea that substances should be arranged according to their density, which would mean that the sea should overlie the land. Thus by the time of the Renaissance, there was considerable curiosity concerning the sea, but very little accurate information, and virtually no measurements.

The development of ocean sounding was discussed in volume 1; in this volume the advances in other aspects of oceanography will be considered (Deacon, 1971). One of the Renaissance works on the sea was that by W. Bourne, published in 1578. He discussed the genesis of coastal morphology, including banks and shoals at the mouths of rivers. He also considered tides, distinguishing between the diurnal tides and the spring–neap cycle, and recognized the relationship with the moon in a somewhat fanciful way. He confused tides and currents (a common misconception at the time), although his account of the major ocean currents was remarkably accurate, indicating that by this time the observations of practical navigators had been fairly widely disseminated.

The theoretical basis of current activity rested on the idea that water moved continuously from east to west, rounding the Cape of Good Hope and flowing into the Atlantic, with some water later moving into the Pacific through the Straits of Magellan, while the rest penetrated along the east coast of South America to the Gulf of Mexico. From here it was thought to emerge through the Straits of Florida to flow along the east coast of North America towards Europe across the north Atlantic. Bourne also described the south equatorial current of the Indian Ocean and the Agulhas stream. The general saltiness of the sea was appreciated by this time and he suggested that this was due to mixing with dissolved minerals.

Bacon assembled some of the knowledge of the sea available at the time he wrote, noting that waves break more readily in shallow water than deep. The cause to which ocean currents were attributed in this period was generally supposed to be related to the greater evaporation in the hot, low latitudes, and the greater precipitation in higher latitudes, which it was argued, would cause the sea surface to slope down towards the equator, thus initiating a current flow in this direction. and then westwards along the equator.

During the seventeenth century oceanography became a more scientific study, there being a greater attempt to explain the empirical knowledge increasingly gained at sea in terms of mathematical and scientific theory. The tides still continued to arouse interest, and the earliest recorded tidal observations were carried out near Liverpool in the years before 1660. The activities of the Royal Society, founded in 1662, assisted the theoretical advancement of tidal and other oceanographical knowledge. R. Boyle was asked by Christopher Wren to make a barometer to test Descarte's tidal theories. This led to the introduction of pressure observations in meteorology, rather than assisting tidal theory.

During the 1660s the Royal Society sponsored oceanographic observations on the depth of the sea, horizontal and vertical variations in salinity, pressure of sea water, tides and currents in the Straits of Gibraltar, and luminescence. A pump type of instrument was advocated for obtaining samples from different depths, while salinity was established by weighing samples and then evaporating them and weighing the salt. The observations, however, never seem to have been made, although suggestions concerning methods do indicate the realization of the need for many accurate observations.

Temperature measurements below the surface were first attempted in the Thames at Greenwich. Some water samples were obtained from a depth of 125 and 145 fathoms in the Bay of Biscay in 1663. Problems with instruments, however, led to several abortive attempts to take soundings and collect water samples at depth. This resulted in Hooke's attempts to improve oceanographic instruments. For water sampling he developed a device through which water could pass as it was lowered, but when it was hauled up the container was closed and the sample was representative of the depth from which it was hauled up.

As a corporate body the Royal Society's interest in oceanography did not last very long, but individual members carried on their work. Data continued to accumulate on tides (which were later of value to Isaac Newton), on currents in the Straits of Gibraltar, and in other fields. After the publication of Newton's *Principia* and on into the eighteenth century, the science of oceanography was neglected in favour of physics, mathematics, and astronomy. Technological problems greatly increased the difficulties of practical oceanography at this time of very elementary technology; without steel and rubber many instruments could not be made accurate enough to give reliable and meaningful results.

By the end of the seventeenth century, however, Newton had explained the tides, Halley had established the cause of the trade winds and the pattern of currents in the Straits of Gibraltar, and Boyle had explained other problems, so that to some there seemed little left to achieve. Indifference, therefore, made oceanographic research difficult at this time. One of the difficulties faced by oceanographers was the fact that one individual could accomplish little on his own.

During the eighteenth century interest in the sea reawakened and a number of different fields of oceanography were being actively explored by the end of the century. Marsigli almost alone bridged the gap between the two centuries. He made the first study of regional oceanography. His work was concentrated in the western Mediterranean. Working near Marseilles, he measured bottom profiles, collected water samples (both at surface and at depth), and noted variations in salinity and density. He also recorded temperatures at different depth and observed tides and currents. As a result of his observations, which included some study of marine life, he wrote the first book to be entirely devoted to oceanography. Another noted worker of the period was S. Hales (1677–1761), who prepared a number of instruments for observations at sea. As the century progressed more routine measurements were carried out at sea on official voyages (when occasion allowed) as well as on private ones. Even when measurements were not made, useful observations were recorded, such as the cold northward-flowing Peru current off western South America. The voyage of James Cook (1728–1779) was the first on which scientific observations were the main purpose of the voyage, although observations of tides and currents had commonly been made along with navigational observations earlier. Little physical oceanography was, however, carried out on Cook's first voyage, although the tides at Tahiti were recorded. The diurnal inequality of the tide on the reefs off eastern Australia was noted when the ship struck a reef. More work was accomplished by the astronomers who accompanied Cook on his second voyage. Their observations included the testing of chronometers and measurement of salinity and temperature at depth, using instruments similar to those developed by Hales.

The general development of science in the eighteenth century led to a greater interest in oceanography. In the later years of the century B. Franklin (1706–1790) revealed the extent of the Gulf Stream, showing that it flowed far out into the Atlantic. Better measurement of longitude also allowed improved mapping of ocean currents. Measurements of temperature at depth had become a challenge, and measurements over time at the surface were now made, revealing the greater warmth of the Gulf Stream than the surrounding water. The stream was now correctly related to the effect of trade winds in piling up water in the Gulf of Mexico. Some attempt to invoke the effect of the earth's rotation in explaining the divergence of the Gulf Stream from the coast was made. The constancy of deep water temperatures compared to those on the surface was appreciated and explained by the effect of heat rising from the ocean floor to maintain the temperature of deep water. Some less complex explanations were also suggested to account for observed temperature fluctuations.

Towards the end of the century it was realized that differences of density could cause internal movements of the water, and the sinking of water cooled in winter was suggested by R. Watson. Differences of density in creating the undercurrent flowing out through the Straits of Gibraltar were recognized by Admiral P. Patton in the 1770s. He showed that Mediterranean water was indeed more dense than Atlantic water. An essay published in 1798 by B. Thompson, Count Rumford, first suggested that the ocean circulation is based on differences in density. He realized the difference between fresh and sea water in its density behaviour, showing that density increases right up to freezing point in sea water. He also suggested that the very low temperatures recorded in deep water in low latitudes must represent the movement of water at depth from the polar regions towards the equator. His conception of ocean circulation was not accepted for 70 years, although A. von Humboldt (1769–1859) supported his views and also showed that the salinity must be such as not to prevent the movement. The first qualitative chemical analyses of sea water were made in the eighteenth century. In 1772 Lavoisier analysed sea water, finding it to contain 1·7 per cent salt, which is half the normal value. The origin of salt in the sea was still being debated. Lavoisier predicted that all substances on earth would be found in sea water, on the assumption that the salt resulted from the action of water over the whole earth. The problem remained unsolved, however. A neglected field of oceanography at this time was the study of sea waves. One idea was that waves were caused by fermentation, which made the water swell. However, the cause of waves was soon recognized to be the wind. Observations on the coast of Sumatra showed that the waves were the result of winds blowing far out in the ocean. The nature of wave motion was also appreciated, such as the apparent forward movement of the wave form.

The first major period of activity in oceanography took place in the seventeenth century, during the general intellectual ferment of the scientific revolution. The realization of the beneficial effects of a better understanding of the oceans was one motive. The second period covered the eighteenth and early nineteenth centuries, and was related to the belief that the oceans might provide clues to some of the important scientific queries of the day. Expansion in the second phase of oceanography reached its peak in the period 1815–1830, following the Napoleonic wars, during which activity was curtailed. A long, slow build-up similar to that which led to the activities of the first period in the 1660s, led to this period of growth. Old problems and queries were attacked by new techniques that had been developed in the interval. J. Rennell (1742–1830) studied currents and distinguished drift currents (caused by wind) and stream currents (resulting from interference by land-masses, gravity being their cause). He combined both types in an explanation of the currents of the Atlantic. His explanation was more successful for the currents of the western side of the ocean, as he misinterpreted the eastern currents. His work on currents was, nevertheless, the best available at the time. During this period Sir J. Banks carried on his interest in oceanography and also encouraged W. Scoresby (1789–1857), who followed his father in the Arctic whaling industry and who also made oceanographic observations, including temperature observations made with a new type of thermometer, which was not very successful. He also studied ice formation and Arctic life. Work in the Antarctic by Bellinghausen between 1819 and 1821 was important on account of his study of plankton, while in 1823–26 work also included the most accurate series of temperature and gravity measurements yet made. Major advances in the chemical analysis of sea water were made during this period and more work was done on the density of sea water and its implications. Problems still arose concerning the effect of pressure on temperature at depth. The work that was achieved in this second period of activity was, unfortunately, never fully written up, and so was not adequately regarded when the next phase for advance came in the later part of the nineteenth century. The third phase had to await government backing in the expensive and complex task of preparing major oceanographic cruises for purely scientific purposes.

A period in which interest was lost followed the second spell of activity, during which activity on tidal work was based on the contributions of Newton, Maclaurin, and Euler, who had established the nature of the tide-producing forces, and the part played by the sun and moon. Bernoulli had calculated the tide-raising forces on a completely water-covered earth, and Maclaurin had examined the effect of the earth's rotation on meridional currents. Laplace worked on tides during the last quarter of the eighteenth and the first of the nineteenth centuries, giving equations for a dynamic theory of tides on a rotating globe. The problem remained, however, of relating the theoretical tide-producing forces, by then well known, to the actual observed tides at different localities. Two names associated with this work were J. W. Lubbock and W. Whewell, who worked together on the problem. Whewell suggested the development of stationary waves from the interference of two progressive waves, and drew the first co-tidal charts for the waters around

Britain, in which the idea of an amphidromic system was first suggested but not finally accepted for a considerable time. In fact it was not until R. A. Harris in 1904 confirmed Whewell's views concerning amphidromic systems that the theory became generally accepted. Harris demonstrated the existence of such systems in the open ocean as well as smaller seas. During the latter half of the nineteenth century the main advance in tidal analysis was the introduction of harmonic analysis.

The third major phase of advance in oceanography took place in the latter half of the nineteenth century. It was realized at this time that much still remained to be learnt about the oceans, and it was felt that the oceans might provide answers to some of the fundamental problems of the day, such as the origin of life and of species. The term 'oceanography' first came to be used in the 1880s, during the third phase of advance, when a number of organizations (both national and international) were set up, such as the Marine Biological Association and the International Council for the Exploration of the Sea in 1902. At this stage more support was given to biological oceanography, the major work in physical oceanography being done at the Tidal Observatory at Liverpool.

At the beginning of the third phase of advance, before the *Challenger* expedition set out, W. B. Carpenter had been observing deep sea temperatures off western Britain and had proposed a model ocean circulation based both on density differences and wind stress. The *Challenger* cruise (1872–1875) was led by C. Wyville Thomson, who had previously worked with Carpenter in the *Porcupine*. It was during this work that hauls of deep sea life finally ended the azoic theory of lifeless ocean depths. The *Challenger*'s work covered both biological and physical oceanography, with an emphasis on the former. During the 3·5 years' cruise, the expedition made observations in all oceans, on types of life at all depths, as well as taking many water samples for analysis and temperature observations at depth. Problems were experienced with deep temperature observations, but nevertheless the results of the *Challenger* expedition were momentous and monumental, as indicated by the scientific publications, which run to 50 volumes, taking five times the length of the expedition to prepare. These results included the first systematic sampling of deep sea sediment, and of life at a variety of depths, as well as routine soundings, temperature and salinity observations. The first volume of results was published in 1880. The *Challenger* carried 144 miles (230 km) of sounding rope and 12·5 miles (20 km) of sounding wire. The biologists discovered 715 new genera and 4417 new species. The two summary volumes were written by Sir John Murray, following Wyville Thomson's death in 1880.

In the early years of the twentieth century, following the final publication of the results of the *Challenger* cruise, money for oceanography was hard to come by, and Murray advocated work in the Antarctic. Two expeditions associated with him took place in 1892–93 and 1902–04. Other countries were also involved in similar work that included marine observations. Nansen's drift in the *Fram* was one of the outstanding events of this period, and revealed much of interest concerning the Arctic Ocean. Germany also took a leading part, with the *Valdivia* expedition of 1898, and oceanography was being actively pursued in Sweden. The First World War stopped most oceanographic work and it was not until the Second World War that oceanography started to advance rapidly, with the aid of many new technological developments. Some interwar work was, however, accomplished, such as the German *Meteor* expedition, and the work of the UK *Discovery* committee. The phase of activity that started in the Second World War has increased in tempo and sophistication from then on, and is now producing an ever-expanding fund of knowledge about the oceans.

The value of oceanography in the world today is being increasingly appreciated and thus more governments and scientists are participating in the development of the subject. In view of the developing sophistication and expense of oceanographic work, an essential element is the availability of government funds and international co-operation. The suggestion of an International Decade for Ocean Exploration stresses the significance of the oceans. This growing importance applies to both pure and applied oceanography, although it is very difficult to separate the two, since research in pure oceanography (in adding to the knowledge of the oceans) also helps in practical problems concerning the use of the oceans.

The aim of the International Decade is to achieve a much fuller coverage of basic data concerning ocean characteristics. The data can now be collected by unmanned buoys, space satellites, aircraft, and submersibles, and many other highly sophisticated pieces of equipment. The data-gathering techniques will provide a world-wide coverage of basic data, much of which is lacking at present. Thus a much sounder knowledge of oceanography will be available with which to plan the optimum use of the ocean resources in the future.

8 Applied oceanography: ocean technology

Modern technology provides both problems and opportunities for oceanography. The opportunities include the availability of advanced instruments with which more numerous and more accurate data can be obtained and processed more rapidly, and stored more readily, by means of advances in computer technology. The problems on the other hand, are concerned with man's increasing ability to modify the ocean environment, often with disadvantageous consequences. The extent to which the oceans can be harmed by deliberate or accidental means is one of the major problems facing man in his exploitation of the marine environment, and it is this that makes the gathering of basic data so essential. The data provide a base against which change can be assessed. A brief introduction to the technology relevant to different uses of the oceans will be given in this section, while some of the points are treated in more detail in chapters 7 and 8.

The oceans provide resources of great variety, both inorganic and organic, but their optimum use depends on developing suitable technological methods and enforcing adequate controls. This entails international co-operation and restraint, based on the best assessment of the state of the resource, its distribution, viability, and potential supply. Inorganic resources include both those substances that can be obtained from the ocean floor, such as sand and gravel in shallow water, and manganese nodules in deeper water, and those substances that can be obtained directly from sea water, which contains most minerals.

The technology for obtaining sand and gravel from shallow water is sufficiently advanced for this resource to be actively exploited, one purpose being the stabilization of beaches by artificial sand filling. Detailed surveys of sand and gravel availability have been made around the coasts of the United States, and similar projects are also being carried out around the coast of Britain. Technological problems concerning the exploitation of deep sea nodules are less easily solved, and although the potential in this field is great, it has yet to be realized in practice.

At present the most intensive activity in the field of inorganic resources from the oceans is the exploration for, and exploitation of, oil and gas. At present the continental shelf is the main area being explored, but eventually technology may enable deeper resources to be utilized. Already the search for oil has led to great advances in the technique of drilling. Some of the work is done from platforms fixed to the sea bed in shallow water, but floating platforms are being developed that can operate in quite deep water. Particularly notable in this field of technology is the development of the deep drilling ship *Glomar Challenger*, which has accomplished very valuable work in pure research in drilling right through the sedimentary layer into the ocean crust in all the world's oceans. Some of the results of this work were considered in volume 1. The technological results of this work have had a direct result on the improvement of drilling methods for oil and gas. The cost of oil and gas production from offshore is high; figures for 1968 for offshore oil production in the US totalled 12·75 billion dollars, of which one-third was for lease and rental payments.

The seas provide a wide range of minerals within the water, some of which are already utilized (for example salt and magnesium) and more could be exploited as technology improves. It is likely, however, that the water itself will eventually be one of the most valuable commodities obtained from the ocean. Fresh water is an essential element of all communities' well being, and one that is likely to become very short in some areas as the standard of living rises. In some areas fresh water is already being derived from sea water by various processes of desalination. Technical advances in this field will make the process increasingly important and commercially feasible. There are several methods by which salt water can be freshened, but some of these are not yet economically viable in competition with more traditional methods of water supply from surface and underground land sources. Since 1900 the USA has increased its water consumption 7 times, and by 1975 the rate of consumption of 1957 will have been doubled. Advances can be made by recycling water, but in some areas desalination of sea water is already providing the bulk of the supply, for example in parts of the West Indies, Israel, Arabia, and Sicily. The energy required to separate the salt ions from the water molecules is 2·8 kw hr/1000 gallons theoretically, but in practice 1000 times this amount is required to operate an effective system of single-stage distillation. Some of the new techniques under study may reduce the amount of energy needed to only 4 or 5 times the theoretical thermodynamic minimum, or about 10–15 kw/hr/1000 gallons. Compression distillation is a method being actively investigated, although the idea is 100 years old. Other techniques include freezing, evaporation, reversed osmosis, and electrodialysis.

Many of the inorganic resources of the oceans are expendable, and once used cannot be replaced, although this does not apply to manganese nodules (which grow at a very slow rate) or the water in the ocean (which is in transit through the hydrological cycle). The other major resource is the organic one, which is self-renewing as long as it is treated with reasonable care. The use of organic resources, therefore, is the most vulnerable field of ocean exploitation, and can only be undertaken successfully in the light of a full understanding of the marine ecosystem to which the organism belongs. There is thus a great need for a more thorough knowledge of biological oceanography, and the complex mutual relationships that form the fragile and easily disturbed natural ecosystems. Until these are thoroughly understood, it is too early to consider modifying the marine environment for human benefit, although this may eventually be possible as it has been on land, where dramatic advances in production are possible under enlightened farming methods.

Despite recent advances in technology, there is still a very long way to go before the biology of the oceans is fully known and can be effectively controlled. The creatures of the sea include a very wide range of both plants and animals, each species having different requirements and being adapted to a particular environment. As already stated, plants form the essential basis of life in the oceans. They consist of two main elements—the seaweeds most of which are attached to shallow, rocky bottoms, and the drifting elements of the phytoplankton in the surface layers of water throughout the ocean. The only exception is the Sargassum weeds which are floating macro-algae. These plants, however, come from attached weeds that are broken loose by wave action and then drift into the Sargasso Sea from which they cannot escape. They grow but do not reproduce, and amount to a total weight of 4–10 million tons, distributed at 1–$1 \cdot 5$ tons/km^2. Some of the plants are very old and provide a habitat for an unique animal population. Exploitation of this large mass of weed could eventually become feasible. The exploitation of the more common attached seaweeds of the coast is considered in more detail in chapter 8. Because of its attached character and rapid growth rate, seaweed provides an element of sea wealth that is particularly suitable for artificial encouragement and exploitation. Technology plays a part in developing suitable harvesting machinery that would facilitate the collection of the weed, for which there are many uses.

Another branch of the marine life that could be relatively easily exploited more fully is the invertebrate population of shellfish and crustaceans that are relatively immobile, at least in the adult phase of their life cycle. Farming of such creatures is already undertaken on a small scale in some areas, notably Japan and France. Oysters are one of the shellfish that can be successfully cultivated, and have been since Roman times. Japan now has an annual production of between 30,000 and 35,000 metric tons/year, the oysters being grown on rafts rather than on the bottom. Other shellfish and crustaceans have long been exploited. The king crab industry in Alaska, for example, expanded from 141,000 lb in 1947 to over 160 million in 1966 (63,800 kg, 72·5 million kg), but thereafter a reduction in catch indicated that the sustainable yield had been reached if not exceeded.

It remains true, however, that at present, despite a considerable amount of experience over a long time, no cultured marine invertebrate is more than a dietary luxury. The only invertebrate to form an item of staple diet is the squid, which is extensively eaten in Japan, where a large effort has been made to extend mariculture of a wide variety of types. Squids for human consumption, however, are still most efficiently obtained by fishing. Apart from their use as food, the control of invertebrates in the marine environment could have other beneficial effects. The control of invertebrates that compete with fish for food, for example, has a beneficial effect in increasing the food available for the fish.

Another example of the problems of marine ecology in which invertebrates are involved is the destruction of coral by the crown of thorns starfish in the Pacific. The cause of the starfish population explosion is not known, nor whether it is a man-induced or a natural phenomenon. The results are nevertheless devastating large areas of reef. Much remains to be learnt concerning ecological problems of this type in the ocean and also in the efficient culture of marine invertebrates for human food. These developments depend both on a knowledge of the life cycle and requirements of the organisms, and on the technology necessary to control their environment optimally.

By far the largest organic exploitation of the sea is derived from the sea fisheries. Technological difficulties in efficient exploitation of these large reserves of high-class protein food are formidable. Fishing is still in the primitive stage of hunting, despite the recent very rapid development of increasingly elaborate and sophisticated fishing gear. Fish farming will be even more difficult than other forms of sea farming because of the mobile nature of fish and their complex life cycles. They will be difficult to rear effectively and to control as adults, although some successful work has been done.

Developments in fishing have been strongly influenced by changes in technology. Changes began to accelerate in the nineteenth century, when from 1880 onwards conversion from sail to steam in fishing vessels gave fishermen much more freedom over the elements, and also added power to develop effective trawling gear. Other important innovations were in mechanized fish handling gear and the development of freezing fish at sea, which allowed longer voyages to be undertaken to the more distant fishing areas, thus enabling increased output. Since the Second World War, fishing has benefited by the great development in electronic equipment. The most useful devices are those concerned with communications and navigation. Developments in fish detection by sonar and similar methods have played a major part in rendering fishing more effective. Schools of fish can be detected by thermometric techniques and underwater sound, the basis of the echo-sounder and similar instruments. Refinements of the equipment are constantly being produced. Fishermen are also being increasingly provided with data on the oceanographic environment that affects the distribution of fish, such as temperature on the surface and at depth. Net materials and handling equipment are also improving. The results of all these improvements in fishing methods are shown in the steadily rising fish catch, which increased from a world total of 19·6 million metric tons in 1948 to 52·4 million metric tons in 1965.

One of the major problems raised by the increasing efficiency of fishing techniques, owing to improved technology, is the danger of over-fishing. The more efficient trawls are too efficient, in that they destroy the bottom across which they are hauled as a habitat for fish (thus reducing the stock), despite some control on the mesh size of the nets. As the more accessible sources of fish become over-exploited, on the continental shelf around the shores of western Europe and North America, the fishery must turn either to more stringent and effective international control, or alternatively to as yet under-exploited stocks, which are likely to be found more in the open ocean.

Harvesting fish in the open sea, especially if they are not of the shoaling species, poses very much greater problems for the fishery technologists. New developments are constantly being produced, so that no doubt problems of location and catching can be overcome if the demand is great enough. Electricity, for example, has been used experimentally for fishing. Light devices have also been used to attract fish, and when they are sufficiently concentrated around the light they can be stunned and pumped aboard. One drawback is the large amount of current needed in salt water compared with fresh, amounting to 500 times more current. The use of bubbles to contain fish has also been tested experimentally. All these methods of catching more fish depend on technological developments.

There is also another very essential element in developing a viable and sustainable fishery, until fish farming is efficient. This element is the human one. Unless adequate national and international control of organic exploitation can be introduced the fate of the blue whale will be that of many other potentially valuable species. Fish have the advantage over marine mammals in that they can multiply extremely rapidly under favourable conditions; they produce an immense quantity of eggs at each spawning, and are thus much more resilient to over-fishing. Nevertheless efficient control measures must be accepted and adequately enforced if the fisheries are to remain productive. The experts on fishery dynamics can calculate the optimum fishing rate to obtain the maximum sustainable yield with the minimum of fishing effort. Agreements should be based on fishing to this intensity, which will provide the maximum benefit. The more mobile the fish or marine animal the more difficult it is to control it effectively, especially if it lives outside territorial waters over which individual nations have jurisdiction. This applies to the whales who inhabit the open ocean, and whose control must depend on international action.

One of the uses to which the oceans have been put since man first made a boat is that of transport. Even now, with the great increase of air transport, the oceans still carry a very large tonnage of heavy goods, of which oil is becoming increasingly important, thus causing problems of navigation and dangers of pollution as ships become ever larger, and thus more vulnerable to grounding and collision in coastal waters. The example of the *Torrey Canyon* is a reminder of this danger and its results, a matter that is considered in chapter 8. The largest tankers are so deep draught that even the sand waves on the sea floor may be a potential hazard and very few ports are available for their docking.

One technological field is concerned with the design and construction of these immense vessels, while another different one, in the field of pure research but with practical applications also, is that of the development in the knowledge of wave propagation and generation. Accurate forecasting of waves can save ocean vessels valuable time and hence money. The problems of wave generation and propagation are being tackled on a number of fronts. There have been elaborate mathematical and theoretical studies of wave generation, but these cannot achieve their maximum value unless they can be tested and justified by field observations. The recording of waves in the open ocean has improved rapidly with the technological development of wave recorders for use in ships and the design of floating buoys that can record the whole three-dimensional wave spectrum and transmit the coded message in digital form suitable for computer analysis by spectral methods. These recorded spectra can then be compared with theoretical ones, based on recorded wind and temperature data.

Kvinge (1969) has described a buoy system for measuring low-frequency variations in hydrographic parameters. The record includes temperature, and current speed and direction, every half hour for a period of 6 months. The instrument is connected to a moored buoy either on or below the surface. It can be used both in deep water up to 2000 m or adapted for shallow water observations near the bottom. On the whole recordings have been made on three-quarters of the days the buoys have been on station, although there were some losses (including some by ship damage) at a point 72 km northwest of Shetland.

Wave analysis is complicated by the very slow decay of ocean waves, making it difficult to find a natural situation that is simple enough to analyse theoretically, because waves coming from considerable distances are mixed with those being actively generated, thus forming a very complex spectrum. The use of microseisms for forecasting ocean waves may eventually become possible in some circumstances, when more is known concerning the refraction and other distortions to which microseisms are subject.

Satellites may also be able to help in the work of wave forecasting, since air and sea temperatures can be recorded by this means over a wide area, and these variables have a considerable effect on wave height. Waves in the open ocean are of more concern to ships carrying goods or people across the sea. Waves close to the shore are also important in several fields, such as coastal protection, offshore drilling for oil and gas in shallow water, and coastal navigation. Advances have also been made in studying important aspects of waves in shallow water. Their generation differs in significant respects from that in the open ocean, and the effects of refraction on the concentration of wave energy at the coast are being studied by means of computer-based refraction diagrams. The effects of surf beat and seiches in harbours have also been investigated, while wave run-up on sea defences is being investigated by coastal engineers, both in hardware models and in the field and theoretically. Coastal engineers, in studying the best type of sea defence work, are increasingly relying on beach replenishment as the most effective method (from both the engineering and amenity points of view) because a natural beach is the best coast defence.

As leisure increases and more people have the necessary resources, so the oceans will become increasingly important in the field of recreation. The seaside holiday has long been a major leisure attraction, and problems arise concerning the optimum land use in coastal areas where the pressure of population is high, and the coastal zone is liable to erosion or other hazards, such as storm surges or hurricane activity. Imaginative planning for these areas must be based on a sound knowledge of coastal processes. Increasingly, however, people are seeking their recreation beneath the sea and on it, rather than merely at its edge. Small boats and yachting have long been popular, and their safe usage depends on a knowledge of tidal and wave currents. A knowledge of waves and surf, and the rip currents that are frequently associated with them, is also desirable for safe bathing and surfing, where rip currents may be potentially dangerous.

Beneath the sea more and more people are exploring the underwater environment by means of the aqualung and SCUBA diving equipment. These innovations give much greater freedom and safety than the traditional hard-hat diving suit. They have more than a recreational value, in that valuable work has been accomplished by their use by marine biologists and geomorphologists, and above all by marine biologists, who for the first time can live in the same world as the creatures they are studying and appreciate the different phases of their life cycle and ecosystem at first hand in their own environment. The return of man to his old environment, according to McKee (1967), has greatly enhanced the possibility of making the optimum use of the shallow water environment, for example, for fish farming. Man's use of the shallow water by direct contact in diving is restricted by the necessity to decompress slowly in ascending to the surface to avoid nitrogen poisoning. This danger can be overcome, and much longer underwater stays are made possible, if use is made of a sea floor installation in which it is possible to live and rest at the pressure of the surrounding water.

Underwater sea laboratories are being increasingly developed and represent a major technological advance. The underwater sea laboratory was pioneered by Cousteau in the Mediterranean, and it has since been elaborated off American shores, where Sea-Lab I was installed off Bermuda in 1964, and Sea-Lab II off California in 1965. Tektite I was in use off the Virgin Islands in 1969. Stays underwater amounting to several months are possible in these installations. These new techniques are greatly increasing direct observations of the marine environment and allowing advances in many fields of oceanography. At the same time indirect observations are being facilitated by developments of deep diving submersibles, and the developments in underwater closed-circuit television, as well as satellite surveys that can cover the entire ocean surface very rapidly. Oceanographical observations can now be made from above, on, and within the oceans.

9 The future of oceanography and ocean exploitation

The knowledge of all aspects of oceanography is expanding rapidly by means of the great advances in underwater technology and other oceanographical techniques. The degree to which the oceans will be required for future resources is also likely to grow quickly with the increasing population and increases in standards of living. These developments will bring new stresses to the fragile ocean environment, and these will need all the accumulated knowledge of oceanographers to work out satisfactory solutions.

One of the current problems under discussion concerns the dangers of ocean pollution (considered in chapter 8). These problems are closely associated with the nature of oceanic circulation and all aspects of water movement in the oceans (examined in chapters 3, 4, and 5). The natural self-purifying capacity of the oceans is dependent on bacterial activity, which is, therefore, an important aspect of biological oceanography. This aspect of biological oceanography must be studied intensively to assess the optimum means of waste disposal in the oceans. Some considerations relevant to this problem are mentioned in chapter 8.

The use of the biological resources of the oceans (chapter 6 and 7) will become increasingly important, and perhaps the greatest need in this field is a wise and just system of human control on the use of these resources, so that they can be managed to the optimum benefit both of the marine communities themselves, and the people who will benefit from their exploitation. Fishery dynamics, legal agreements, international co-operation, and fishing technology are all involved in this problem, one of the most difficult and complex to solve. It will be a long time before the seas are farmed as efficiently as the land, but this is a goal towards which it is well worth while to aim. It will involve the co-operation of specialists in many fields and a thorough understanding of the ocean environment in all its aspects, both physical, chemical, biological, and social.

Further reading

BRATZ, J. F. (editor). 1968: *Ocean engineering*. New York: Wiley. 720 pp.
BRETSCHNEIDER, C. L. (editor). 1969: *Topics in ocean engineering*. Houston: Gulf Publishing. 420 pp.

DEACON, M. 1971: *Scientists and the sea.* London: Academic Press. 445 pp.

FIRTH, F. E. (editor). 1969: *Encyclopedia of marine resources.* New York: Van Nostrand Reinhold. 740 pp.

HOOD, D. W. (editor). 1971: *Impingement of man on the oceans.* New York: Wiley–Interscience. 738 pp.

LAWRENCE, L. G. 1967: *Electronics in oceanography.* Indianapolis: H. W. Sams. 288 pp.

MCKEE, A. 1967: *Farming the sea.* London: Souvenir Press. 314 pp.

MARX, W. 1967: *The frail ocean.* New York: Ballantine Books. 274 pp.

MEYERS, J. J., HOLM, C. H., and MCALLISTER, R. F. 1969: *Handbook of ocean and underwater engineering.* New York: McGraw-Hill.

MOORE, J. R. 1971: *Oceanography.* Readings from *Scientific American,* with introductions by J. R. Moore. San Francisco: Freeman. 417 pp.

2 The water of the ocean

1 Oceanographic instruments

2 The character of sea water: 2.1 Temperature of ocean water; 2.2 Salinity of surface water; 2.3 Oxygen content of ocean water

3 Water-masses: 3.1 Definition and concept; 3.2 Method of formation and character of the major surface water-masses, *3.2a North Atlantic central water, 3.2b South Atlantic central water, 3.2c Arctic surface water, 3.2d Subantarctic water, 3.2e Subarctic water 3.2f The central water-masses of the Pacific, 3.2g Pacific equatorial water, 3.2h Subantarctic water-mass, 3.2i Indian Ocean central water, 3.2j Indian Ocean equatorial water, 3.2k The Antarctic circumpolar water;* 3.3 Deeper water-masses, *3.3a Antarctic intermediate water, 3.3b North Atlantic intermediate water, 3.3c Pacific intermediate water, 3.3d Mediterranean water, 3.3e North Atlantic deep water, 3.3f Deep water of the Pacific, 3.3g Antarctic bottom water*

4 Ice in the ocean: 4.1 Sea-ice; 4.2 Icebergs; 4.3 Ice islands

The oceans contain about 97 per cent of all the water on earth. They are the great store of water which is always available for evaporation into the atmosphere and precipitation onto the land. Water as a substance has some interesting properties. It is the only natural substance to be found in all three states (solid, liquid, and gaseous), while it is one of the very few inorganic materials in nature that is liquid at normal temperature and pressure. Its other special properties include its very great solvent power, which makes it indispensable for life. This property is due to its molecular structure, in which the two hydrogen atoms are linked to the oxygen at an angle of 105 degrees. It also has a higher surface tension than any other liquid and has an exceptionally great capacity for absorbing and conducting heat. It requires a large amount of heat to change state from solid, through liquid to gas. Water heats and cools more slowly than other substances and as a result, it exerts a very important modifying effect on the earth's temperatures. Heat used for evaporation also exerts a strong influence on temperature extremes in some areas.

The maximum density of pure water is at 4°C, but in sea water the density increases steadily to its freezing point, which is lowered as salinity increases. Thus sea water can reach its greatest density as it is about to freeze at temperatures around −2°C. This means that very cold sea water can be formed, which is also very dense and can, therefore, sink to the bottom of the ocean basins. Another property of interest is the compressibility of sea water. It is small, but it is also enough to make sea level 30 m lower than it would be if it were completely incompressible.

Water is the agent by which almost all transport of solid and dissolved matter is accomplished in the world on land or in the sea, but on land the air plays a minor part. The ultimate source of all the energy on which the movement of ocean water depends is the sun, though the moon plays an important part in the generation of tidal energy. The energy of the sun is imparted to the oceans through the atmosphere, some being in the form of heat. This causes evaporation, which carries the ocean water into the atmosphere, from where it can be transferred to the land. The atmosphere also affects the movement of the ocean waters via the wind, which exerts a direct stress on the water surface, creating both ocean currents and waves.

Calculations of the age of the oceans from salt input by rivers provide various possible ages, such as 18 m years, using total salt in the sea of 48×10^{21} g. and an annual input by rivers of $2 \cdot 7 \times 10^{15}$ g/year. Joly obtained 80–90 m years, using amounts of sodium. Further refinements suggest 200 m years, when corrections for the sodium cycle are taken into account. An age of 700–800 m years is obtained if allowance is made for the concentration of dissolved solids to be lowered in time of flood. None of the estimates allow for the effects of sedimentation and cycling. The main objection to the Halley model is that it is a static and not a dynamic one. The age of the ocean as calculated for different substances is varied, because each has a particular cyclic period. The 'age of the ocean' for magnesium, calcium, sodium, and potassium are respectively 23, 1·3, 120, and 10 m years.

The properties of the new dynamic model of the igneous rock–sea water–sediment system can be summarized as follows: 1) The system is dynamic, both physically and chemically, changes taking place during transit in the river, in the sea, and through ageing of sediments. 2) The system is open. Weathering of igneous rocks is not the only source of chemical compounds; there are the excess-volatiles, including water, sulphur compounds, halogens, boron, and carbon, coming from the earth's interior and some from outer space. 3) The system is in a steady state, in that material added by the rivers is removed by

Table 2.1 Residence times of elements in sea water

Element	From weathered igneous rock	Present in sedimentary rock	Present in ocean	Residence time in the ocean years
Si	5820	5820	$4 \cdot 2 \times 10^{-2}$	$8 \cdot 0 \times 10^3$
Al	1620	1620	$1 \cdot 4 \times 10^{-4}$	$1 \cdot 0 \times 10^2$
Fe	862	862	$1 \cdot 4 \times 10^{-4}$	$1 \cdot 4 \times 10^2$
Ca	739	733	5·61	$8 \cdot 0 \times 10^6$
Na	572	425	147	$2 \cdot 6 \times 10^8$
K	524	519	5·33	$1 \cdot 1 \times 10^7$
Mg	358	340	18·2	$4 \cdot 5 \times 10^7$
Ti	98·6	98·6	$1 \cdot 4 \times 10^{-5}$	$1 \cdot 6 \times 10^2$

processes of sedimentation. 4) A significant part of the material taken as salt to the sea via the rivers is cyclic salt that has been blown in spray from the ocean surface and precipitated on land with rain and snow. River composition is, therefore, not a measure of erosion and weathering. 5) The degree of chemical reaction determines the period of solution in the sea, and thus the apparent age of the ocean as measured by that particular substance. Thus residence time has now replaced the idea of the 'age of the ocean'. Some of the residence times are given in table 2.1, derived from Horn and Adams (1966). See also Carrit (1971).

1 Oceanographic instruments

The density is of great significance in the circulation of ocean water and depends on water temperature and salinity. Temperature at the surface can be measured by using ordinary thermometers, but for subsurface measurements, other instruments must be used. A sample may be brought to the surface in an insulated bottle for measurement, or a reversing thermometer may be used. This instrument turns upside down at the point of observation, breaking the mercury at a constriction in the tube. The thermometer is reversed by the action of a weight sent down the wire cable on which the sampling bottles and thermometers are fixed.

Salinity is measured in the laboratory and for this purpose a sample of water is brought up from a specific depth. Temperature and salinity must be measured in the same position, in order to obtain the density. Reversing thermometers are, therefore, usually attached to the sample bottles. The weight which reverses the thermometer at the same time automatically closes the sampling bottle by reversing it. Two thermometers are usually used, one of which responds to the pressure and the other which does not. The protected thermometer eliminates the effect of hydrostatic pressure, while the unprotected one gives a fictitious temperature, due to the pressure of the overlying water causing a higher reading. This is caused by the contraction of the thermometer. The difference between the two temperatures recorded provides a determination of the depth, which can be checked against the length of wire paid out on which the thermometers were fixed. It is useful to be able to measure the water character at several depths in any one position. In order to accomplish this a series of sampling bottles and thermometers are fixed to the wire. When one is reversed by the weight, it releases another weight, which reverses the next bottle beneath it, and so on.

Sea water characteristics play an important part in its subsequent movement and in the fertility of the water for the production of the basic forms of marine life, on which larger creatures depend. The plankton in turn may assist in the determination of the water type.

2 The character of sea water

2.1 Temperature of ocean water

The temperature of sea water is measured in degrees centigrade to an accuracy of $\pm 0.02\,°c$, the range of oceanic temperature falling between $-2\,°c$ and $+30\,°c$; at the lower limit ice forms. The conduction of heat through sea water is complicated by the fact that the motion is nearly always turbulent. The specific heat of sea water has been determined by Cox and Smith (1959). The value gives the number of calories required to increase the temperature of 1 g of water 1 °c. Their value indicates that there is greater adiabatic cooling (as water is raised to the surface) for low temperatures than previously estimated.

The specific heat decreases with increasing salinity. Adiabatic heating may lead to temperature increases at the bottom of deep isolated basins, owing to the low compressibility of water. This process is less important (because of this property) in water than in the atmosphere.

The potential temperature of sea water is defined as the temperature that it would have if it were raised adiabatically to the surface. The freezing point of sea water depends on the salinity, falling from $-0.5°$C at a salinity of $10‰$ to just below $-2°$C at a salinity of about $36‰$ (‰ equals parts per thousand).

The distribution of temperature in the surface layers of the ocean reflects the general distribution of heat supply from the sun. The value of heat received from the sun is greatest at the equator and falls off towards the poles. Total heat loss falls off similarly from equator to poles, but not at the same rate. There is, therefore, a surplus of heat received from the sun between the equator and $30°$ latitude, and a net deficit from there to the poles. The transfer of heat from low to high latitudes is achieved by air currents and water movements, and reaches a maximum around $40°$ latitude. Most of the heat transfer probably takes place in the atmosphere, but at least in some places the oceans play a significant part.

Table 2.2 Surface temperature in the Atlantic Ocean

North latitude			South latitude		
70–60°		5·60°C	70–60°		−1·30°C
60–50		8·66	60–50		1·76
50–40		13·16	50–40		8·68
40–30		20·40	40–30		16·90
30–20		24·16	30–20		21·20
20–10		25·81	20–10		23·16
10–00		26·66	10–00		25·18

For example in the north Atlantic at about $55°$N 0.3×10^{16} g cal/min is carried by the ocean which is one-tenth of the total transport of heat in this latitude.

The heat budget of the oceans can be given as $Q_s - Q_b - Q_h - Q_e = 0$. Q_s is the heat received and equals 0.221 g cal/cm^2/min. It is transmitted by radiation from the sun and sky. Q_b is radiation back from the sea surface and equals 0.090. Q_h is the convection of sensible heat to the atmosphere and equals 0.013 in the same units, and Q_e is the heat of evaporation, equalling 0.118. These three together equal 0.221, and thus the budget is balanced. The budget applies between $70°$N and s, according to Mosby (1936). Heat can also be added to the oceans in various minor ways, such as convection of heat through the ocean floor, transformation of kinetic energy to heat, heating due to chemical processes, convection of sensible heat from the atmosphere, and condensation of water vapour.

Only a generalized picture of the surface temperature of the oceans can be given, because of the variability of the large number of factors which affect the temperature of the oceans. Table 2.2 reveals some interesting features of the Atlantic Ocean, given by Böhnecke (1938). The values are the average surface temperatures for each $10°$ latitude. One of the important points, which applies not only to the Atlantic Ocean, is the fact that the maximum temperatures occur north of the equator. The oceanic thermal equator lies north of the geographical equator. This fact has repercussions on the movement of water and other factors in the equatorial zone. The zone of maximum temperature moves with the seasons, but is rarely south of the equator. The temperatures in the southern hemisphere are considerably lower than those in the north. The large mass of the ice-covered Antarctic continent is an important source of cold in the south. The much smaller amount of land in the southern hemisphere is also significant in this respect.

The difference between the sea surface temperature and the air temperature over it is important. It influences the generation of waves, for example. In general heat is given off from the oceans to the atmosphere. The sea surface temperature is, therefore, on average warmer than the atmosphere above it. Between 20°N and 55°S in the Atlantic, at a height of 8 m above the sea surface, the air temperature is on the average 0·8°C cooler than the sea. There is, therefore, a fall in temperature in the layers immediately above the water surface. The heat given off by the oceans, however, varies greatly in winter

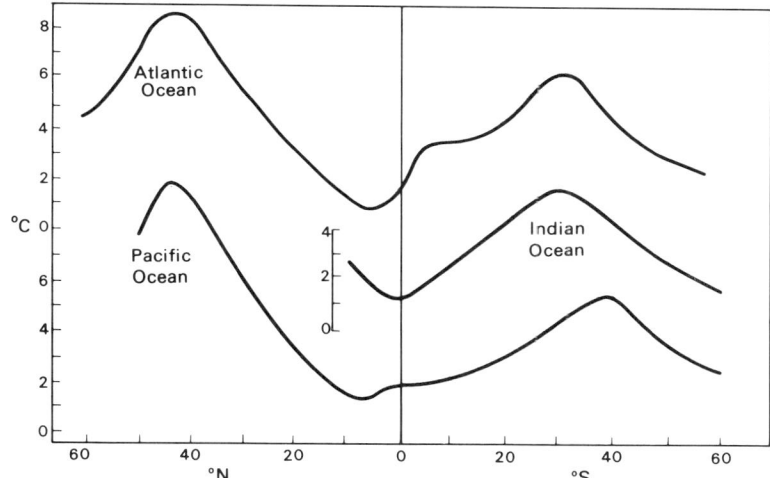

Figure 2.1 Average annual range of surface temperature with variation at different latitudes in the different oceans. (*After Sverdrup* et al., *1946*.)

and in summer in middle latitudes. It is this factor that causes the great difference of air temperature over the oceans and continents during the winter season in particular. The temperature for January is 22·2°C higher over the ocean between 20° and 80°N, while in July it is 4·8°C lower. The mean annual temperature is 7°C higher over the water meridian. This illustrates the important modifying effect the oceans have on the surface air temperatures over them.

Although sea temperatures do not vary so greatly as on land throughout the year, there is nevertheless an important seasonal fluctuation of surface temperature in the oceans. This varies with a number of variables, such as the ocean currents, prevailing winds, and so forth. It is generally true that zones of maximum variation occur in middle latitudes (figure 2.1). In the Atlantic Ocean the zone of maximum range of over 8°C occurs around latitude 40°N. From this position it falls off steadily to north and south. In the latter direction it reaches a minimum value of less than 1°C about 5°N. A smaller maximum range is reached in the southern hemisphere at 30°S.

In the north Pacific the maximum range is sharper and higher, reaching 10°C in latitude

40°N; the southern maximum range occurs further south at 40°s. These values are derived from the observations of Böhnecke (1938) for the Atlantic and Schott (1935) for the Pacific. The cold, continental winter winds of the northern hemisphere are an important factor in modifying the winter temperatures of the water on the east side of the oceans. The larger amount of land in the northern hemisphere is an important influence in this respect.

The relatively great depth to which heat is distributed in the oceans is one of the processes that makes them such an important source of heat. This is reflected in the annual variation of temperature with depth in the ocean. The change of temperature with depth depends on four factors: 1) the variation in the amount of heat absorbed, 2) the effect of heat conduction, 3) lateral displacement of the water by currents, and 4) vertical motion of the water. The variation in these four factors makes generalization impossible.

One example will illustrate the effect of some of the variables. Fig. 2.2A shows water

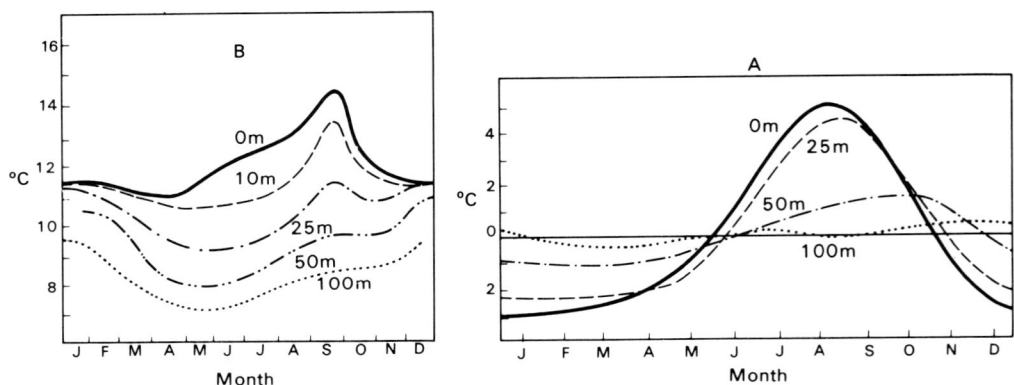

Figure 2.2 Annual variation of temperature with depth. **A** Off the Bay of Biscay in 47°N, 12°W. **B** In Monterey Bay, California. (*After Sverdrup* et al., *1946.*)

in the Bay of Biscay, studied by Helland-Hansen. The conduction of heat downwards is illustrated. The area was chosen because the type of water remained the same throughout the year. The surface layers at a position 47°N, 12°w showed a range of 3°c below the annual mean in January to nearly 5°c above the mean in August. At a depth of 25 m the maximum occurred three weeks later and was about 0·5°c less, while the minimum remained at about 2° below the mean from mid-December to April. At 50 m depth the maximum was reduced to just over 1°c above the mean and occurred between September and October, while the minimum was about 1°c below the mean and was still at about the same time. At 100 m depth the whole range about the annual mean was reduced to less than 0·5°c, the maximum taking place in early December and the minimum about March to April.

Other places show a greater range at 100 m depth owing to different circumstances, such as cold winter winds, while some places do not show a systematic variation relating to conduction, being influenced by different waters at different seasons of the year. For example off California, as shown in figure 2.2B, the range at the surface is about 4·5°c, while at 100 m it is still as much as 3·5°c during the year.

The relationship between air and sea temperature is responsible for sea-fog formation. Sea-fogs are formed when warm damp air passes over a cold sea surface. The sea surface temperature must be lower than the dew-point of the air, which is cooled below its dew-point, and fog is formed. Owing to the small diurnal variation of sea surface temperatures, sea-fog is not dependent on daily heating and cooling, and can occur with strong winds. Sea-fogs occur most frequently in spring and early summer, when air passing off the land may be warm, but the sea is still very cold. They are most common in high latitudes in summer. The Grand Banks of Newfoundland are notorious for their fogs, which result from the cold water of the Labrador current flowing alongside the heated land-mass and towards the warmer waters of the northwest Atlantic. Only north winds, which blow over the cold water, can prevent fog, which occurs about 50 per cent of days in summer. Similar conditions occur in the northwest Pacific, where the cold Kamchatka Current acts in the same way.

Coastal fogs also form where offshore warm winds from the land cause upwelling of colder water. Such coasts lie within the trade-wind belts, on the western side of land-masses, and are found along the coasts of California, Chile, southwest Africa, and Morocco. Around the British Isles fogs form when warm winds blow from the south or off the heated continent in spring and summer particularly. Fog is rare in tropical seas and does not often occur in the Mediterranean.

2.2 Salinity of surface water

The normal chemical content of sea water with a total salinity of 35‰ is given in table 2.3. Silver, gold, and radium also occur in sea water, but in minute proportions, which are

Table 2.3 Chemical content of sea water in g kg

Chloride	19·353
Sodium	10·760
Sulphate	2·712
Magnesium	1·294
Calcium	0·413
Potassium	0·387
Bicarbonate	0·142
Bromide	0·067
Strontium	0·008
Boron	0·004
Fluoride	0·001

respectively 0·3, 0·006, and 0·000,000,2 mg/metric ton or parts per thousand million. The proportion of the various elements remains very constant in sea water from place to place even when the total salinity differs.

Some of the important elements in sea water are the nutrients used by living organisms for growth and reproduction. They vary inversely in quantity with the abundance of living organisms at any place. These elements include silicon, nitrogen, and phosphorus, as well as arsenic, iron, manganese, and copper in smaller quantities. Where the nitrates and silicates and dissolved oxygen necessary for life in the sea cannot be replaced as they are used up, the sea is relatively barren, but where they are replaced by upwelling of water from below, the sea is fertile. The zones of fertile water are greenish blue, on account of the large number of small organisms in the water, where the deep blue of some tropical waters indicates paucity of organic matter and marine desert conditions.

In some areas the sea supplies the needs of common salt and other elements are becoming important. Magnesium is one of these, the oceans holding enough of this element to cover the dry surface of the earth to a depth of 60 m, while the salt would cover the dry earth's surface to a depth of over 150 m. Bromine is also extracted from sea water. Another element which would be very valuable as a fertilizer, if it could be extracted from the sea, is potassium.

Total salinity is used to describe sea water. The technique of measuring this has been speeded up and made more accurate. An electric salinity meter has been developed at the National Institute of Oceanography and in the United States. The former can determine the salinity to an accuracy of $\pm 0.003\%$ and about 200 determinations can be made in a day. Salinity is expressed as $0.03 + 1.805 \times$ chlorinity in parts per thousand. In the open ocean the salinity normally varies between 33% and 37%, with a mean of 35%. The value is lower where large rivers enter the sea, but in areas where the influx from the land is negligible, and where surface evaporation is great, as for example in the Red Sea, it may exceed 40%. At depth salinities range between 34.6% and 35%.

The surface salinity is closely related to the process of evaporation, by which the salts are concentrated. Thus conditions in which evaporation is great will lead to areas of high surface salinity. The salinity itself will exert some influence on evaporation, but the relationship between the sea and air temperature is much more important. When the water is warmer than the air evaporation will be facilitated by the instability of the lower layers of air, brought about by the rapid lapse rate near the surface. Evaporation will be greatest where the cold air flows over warm water. However, when the air is very much colder than the water it can contain relatively little moisture, and condensation takes place to form sea-smoke. This process is limited to inland and coastal areas, because only in these can the necessary extremes occur. When the sea is colder than the air, turbulence is reduced and evaporation is only possible if the air is not saturated.

The average annual rate of evaporation in the Atlantic falls off rapidly north of $40°N$, where it is 94 cm/year. It increases to 149 cm/year at $20°N$, falling off again at the thermal equator to 105 cm/year at $5°N$. It increases again to 143 cm/year at $10°s$, but falls off rapidly further south to 43 cm/year at $50°s$, according to Wüst (1935). Similar but rather smaller values hold for the Pacific Ocean. The values are related to the general climatic conditions. Subtropical high-pressure belts and trade-wind zones allow rapid evaporation, while the cloudy skies and calmer conditions of the equatorial zone diminish evaporation.

The difference between evaporation and precipitation shows a linear correlation with surface salinity. Precipitation is on the whole inverse to evaporation. Between $40°N$ and $35°s$ evaporation exceeds precipitation, except for a narrow belt around $5°N$. The curve for surface salinity (shown in figure 2.3 for all oceans) follows very closely the curve for net evaporation. It reaches a maximum at $25°N$ of 35.79%, and a minimum occurs at $5°N$ of 34.54%. A secondary maximum occurs in the southern hemisphere at $20–25°s$ of 35.69%. The values fall off rapidly to higher latitudes. Mixing with subjacent water also affects the surface salinity. The mean salinity at depths of 400–600 m is 34.60%.

The annual variation of salinity in the north Atlantic shows a maximum in March and a minimum in November, the values being 36·70 and 36·59‰ respectively. They correlate fairly well with changes in evaporation, confirming the close relationship between evaporation and surface salinity. The salinity of coastal waters in the southern North Sea is much higher in the water that has penetrated from the Atlantic, from the English Channel or around the north of Britain, than it is in the coastal water. This coastal water forms a belt of low salinity along the Dutch coast and across the southern North Sea from the Netherlands to East Anglia. The salinity in this coastal water falls below 34·2‰. The difference is due to the dilution of sea water by river water.

A good example of the effect of evaporation on salinity is found in the Mediterranean Sea. The water flowing into the Mediterranean is completely altered by external processes and flows out of the Mediterranean as a different water type. At the Straits of Gibraltar

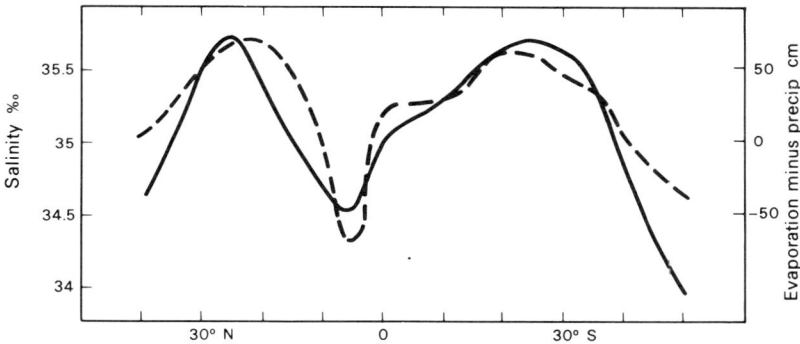

Figure 2.3 Average values of surface salinity (full line) and evaporation minus precipitation (dashed line) for all oceans. (*After Sverdrup* et al., *1946.*)

relatively fresh, low-temperature Atlantic water flows into the Mediterranean, while very saline, warm water flows out beneath the inflowing water. The increase in salinity more than counteracts the increase in temperature, thereby increasing its density relative to the inflowing water. These changes are due to the great increase of evaporation and the relatively low rainfall and runoff in the Mediterranean area, characteristics that become progressively more effective towards the eastern end of the basin. This is reflected in the increasing salinity and temperature eastwards.

The reverse occurs in the Black Sea. Rainfall and runoff exceed evaporation and the surface water becomes less saline and its density is decreased as a result. The lower layers of water become stagnant, as the surface water is never dense enough to sink to the bottom. The bottom water, therefore, lacks oxygen, but instead contains hydrogen sulphide, prohibiting normal life in the deeper waters of the Black Sea.

2.3 Oxygen content of ocean water

Oxygen in sea water is one of the essential elements for marine life. It also acts as a pointer to the movement of the water. It is acquired by sea water when it is at the surface, and becomes slowly reduced in amount with time since the water was at the surface. Oxygen and carbon dioxide are the two most important dissolved gases in the ocean. The amount of oxygen in water is much less than that in air, the values being respectively 9 ml/litre and 200 ml/litre. A study of the oxygen content of the deep waters of the oceans gives useful information concerning the rapidity of deep water circulation and the renewal of water. Table 2.4 gives oxygen values, obtained by various expeditions at different dates, for the Atlantic and Pacific Oceans. At depths between 2000 and 4000 m there is a general diminution of oxygen content southwards in the Atlantic Ocean. The Pacific values change in the reverse direction. Only in a latitude of nearly 60°S are the Pacific values higher than those of the Atlantic. The figures show that on the whole the oxygen content of the deep Pacific waters is considerably lower than that of the Atlantic, particularly in the northern hemisphere

Table 2.4 Oxygen content in Atlantic and Pacific Oceans

Depth, m	Atlantic Ocean					
	50°27·5′N 40°14·5′W	33°19′N 68°18′W	19°16′N 27°40′W	32°49′S 40°01′W	39°46′S 22°12′E	58°53′S 4°54′E
2000	6·30	6·08	5·07	4·71	4·70	3·37
2500	6·26	6·04	5·30	5·53	4·99	3·52
3000	6·17	5·99	5·27	5·65	5·14	3·59
3500	6·28	6·03	5·32	5·46	5·04	3·67
4000	6·34	6·06	5·42	4·88	4·97	3·79

Depth, m	Pacific Ocean					
	50°30′N 175°16′W	28°02′N 122°08′W	3°18′N 129°02′E	4°20′S 116°46′W	31°25′S 176°25′W	59°05′S 163°46′W
2000	1·64	1·89	2·56	2·53	3·32	4·27
2500	2·20	2·44	2·82	2·75	3·25	4·33
3000	2·58	2·72	3·15	2·86	3·75	4·37
3500	3·00	3·00	3·26	2·96	4·27	4·20
4000	—	—	3·27	3·00	4·52	4·06

The pattern can be partly explained by the sources of deep water and the general circulation of the oceans (discussed in chapter 3). The oxygen content is related to the age of the ocean water. A method of estimating the age of ocean water is by carbon 14 analysis; some of the results of this method are presented in table 2.5. On the surface there is a nearly uniform distribution of carbon 14/carbon 12 ratio between 50°N and 50°S, but the ratio is significantly lower in the Antarctic and slightly higher in the northern north Atlantic. In the deeper waters of the Atlantic the ratio of carbon 14/carbon 12 is 10 per cent lower than that in the atmosphere, except for three water types. The north Atlantic central water has a value intermediate between that of the deeper and the surface water. A tongue of western north Atlantic water has a 3 per cent higher value, and the deepest water in the eastern basin of the Atlantic is about 2·5 per cent lower in carbon 14 than the overlying north Atlantic deep water and the water on the other side of the ridge at the same depth. The Antarctic intermediate water and the Antarctic bottom water are both

Table 2.5 Age of ocean water

Surface water		Years corrected for bomb carbon 14 to 1854
North Atlantic	60–80°N	38
,, ,,	15–40°N	52
Caribbean		56
South Atlantic	0–40°S	63
Southwest Pacific	15–42°S	49
East Pacific	20°N–30°S	56
Average surface oceans		58 ± 10

Atlantic subsurface	Depth, m	Apparent age, years	Average carbon 14
North Atlantic central water	200–400	160 ± 70	−71
South Atlantic central water	200–400	less than 80	−61
North Atlantic intermediate	800–1100	400 ± 90	−99
Antarctic intermediate	500–1200	less than 350	−118
Upper north Atlantic deep	1200–2400	350 ± 160	−72
Upper north Atlantic deep	1400–2000	500 ± 90	−104
South Atlantic deep	1500–2200		−130
North Atlantic deep	2500–4000	500 ± 90	−105
Antarctic bottom	more than 4000	less than 250	−144

Pacific subsurface			
5–25°S, 170–180°W	200–500		−90
5–30°N, 115–130°W	300–600		−130
5–40°S, 170–180°W	700–1000		−115
5–25°S, 170–180°W	2000–3000	less than 350	−176
25–40°S, 80–145°W	3500	425 ± 150	−195
15–30°N, 115–135°W	2000–3500	925 ± 150	−233

lower in carbon 14 than the deep waters from the north. The conclusion from carbon 14 studies is that residence times in the Pacific may be of the order of 800 years and in the Atlantic of 500 years. There seems to have been a steady state during the last 2000 years. The use of radium for tracing movement of ocean water-masses may prove of considerable value, particularly in the relation between the age of surface and deep waters.

Another use that can be made of oxygen values is related to the O^{18}/O^{16} ratio. Oxygen occurs both as O^{18} and O^{16}, and the ratio between the isotopes in calcium carbonate foraminifera has been used to establish temperature fluctuations by Emiliani (1957), as mentioned in volume 1. Such studies can be carried back through long periods of geological time in cores of deep sea sediments.

3 Water-masses

3.1 Definition and concept

Just as the concept of air-masses in meteorology has greatly assisted the description and analysis of the weather, so in the oceans the concept of water-masses has proved fruitful. A water-mass is a relatively homogeneous body of water which can be described by its characteristic temperature and salinity determining its density. The density is usually expressed in terms of sigma$_t$, which refers to the density at atmospheric pressure and the temperature at which the sample was collected. The density of pure water at 4°c is unity, but because of the salt in sea water its density exceeds that of pure water. An average density is about 1·02575. This value is given in the form, density -1×1000 or $\sigma = (\rho - 1) \times 1000$, as 25·75. The density depends on the temperature, salinity, and pressure. Sigma$_0$ refers to the density the water would have at 0°c; in this form the density is only a function of salinity.

In order to describe the water-masses in terms of their temperature and salinity, Helland–Hansen (1916) devised the T–S diagram, relating these two variables. Characteristics of temperature and salinity may be plotted separately against depth, or as in the T–S diagram, they may be plotted against each other. The points so plotted usually fall on a

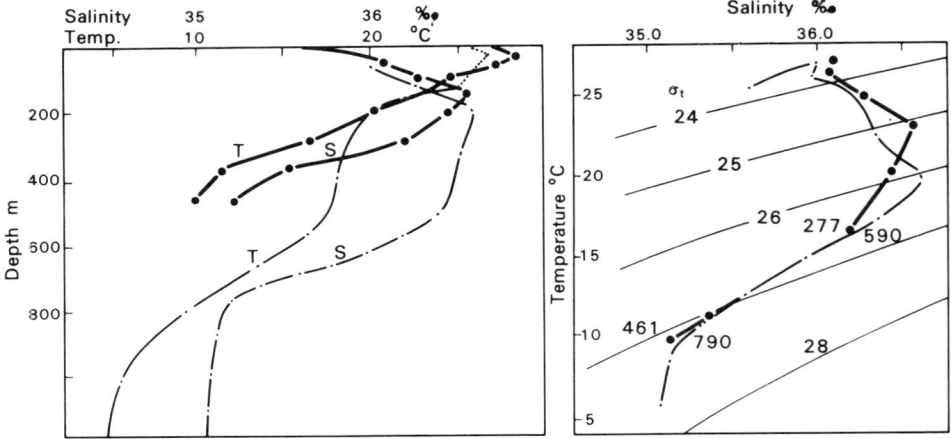

Figure 2.4 The left-hand diagram shows temperature and salinity observations plotted against depth in the Gulf Stream off Onslow Bay, and the right-hand diagram shows the same data plotted as a T–S diagram. The density lines are shown. (*After Sverdrup* et al., *1946.*)

well-defined curve, in which temperature and salinity decrease together. There is, therefore, only a slow increase of density downwards, as the two variables causing an increase of density are working in opposite directions. Sigma$_t$ values can be entered on the T–S graph as lines that slope gently across the diagram from right to left (figure 2.4). The curves show that the same water-mass is present at different depths, while the stability of the stratification can be seen from the relationship between the angle of the sigmat lines and the slope of the temperature–salinity figures.

A water-mass is defined by its characteristic T–S curve. If the water were entirely homogeneous, it would be possible to define it by one temperature and one salinity value. However, water-masses or water types are usually defined by a T–S curve. A completely homogeneous water-mass only occurs occasionally, for example in a basin where one water is found over a considerable depth range. A straight line T–S relationship results from the mixing of two different water-masses, each of which can be defined as a point on the T–S graph.

In considering the formation of water-masses the conditions at the surface are important, as the water attained its original characteristics of temperature and salinity at the surface. Modifications of the original T–S relationships come about by mixing with adjacent water-masses. Mixing takes place either laterally or vertically. In the former movement is along the sigma$_t$ surfaces, in the latter across them. The movement of one water-mass through the ocean can be traced by noting the changing extremes. The rate of change of its characteristics gives an indication of the amount of dilution a water-mass undergoes by mixing with other water-masses. The method of tracing water-masses by their extreme values of temperature and salinity is called the core method by Wüst (1935).

Cochrane, Pollak, and Montgomery have described the distribution of water in terms of the frequency of temperature and salinity relationships. Cochrane (1956, 1958) deals with the waters of the Pacific Ocean, Pollak (1958) discusses the Indian Ocean, and Montgomery (1958) considers the Atlantic and world oceans. The technique used is to work out the number of observations of temperature and salinity which fall within certain divisions of a T–S diagram. The ranges used were 2°c for temperature and 0·4‰ for salinity for the coarse scale graph, and 0·5°c and 0·1‰ for the fine scale graph. The results are worked out for different seasons of the year. Isopleths are drawn round the frequency values, which indicate clearly the most common values of temperature and salinity relationships at different seasons and in different areas.

There are many more observations of high temperatures on the surface of the Pacific Ocean in the north than the south. This result is partly due to the great width of the north Pacific in low latitudes. The inclusion of the Antarctic region accounts for the concentration of low temperatures in the south Pacific. In the north Pacific the salinities are concentrated in the higher values, with a secondary concentration at lower values, while in the south Pacific the salinity range is smaller. On the whole the densities of the north Pacific exceed those of the south Pacific.

The deeper waters of the Pacific Ocean have much lower temperatures. T–S values concentrate about 1·5°c and 34·65‰ salinity, 33 per cent of the observations fall between 1 and and 2°c and between 34·6 and 34·7‰. The 50 per cent boundary does not cover a much wider range. Water having these characteristics occupies much of the Pacific basin below a few hundred metres in the Antarctic and below 2000 m elsewhere. It is designated Pacific deep water. The bulk of the water in the Pacific is very homogeneous. A secondary peak occurs within the 75 per cent boundary around the T–S value of 4·25°c and 34·55‰. This water is found across the tropics at depths of 700–1100 m, and is called tropical water. A third mode occurs with a temperature of 0·25°c and 34.05‰ in water in the Sea of Japan. It is one of the very few definitely Pacific deep waters and fills the Sea of Japan below 400 m to the bottom. It results from cooling in cold winters and attains great depths because the main Pacific deep water cannot enter the Sea of Japan.

Pollak has shown that there is a similar homogeneity in the deep water of the Indian Ocean. Almost one-third of the total volume lies between 0·5 and 2·0°c and 34·7 and 34.8‰, a secondary maximum occurring with a similar density but lower temperature and salinity. The former is the main deep water of the Indian Ocean, and similar T–S relationships are found over a range of latitude of 66°. The Red Sea and Persian Gulf add a small quantity of very saline water, ranging from 38·5 to nearly 44‰.

Montgomery's results for the Atlantic include all the adjacent seas, which show up on the graph as separate entities. They have very distinctive salinity characteristics—high in the Mediterranean and low in the Black Sea. The 50 per cent line in the finer scale diagram shows two maximums. The major one, at 2·25°c and 34·95‰, represents the north Atlantic deep water; the other at 0°c and 34·65‰ represents south Atlantic water, including Antarctic bottom water.

The relative homogeneity of the three oceans may be compared by means of the number of fine classes that cover 50 per cent and 75 per cent of the total observations. The values are as follows:

	50%		75%	
Pacific Ocean		5 classes		20 classes
Indian Ocean		8		33
Atlantic Ocean		11		44
World ocean		12		43

The figures show that the Pacific is the most homogeneous and the Atlantic the least. The reason for this will become apparent when the formation of the different water-masses is discussed. Calculations for the world ocean show that for the coarse scale class, 75 per cent of the entire ocean falls into 3 classes, from 0 to 6°c and 34 to 35‰. This is a small range considering that the total range of the ocean lies between −2 and 36°c and from 0 to 40‰. Most of the ocean water is remarkably homogeneous. Within this general homogeneity, however, a number of separate water-masses can be differentiated and these begin to emerge when the modes for the 50 per cent boundary for the fine scale diagram are analysed. The mean density is greatest in the Atlantic, intermediate in the Indian Ocean, and least in the Pacific Ocean, while the world ocean falls between the Pacific and Indian Oceans.

Table 2.6 Surface and deep water-masses

Surface water-masses	Arctic surface water
	North Atlantic central water
	South Atlantic central water
	Subantarctic water: all oceans
	Subarctic Pacific water
	Western north Pacific central water
	Eastern north Pacific central water
	Pacific equatorial water
	Western south Pacific central water
	Eastern south Pacific central water
	Indian equatorial water
	Indian central water
Surface and deep water	Antarctic circumpolar water: all southern oceans
Deep water-masses	Antarctic intermediate water
	Intermediate water: north Atlantic
	Intermediate water: north Pacific
	Mediterranean water
	North Atlantic deep water
	Arctic deep water
	Antarctic bottom water

3.2 Method of formation and character of the major surface water-masses

Water-masses are stratified according to their density, the densest sinking to the greatest depth. Only in high latitudes is it cold enough for deep waters to form. These dense waters occupy most of the ocean basins, but on the surface, floating above the deep water, are the shallow surface water-masses. Some of the water-masses have been given different names by different workers, but the better-known names, as given by Sverdrup, Johnson, and Flemming (1946), will be used. The water-masses are listed in table 2.6.

Distribution of the sources of deeper surface water-masses is shown in figure 2.5. The central water-masses cover a wide area of all the oceans. In the Atlantic Ocean, northern and southern central water-masses can be differentiated. The difference between them reflects the difference of surface temperature and salinity.

3.2a North Atlantic central water

This water covers a wide area of the northern part of the ocean, but is restricted to shallow depths. The T–S relationship can be described by an almost straight line on the T–S graph from T $8°c$ and S $35 \cdot 10\%_0$ to T $19°c$ and S $36 \cdot 70\%_0$ shown in figure 2.6. Towards the equator this water is replaced by water that has crossed from the south Atlantic. It is similar to the west of the Bay of Biscay and east of Cape Hatteras on either side of the Atlantic. This large body of water, formed in the north Atlantic, attained its characteristics at the surface. The vertical arrangement of temperature and salinity closely resembles the northward decrease of these variables in the horizontal plane. This water probably originated by sinking along the correct density plane during winter cooling. Some lateral and vertical mixing may also have taken place. The thickness of the water-mass depends partly on the currents, which are all confined within it on the surface. It is thicker on the right-hand side of the currents and thinner on the left of their direction of flow, owing to the earth's rotation. Its maximum thickness of 900 m is reached in the Sargasso Sea, but over much of its extent the thickness does not greatly exceed 200 m.

The north Atlantic central water-mass is formed between latitudes 30 and 40°N at the subtropical convergence. This is a wide zone in which currents tend to converge owing to the deflection of wind-driven currents to the right in the northern hemisphere. Currents generated by the northeast trade winds are deflected northwest, where they meet water directed southeastwards in the west-wind zone north of the trade winds. At the subtropical convergence, the surface water increases rapidly in density towards the north. This zone of convergence (being primarily dependent on the winds) is much more variable with the seasons than the Antarctic convergence.

3.2b South Atlantic central water

This water also covers a wide area. In general it is rather similar in its T–S relationships to its northern counterpart, but it is cooler and less saline than the northern one. It is defined by a nearly straight T–S curve with values of T $6°c$ and S $34 \cdot 5\%_0$ to T $18°c$ and S $36 \cdot 0\%_0$. There is a similar range between the vertical and horizontal change of temperature and salinity between 30 and 40°s. This is the region of the southern subtropical convergence. The water sinks at this convergence and spreads northwards along its correct density surface. The water-mass extends from coast to coast in the south Atlantic, and also reaches north of the equator, because the thermal equator lies at 5°N. Like its northern

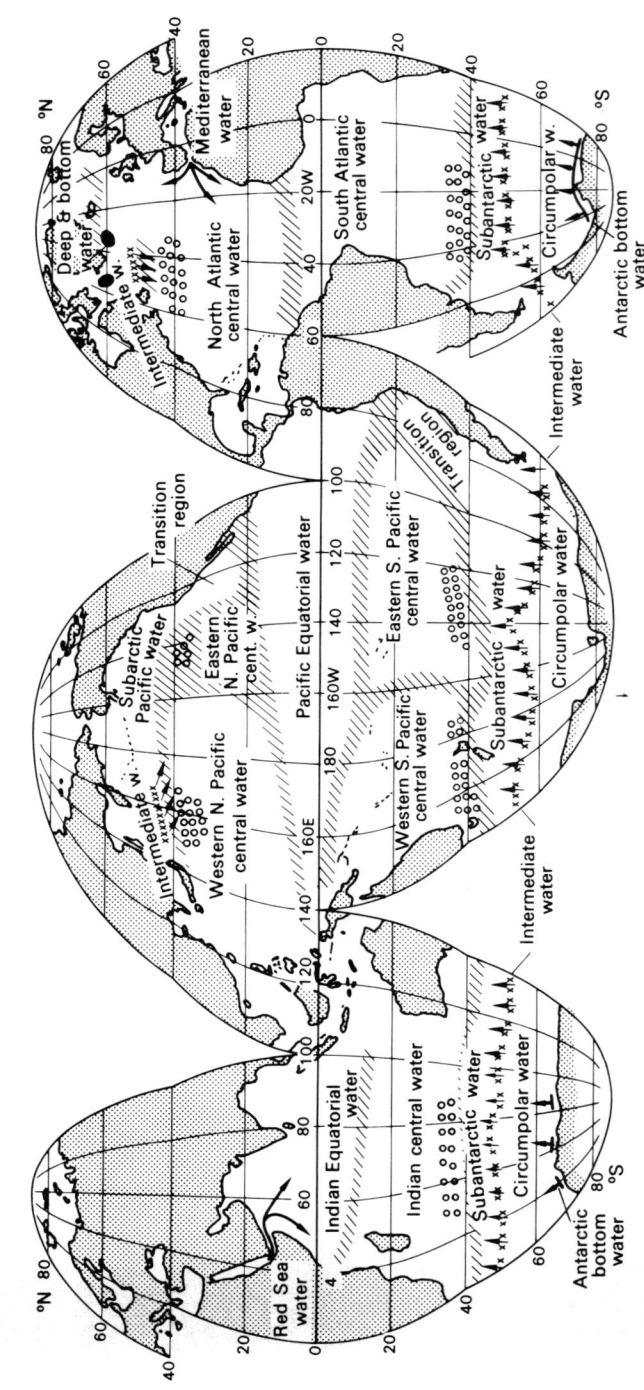

Figure 2.5 Map to show the distribution of surface-water masses and the source of some of the deep water masses. The circles indicate regions of Central Water formation, and the regions of Antarctic and Arctic Intermediate Water formation are shown by crosses. (*After Sverdrup* et al., *1946*.)

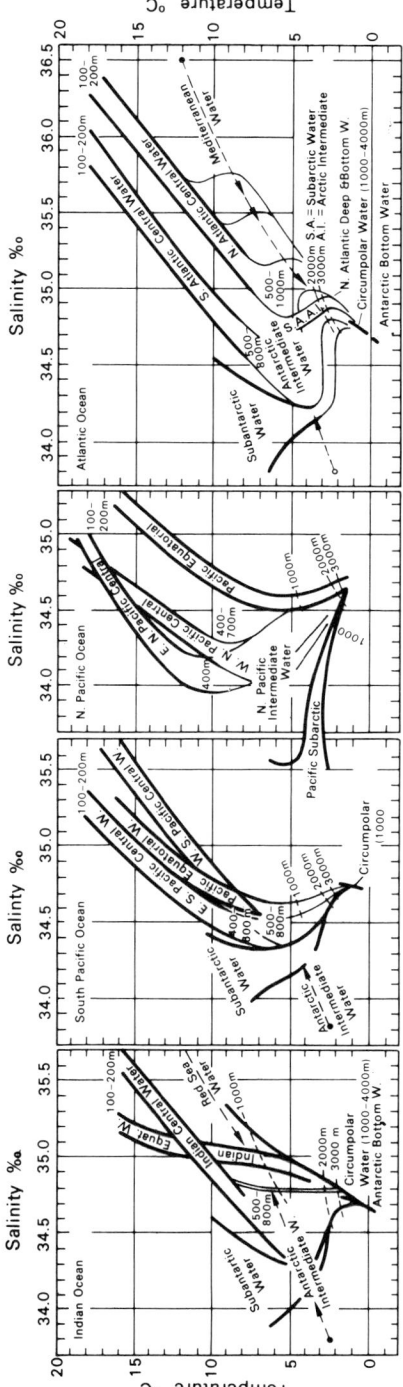

Figure 2.6 Characteristics of temperature-salinity relationships for the principal water masses. (*After Sverdrup et al., 1946.*)

counterpart, the water is shallow, rarely exceeding 600 m in thickness. The main surface currents flow in this surface layer.

North and south of the central water-masses the character of the water on the surface differs considerably from that found in lower latitudes. The area has been termed the Arctic Mediterranean sea in the zone of the Norwegian Sea north of the Wyville Thomson Ridge. It adjoins the Barents Sea and north polar basin. The water flowing into it from the south is relatively warm and saline, compared with the water flowing out, the mean values being 35·30‰ and 32‰ respectively. Excessive precipitation, runoff, and melting ice all help to cool and freshen the water, changes opposite to those taking place in the Mediterranean proper.

3.2c   Arctic surface water

In the North Polar Sea there are three water-masses. The Arctic surface water, the Atlantic water, and the Arctic deep water are found in that order in depth. The main character of the surface water is its low salinity, particularly near the mouths of the large Siberian rivers, where it is very low. To the north of Svalbard it attains 32–33‰. On the north Siberian shelf the salinity is 29·67‰ in May (its maximum value), but as ice continues to melt during the summer a layer of almost fresh, cold water overlies the more saline water. This cold, fresh water rests on top of water of Atlantic origin, which occurs at depths between 75 and 400 m. The salinity of the latter water is 35·10‰ and it has a temperature of 3–4°c.

3.2d   Subantarctic water

This water is found at the surface to the south of the subtropical convergence in the south Atlantic. It lies between this convergence and the Antarctic convergence, at about 52–53°s. The water-mass is transitional between the central water to the north and the circumpolar water to the south. A similar water is only developed to a very minor degree in the north Atlantic. The subantarctic water has a relatively low salinity and mixing with vertical circulation plays a part in its formation. It can be traced all round the Antarctic continent and has very similar characteristics in all the southern oceans.

3.2e   Subarctic water

The Pacific is much larger than the Atlantic, and its central water-masses are correspondingly diverse, falling into two classes in each hemisphere. In addition, there is a large area of equatorial water. The subarctic water-mass of the north Pacific is much better developed than its Atlantic counterpart, and is one of the important water-masses of the Pacific. Its average temperature in latitude 50°N is between 2 and 4°c and it has a low salinity of about 32‰ at the surface, increasing to about 34‰ at depths of a few hundred metres. As it is traced eastwards towards the coast of California, excess evaporation and heating raise its salinity and temperature, forming a transitional water.

3.2f   The central water-masses of the Pacific

These form a western and eastern type, each of which has a rather similar range to its Atlantic counterpart, but with a lower salinity. The western type is more extensive than the eastern. Table 2.7 gives the salinities at a temperature of 16°c in the different water-

Table 2.7 Salinity of Pacific, Atlantic, and Indian Ocean surface water-masses at 16°C

	‰
North Atlantic	36·12
Western north Pacific	34·67
Eastern north Pacific	34·62
South Atlantic	35·64
Western south Pacific	35·55
Eastern south Pacific	35·08
Indian Ocean	35·62

masses. The lowest values for salinity occur in the central masses of the north Pacific. These values can be correlated with the distribution of surface temperature and salinity.

3.2g Pacific equatorial water

This water forms a wedge between the central water-masses. It is found at the surface in a broad tongue pointing to the west and covers the zone from 20°N to 20°s in the eastern part of the ocean, tapering to a width of only a degree or two north of the equator off New Guinea. The water-mass is very uniform. The T–S curve is almost straight, ranging from T 15°C and S 35·15‰ to T 8°C and S 34·6‰. At a depth of 800 m the temperature is 5·5°C. The equatorial current system takes place within this water-mass.

The south Pacific central water-masses also split into an eastern and western type. These two waters cover roughly similar areas of the southern part of the ocean. The western central mass is practically identical with that of the Indian Ocean central water-mass. The T–S relationship again forms a nearly straight line, with the values ranging from T 15°C and S 35·50‰ to T 8°C and S 34·60‰. In the eastern south Pacific central water temperature lies between 10 and 18°C, and the salinity is 0·50‰ lower than in the west. These central water-masses form at the subtropical convergence in the same way as those of the Atlantic.

3.2h Subantarctic water-mass

This water in the south Pacific lies in a broad belt south of the central water-masses. The belt is continuous with that in the Atlantic and Indian Oceans south of 40°s. In its upper layers the salinity is 34·20‰ to 34·40‰, and the temperature is between 4 and 8°C. Like its northern counterpart, it extends equatorwards along the Pacific coast of South America, where its temperature and salinity are similarly increased by heating and excess evaporation. In this transitional form it reaches a little north of 20°s. It can be followed westwards into the Indian Ocean at 40°s and 50°s between the subtropical and subantarctic convergences. Its character is very similar to the same water-mass in the other oceans.

3.2i Indian Ocean central water

This water lies to the north on the surface of the subantarctic water-mass. Its character has already been mentioned and is similar to that of the western south Pacific central water. It is very uniform throughout the width of the Indian Ocean. There is, however, a rapid change at the surface at the southern boundary of the water-mass near the subtropical convergence. The water is formed by sinking at this convergence. Its vertical characteristics to a depth of 800 m resemble closely the horizontal variation at the convergence. Its northern limit is less clearly defined as it grades into the Indian equatorial water, which occupies the upper layer of the Indian Ocean, north of about 10°s.

3.2j   Indian Ocean equatorial water

This water is reasonably homogeneous. The T–S graph is almost straight from T 4°c and S 34·90‰ to T 17°c and S 35·25‰. Equatorial water has a higher surface temperature than central water. In the northern Indian Ocean the water is slightly modified at depth by the addition of very saline water from the Red Sea. Evaporation greatly exceeds runoff from the land and precipitation in the Red Sea, producing highly saline, warm water, with a T value of 21·5 to 22°c and S 40·5 to 41‰ at a depth of 750 m. On the whole, the Indian equatorial water covers a much smaller range of salinity over the same temperature range than does the central water.

Warren, Stommel, and Swallow (1966) studied the water-masses in the Somali basin during the southwest monsoon of 1964. Their results are summarized in table 2.8.

Table 2.8 Somali basin water-masses

Water layer	Layer thickness, m	Volume transport 10⁶ m³/sec	Renewal time days Boundary current	Whole region	Boundary current width, km
Surface	200	30	20	86	150
Subtropical subsurface water	300	5	120	778	150
Red Sea water	300	5	120	778	150
Deep water	2000	25	158	1037	100
Full depth	4000	35	332	1736	

(Seasonal change)

The water-masses that have been discussed so far are those which form at, and remain fairly near, the surface of the ocean, owing to their relatively high temperature compared with the waters beneath them. Their low density prevents them sinking to greater depth. However, in a few special circumstances deep water-masses of great density (which can sink to the bottom of the ocean) are formed. Because deep water-masses can form in some areas, there must be upwelling in others. Although upwelling is probably widely distributed throughout the oceans on a small scale, it is also sufficiently concentrated in one zone to give rise to a definite water-mass. This upwelling water forms the only major surface water-mass yet to be mentioned.

3.2k   The Antarctic circumpolar water

This water is found in a belt encircling the Antarctic continent in high southern latitudes. It lies between the Antarctic convergence at about 52–53°s and the Antarctic shoreline.

The water is found a short distance below the surface, and it has a maximum temperature at a depth of 500–600 m. Its temperature is normally above 0·57°c and its salinity slightly above 34·7‰. The salinity maximum varies from 700 to 1300 m depth. The water is very uniform both horizontally and vertically, slowly increasing in temperature and salinity downwards. Throughout the water-mass the temperature does not often exceed 0–2°c, while below 800 m the differences in salinity fall within a range of 0·1‰. The uniformity with depth makes it possible for the upwards movement. In the Atlantic, at depths between 1000 and 4000 m, the salinity is a little above 34·70‰, increasing rapidly east of 20°E, where the water mixes with saline water from the north Atlantic. The relatively high salinity can be traced into the Indian Ocean, giving a value of 34·76‰ south of Australia at 140°E. In the Pacific, however, the salinity is again decreased due to the admixture of Pacific deep water, which has a low salinity.

The temperature behaves in a similar way. Water moving into the Atlantic from the west has a lower temperature, but it increases at about 10°E. It exceeds 2·5°c between 60 and 120°E, due to admixture of fairly warm Atlantic and Indian Ocean water. The cooler deep Pacific water causes a fall of temperature as the water is traced into the Pacific.

This water-mass incorporates parts of several different water types, which come together and mix in high southern latitudes. They move slowly towards the surface, thus accounting for the homogeneity with depth. The oxygen values show that the water cannot travel around the Antarctic continent without renewal. The water entering the Atlantic from the west has an oxygen content of 3·7 ml/l at about 1200 m. The amount increases eastwards to 4·8 ml/l east of the South Antilles arc. There is then a slow decrease at most depths further eastwards. The Antarctic circumpolar water has a rather higher oxygen content than the deep water further north. This is due to mixing with more aerated waters, which are nearer their surface sources. The Antarctic bottom water is one of these, as this has a high oxygen content above 4·6 ml/l. Circumpolar water is the only major water-mass which is not formed by sinking, but instead is the result of subsurface mixing and subsequent rising.

3.3 Deeper water-masses

Deeper water-masses can be divided into two main classes—the intermediate waters, and the deep and bottom waters. The former are not as dense as the latter and therefore only sink to intermediate depths. The intermediate waters are formed in high latitudes in all oceans, in varying amounts and having different properties.

3.3a Antarctic intermediate water

This is the best known of the intermediate water-masses. It is formed at the Antarctic convergence. This area of convergence, at about 50°s, is one in which there is a rapid increase of surface temperature away from the Antarctic continent, values rising 1–3°c in winter and 4–6°c in summer.

The convergence, or polar front as it is sometimes called, is situated in the zone where the deep warm water climbs southwards towards the surface, while lighter, colder Antarctic surface water sinks northwards above it. The position of the front according to Deacon (1959) depends on the volume of northward-flowing Antarctic bottom water. The position of the convergence (figure 2.7) is plotted from surface observations of temperature; it also

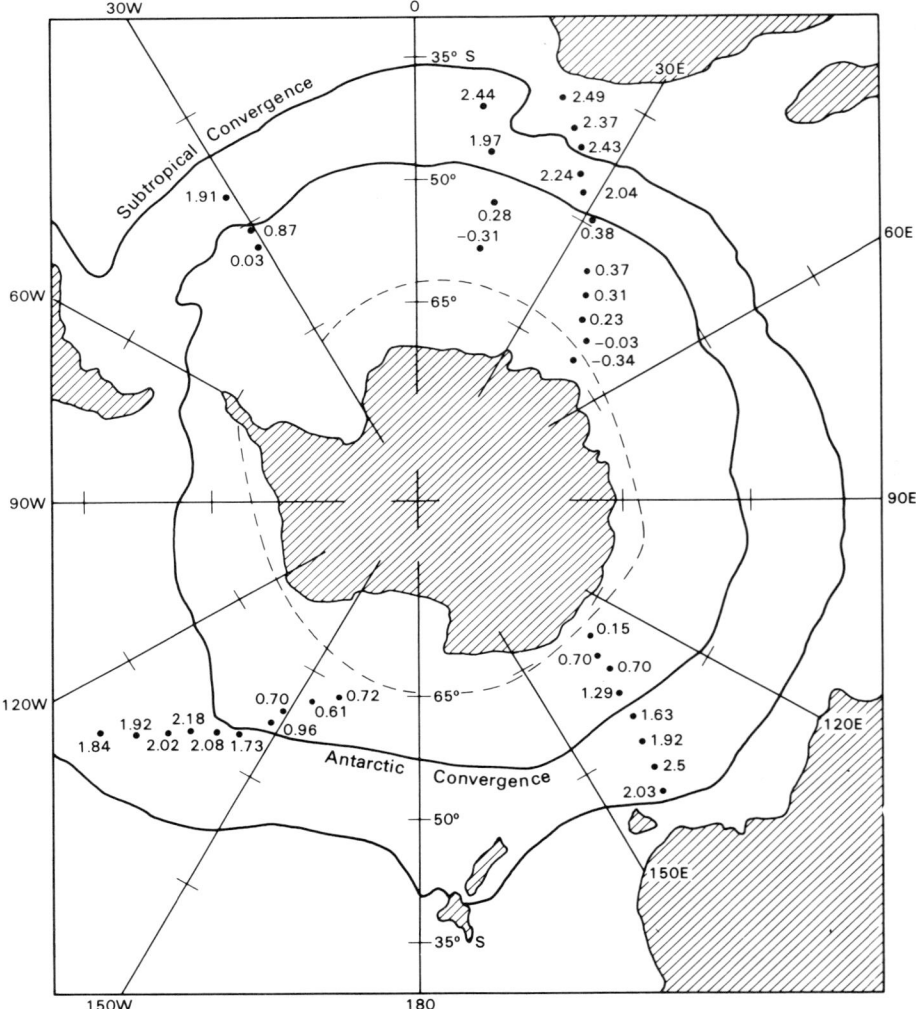

Figure 2.7 Map to show the southern hemisphere convergences. The dashed line indicates the boundary between the easterly and westerly drifts around the Antarctic continent. The figures give the temperature in °C at a depth of 2500 m. (*After Deacon, 1959.*)

agrees closely with temperature change at 2500 m (shown in the figure). These values suggest a steep slope above the bottom water in this zone. The change is sharpest where the greatest volume of bottom flow is indicated in the Atlantic. The greater volume of bottom water in the Atlantic may help to explain the more northerly position, by about 10° latitude, of the position of the convergence in the Atlantic.

Another line of argument relates the convergence to the surface currents and winds. The meridional gradient of atmospheric pressure is greatest at the convergence; which leads to a convergence of surface water in this zone. It can be argued, however, that it is the influence of the surface temperature that causes the greater atmospheric pressure gradient. This in turn could be due to the amount of bottom water moving north, the formation of which is greatest in the favourable conditions of the Weddell Sea in the Atlantic.

The convergence is remarkably stationary, both seasonally and for longer periods (see chapter 3). Its position would be expected to change seasonally if it were related to the surface wind. It does, however, exert a strong influence on the climate in the vicinity. The climate is very much colder in South Georgia, in latitude 54°s, which is in nearly the same latitude as Staten Island, off Tierra del Fuego. The latter is north of the convergence, while the former is south of it. Staten Island is clothed with luxurious vegetation, while South Georgia is much more barren. The tree-line of Tierra del Fuego is as low as the snow line of South Georgia. The position of the Antarctic convergence also affects the marine life of areas on either side of it.

Water sinks in this zone of convergence to form the Antarctic intermediate water-mass, which has a temperature between 2·2 and 7°c and a low salinity of 33·8‰, giving it a density of $\text{sigma}_t = 27\cdot0$. Its essential character (by which its movements can be traced as it sinks and moves northwards) is its low salinity. In the south Atlantic it lies below the central water, and can be recognized by its low salinity at a depth of about 800 m. It is better developed on the western side of the ocean in the south and has been traced into the north Atlantic, where its density is still about 27·4 just north of the equator. It does, however, show evidence of mixing with waters of roughly the same density. Some evidence of the corresponding water-mass of the north Atlantic is also found.

3.3b   North Atlantic intermediate water

This mass is only formed in small quantities. It has a salinity of about 34·88‰ and a temperature of 3·5°c. This water does not spread as far as its southern counterpart.

3.3c   Pacific intermediate water

This mass, however, forms a large body of water. It does not form in such high latitudes as the Antarctic intermediate water, nor does it have the high oxygen content that characterizes this water and the north Atlantic intermediate water. The Pacific intermediate water is formed about latitude 40°N and is found below the central waters of the north Pacific, being characterized by a salinity minimum. It is found at depths of 600–800 m in the northwestern part of the ocean; but nearer the equator in the eastern Pacific, where it divides, one layer is at 200 m and the other at 900 m. Low oxygen values, particularly off the American coast, are found above 1000 m at 400–500 m. This water, low in oxygen, sometimes rises almost to the surface, where the oxygen value may be less than 0·1 ml/l. In this area the circulation at this depth seems to be very slow.

3.3d   Mediterranean water

Because of its special quality, this water can be traced for great distances. Reasons for its characteristics have already been mentioned. The water-mass is one of the denser ones owing to its very high salinity. As it leaves the Straits of Gibraltar it has a salinity of over 37·00‰ and a temperature of 13°c. Intense mixing takes place in the Straits, as these values are considerably modified from those characteristic of the deep water of the Mediterranean itself.

The Mediterranean Sea can be divided into four basins, each of which has its own characteristic type of deep water. There is also a recognizable surface layer, intermediate layer, and a transition layer. The intermediate layer is lacking where the deep water is formed. Where it is present it has a salinity maximum and is composed of the surface water of the eastern Mediterranean, flowing westwards beneath the fresher, inflowing Atlantic water. Its salinity is obtained in the eastern part of the sea, where the excess of evaporation over runoff and rainfall reaches a maximum. This high salinity, despite its high temperature, enables the water to sink below the surface water. The transition layer between 600 and 1500 m links the intermediate and deep water. It is a zone in which the temperature falls off more rapidly than the salinity.

Deep waters are formed in the northern part of the sea, where cold winter winds (such as the mistral in south France and the bora at the head of the Adriatic) lower the surface temperature sufficiently in winter for the water density to be increased enough for it to

Table 2.9 Mediterranean water characteristics in the deep basins

Basins		Temperature °C	Salinity ‰
Algiers–Provençal	West	13·00	38·39
Tyrrhenian		13·10	38·44
Ionian	East	13·57	38·65
Levantine		13·62	38·66

sink to the deep basins. It is probably only in extra cold seasons that the temperature is lowered sufficiently for bottom water to form.

The oxygen content and the amount of inflow and outflow through the Straits of Gibraltar indicate that renewal takes place at such a rate that on average the whole water of the Mediterranean is renewed every 75 years. The oxygen decreases from east to west in the intermediate water, which confirms its westwards travel. Values of temperature and salinity in the deep water suggest that this water forms most readily in the basin west of Corsica and Sardinia. The character of the deep water (table 2.9) for the four basins reflects the external circumstances which govern the formation of this very distinctive water-mass. The further east the basin is, the warmer and more saline its water is, and the greatest difference being between the two western and the two eastern basins, owing to the shallow straits between them southwest of Sicily.

The density of the very saline Mediterranean water is high. It therefore takes part in the deep water circulation of the Atlantic and can be readily recognized by its high salinity, which can be traced as far as the tip of South Africa at least. The amount of water which enters the Atlantic from the Mediterranean has been estimated as 1,680,000 m^3/sec. This is an appreciable quantity of the total deep water formation in the north Atlantic Ocean, estimates of this proportion varying between one-third and one-tenth of the whole. Thus the Mediterranean plays a greater part in the general oceanic circulation than other adjacent seas. In the Atlantic, Mediterranean water mixes with waters whose densities lie between sigma$_t$ 27·5 and 27·7. Its density is nearest to that of the north Atlantic deep water and it lies below the Antarctic intermediate water.

3.3e North Atlantic deep water

This water occupies the bottom of the north Atlantic Ocean basin and is the densest water in this area. In the south Atlantic it is found as a deep water above the even denser Antarctic bottom water. North Atlantic deep water is formed in the north Atlantic in the region on either side of Cape Farewell, south Greenland, where the warm, saline waters of the north Atlantic drift come into contact with the cold currents flowing south from the Arctic Ocean on either side of Greenland. This surface mixture is cooled in winter, increasing its density, and allowing it to descend to depths greater than 1000 m. Its sigma$_t$ value reaches a maximum of 27·88, which is rather greater than that of the densest Mediterranean water. Its temperature lies between 2·8 and 3·3°c and its salinity is in the range 34.90 to 34·96‰.

Corresponding deep water is not formed in the north Pacific, owing to the absence here of cold currents flowing out of the Arctic Ocean as a result of the very narrow and shallow Bering Straits. This has important repercussions on the general circulation of the oceans. It has been estimated that on the average, north Atlantic deep water forms at the rate of about 2 million m^3/sec on either side of Greenland. More recent views, however, suggest that the amount of deep water forming in this area may be as much as 20 million m^3/sec. Whatever the exact amount of water formed, it is generally agreed that this (with the Mediterranean water) is one of the most important sources of deep water formation in the whole world ocean.

The rate of formation given for the deep water is an average. It only forms in winter. Its formation at sufficient density to sink to the bottom of the north Atlantic has been discussed by Worthington (1954a). He has come to the conclusion that the water that now fills the deep north Atlantic basin was formed a long time ago. A study of the oxygen content showed that values at 2500 m obtained in 1921–22 and 1931–33 when compared with those obtained during 1947–54 were lower by 0·3 ml/l. Profiles throughout the ocean showed a systematic drop in the value of the oxygen between 1922 and 1953 at depths of 1500–5000 m. This loss of oxygen also appears to occur in the eastern Atlantic off the Mediterranean at the 2000 m level, which is the only one for which earlier values are available. The rate of oxygen loss suggested by these data is 0·015 ml/l each year. Assuming that this loss has continued at the same rate since the water was saturated at the surface, the data can be extrapolated backwards in time. The results suggest that the water now found in the deep basins of the Atlantic was at the surface about 1810. The first decade of the nineteenth century was abnormally cold in northern Europe, and the northeastern United States was also affected. It seems likely, therefore, that the present north Atlantic deep and bottom water was formed during this cold period. Never since has water of sufficient density been formed to replace the bottom water. This one water-mass occupies about half the Atlantic Ocean. It must have formed by the mixing of water of different origins to form a new and homogeneous water-mass. Its movement can be traced partly by the very saline Mediterranean water, which mixes with it to give it a salinity maximum south of the Straits of Gibraltar.

Water of north Atlantic origin also finds its way into the deep basin of the Arctic Ocean over the sills between the Atlantic and the Norwegian Sea and the Arctic Ocean. There is no means whereby water with the characteristics of deeper polar water could form within the Arctic Ocean, but this Arctic water is very similar to the deep water of the Norwegian Sea. Its temperature is $-0.85°c$ and its salinity is $34.93‰$; it therefore probably is derived from the Norwegian Sea.

Lee and Ellett (1967) discuss the water-masses of the northwest Atlantic Ocean by means of potential temperature and salinity diagrams. There are three water-masses below 750 m. The Labrador Sea water overlies the northeast Atlantic deep water, which comes from the overflow across the Scotland–Icelandic Ridge; the northwest Atlantic water is at the bottom. This lowest water-mass is derived from the mixing of the water overflowing the Iceland–Greenland Ridge, the Atlantic water, and modified Labrador Sea water. The two deeper layers mix to give the north Atlantic deep water. The bulk of the water is overflow water from the deep basins in the Arctic, flowing across shallow sills and sinking to great depths beyond. The contribution from northeast Atlantic water has also been estimated by Worthington and Volkmann (1965), using Swallow floats. They suggest that the westward flow at 53°N 36°w is 4.6 million m³/sec, moving from the Atlantic to the Labrador Sea. The current is irregular and volumes vary between 2.0 and 7.6 million m³/sec. This water is found at depths below 1700 m in the northwest Atlantic and Labrador Sea. In the Irminger Sea it is between 1750 and 2470 m; in the Labrador Sea it is at 2500–3000 m; and in the southern part of the northwest basin it is at 2200–3500 m.

3.3f   Deep water of the Pacific

This water is derived from distant sources, hence its relatively low oxygen content. In the south Pacific the deep water shows a slow increase of salinity, but it remains constant below 2500–3000 m. A similar zone of constant salinity is found in the lower 1000 m of the north Pacific. The oxygen is usually lower in the north Pacific than the south in the deeper layers, suggesting that the northern water has travelled further and longer from its source, either in the north Atlantic, Mediterranean, or the Antarctic.

3.3g   Antarctic bottom water

This is the densest of all water-masses and is formed mainly in the Weddell Sea off the Antarctic continent from very cold surface water, the density of which is further increased by an increase of salinity due to freezing. It has an equal mixture of circumpolar water and shelf water. The latter has a temperature of $-1.9°c$ and a salinity of about $34.62‰$, while the former has a temperature of $0.5°c$ and a salinity of $34.68‰$. The resulting water, at a level below 4000 m, has a temperature of $-0.4°c$ and a salinity of $34.66‰$, where its density is 27.86. The shelf water, which sinks along the continental slope, has a higher density of 27.89, but this is slightly decreased as it mixes with the circumpolar water of density 27.84.

A small amount of a similar water may form off the Antarctic in the Indian Ocean sector, but there is little evidence that similar bottom water is formed in the Pacific sector. The amount forming in the Weddell Sea has been estimated by Stommel to be about the same volume as that forming in the north Atlantic, which is about 20 million m³/sec. In general the water of the Antarctic bottom water-mass becomes more saline and warmer as it is traced eastwards around the Antarctic continent.

4   Ice in the ocean

Ice in the sea is of importance in several respects. The danger to shipping of floating icebergs has led to the initiation of a system of iceberg warning in the seas around Newfoundland, where they are particularly menacing on account of the frequency of fogs. The formation of sea-ice is a limiting factor in the use of many northern sea routes and ports. The distribution of sea-ice is closely related to climatological processes.

Ice in the sea is divisible into two main types: icebergs (originating on the land) and sea-ice or pack-ice (which forms directly by the freezing of sea water). Icebergs are a major menace to shipping as they can reach fairly low latitudes at times, while pack-ice freezes up harbours in winter.

4.1   Sea-ice

If the sea were to freeze very slowly indeed, the resulting ice would be of very low salinity, as the crystal lattice that formed would be selective, accepting only hydrogen and oxygen. Usually, however, sea water freezes too quickly for this to occur. Newly formed sea-ice can have a salinity as high as 20‰ if freezing takes place at -30 to $-40°C$. Annual sea-ice normally has a salinity of about 4‰, 90 per cent of the salt being rejected on freezing. Ice forms when the temperature of sea water falls below $-2\cdot1°C$. As freezing takes place there is a unique temperature at which the ice and brine of a given concentration are in equilibrium. At a temperature of $-8\cdot5°C$ the brine has a salinity of 125‰. As temperature is lowered still further a limit is reached at the eutectic point, which is at $-21\cdot1°C$ for salt; the salinity at this point reaches a value of 233‰ and the whole mass becomes frozen solid, including the brine. As freezing takes place in the sea, salt may become enclosed in pockets of high salinity, or it may drain out below, mixing with the underlying water to increase the salinity values. In general it may be assumed that the salt enclosed in sea-ice has the same chemical composition as in sea water.

As sea-ice begins to freeze, frazil ice is formed first. This consists of many platelets or small discs of pure ice in the top few centimetres of water. The c-axis of the platelet is perpendicular to its plane surface, which floats horizontally in calm water. Any disturbance of the water will, however, force the platelets together and as they freeze into a mass, they will tend to have a variety of angles and orientation. The ice when first formed is very flexible and waves of several centimetres amplitude can travel through it. The azimuth of the crystal angles is random, but their polar or vertical angles become arranged so that the c-axes are almost horizontal at a depth of 20 cm or so, while near the surface they tend to be nearer vertical. This is due to the preferential crystal growth as the ice grows downwards. In the lower 1–2 cm there are pure ice platelets, with layers of brine between, the platelets usually being arranged vertically and parallel to each other. Ice bridges develop between the platelets, giving strength to the structure. The average thickness of the platelets in annual sea-ice is 0·5 mm, and the brine pockets are about 0·05 mm diameter and up to 3 cm long. They tend to work their way down through the ice, by freezing above and melting below, thus reducing the salinity of the ice. Brine can also drain out by gravity when the cells are interconnected, as well as by the process of migration along the temperature gradient, which is normally downwards in sea-ice, as it is colder on the surface than at the lower ice–water contact. Loss of salinity is particularly rapid in spring and summer when melting allows the brine cells to connect. Sea-ice after one season is much fresher than annual ice and is called polar ice. Annual sea-ice has typically a salinity of 4‰, while that of ice 1·5 years old varies between 1 and 0·5‰.

Polar ice appears to be pale blue, while annual ice is greyish white. Polar ice-floes have undulating surfaces with waves of amplitude of 1 m and lengths of 30–40 m. Annual ice grows to about 2–3 m in the first winter, and during the following summer it loses salt to become fresh polar ice. Pools will collect on its uneven surface, which because of lower albedo will melt preferentially, to produce the hummocky surface typical of polar ice. Fresh water may flow down under the ice, where it reduces the salinity and freezes when the next winter comes. There is thus a cyclic process of surface thawing in summer and freezing from below in winter. Where ice never melts entirely, it reaches an equilibrium thickness of 3–4 m in a few years. Any particular particle of ice gradually moves up through the flow. In this process the platelet structure may be lost. The resulting polar ice is extremely hard, and cannot be broken by ice-breakers as its compressional strength is very high. Greater ice thickness can be achieved by pressure ridge formation, which results from the movement of the ice.

From the geographical and general oceanographical point of view, the drift of pack-ice and icebergs is an important aspect of sea-ice; there is a great contrast between conditions in the Arctic and the Antarctic. The Arctic basin is sea and has much less extreme temperatures than the Antarctic, which is a high continent. Minimum temperatures are around $-47°$c in the Arctic, but can reach $-74°$c in the Antarctic. In the Arctic the whole ocean is covered with pack-ice, which is constantly moving. Four to five pressure ridges normally occur each kilometre, but the number can reach 30 in places. Knowledge of the drift of Arctic ice is provided by drifting stations that have been occupied in increasing numbers since the first successful drift of the *Fram* in 1893–96, led by Nansen. The ship froze in near the New Siberian Islands after passing through the Bering Straits and drifted across the Eurasian side of the basin until released near Svalbard three years later.

The polar pack drifts in two main circulations. There is an east to west drift on the USSR side of the north pole and a clockwise circulation, the Beaufort gyral, in the area between Alaska, Canada, and the pole, as shown in figure 2.8. The ice drifts at a rate of 1·4 per cent of the windspeed at an angle of 15° to the wind in winter, when it is tightly packed, but when open leads occur in summer the figures are 2·4 per cent and 49° on the north Siberian shelf. The time taken to drift across the entire Arctic Ocean is about 4–5 years, as shown by the movement of driftwood from the Siberian rivers to Svalbard, Jan Mayen, and Greenland. The pack-ice enters the Atlantic via east Greenland and east Canada, where cold currents flow to the south.

Local circulation exist in the Laptev, Kara, and Barents Seas on the Eurasian side of the Arctic basin. Ice may circulate indefinitely in the Beaufort gyral, taking 10 years to make the circuit on the outer edge and 3–4 in the inner part. Some is removed via the westward flow into the Atlantic. The movement of the pack-ice is, however, very erratic and circulation can only be given as a long-term average. An ice-flow is subjected to four forces: the wind drag, the water drag, the Coriolis force, and the lateral forces due to pressure of surrounding ice, the last being the most difficult to predict. Predictions of ice movement are now based on empirical equations, and it is suggested that ice drifts along the direction

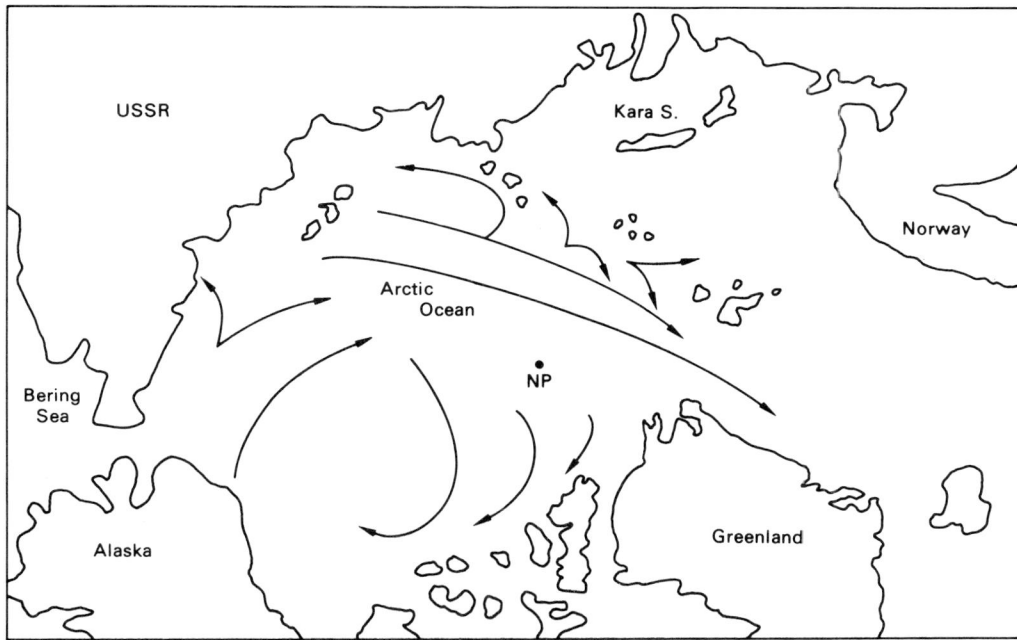

Figure 2.8 Map to show the pattern of ice drift in the Arctic Ocean. (*After Pounder, 1965.*)

of the isobars on the weather map at a speed proportional to the pressure gradient. The Coriolis effect is balanced by the difference between the geostrophic and the surface wind, as the atmospheric circulation is anticyclonic in the Arctic as a rule. Water currents seem to exert only a minor effect on ice movement, although air, ice, and water circulation appear to coincide fairly closely.

The annual inflow into the Arctic is 86,000 km³ of water—61,000 coming from the Atlantic, 20,000 from the Pacific, and 5000 from rivers and precipitation. The Atlantic water enters at depth and the east Greenland current carries 80,000 km³ of water and 10,000 km³ of ice out of the Arctic annually on the surface.

In the Antarctic, ice is fast to the shore for 160 or more km from the coast, and beyond lies the sea-ice, bits of shorefast ice being incorporated in the sea-ice when it breaks off. In winter and spring the pack-ice covers 23 million km² and in autumn it reaches a minimum of 4 million km², although the annual variation is considerable. The pack-ice drifts somewhat west of north under the influence of the northerly winds, deflected west by the Coriolis effect. The increasing width of the ocean northwards allows the pack to become open and thus navigation is possible. The northern boundary is difficult to penetrate because in this zone the ice is affected by the strong westerly winds, which blow from 40 to 60°s. Its northwards movement is slowed down and ice converges in this zone.

Ice first forms in sheltered bays and where the water is not so saline, as off Siberia. This ice is called bay ice, land ice, or shore ice. It may extend out for 430 km from the Siberian coast and is widespread among the islands off northern Canada. The pressure in the pack-ice has caused the loss of many ships, for example the crushing of the *Endurance* in

c

Plate 1 Ice in the sea (*Institute of Oceanographic Sciences*)
a (*above*) Bergs and pack-ice in the Bellingshausen Sea. The ice-floes are separated by leads full of
 brash-ice, and some have upturned edges typical of pancake ice. Note the ski-track for scale.
b (*below*) Young pancake ice, showing upturned edges of the floes.

1914 in the Weddell Sea, when under the command of Shackleton. Drift-ice has been estimated to cover over 22·9 per cent of the total ocean area. In the Antarctic both icebergs and sea-ice are found, but in the Arctic sea-ice covers by far the larger area. In the Davis Strait–Baffin Bay area, where icebergs are most numerous, they only amount to 2 per cent of the total sea-ice. In the Antarctic, the sea-ice is associated with the area of the east wind belt, which extends to about 65°s, but there are various 'bays' in the pack-ice, due to warmer currents. The ice extends further north in the south Atlantic and Indian Oceans than in the Pacific. The mean total area occupied by sea-ice is about 22 million km², but this varies greatly with the seasons, being at a maximum in October in the southern hemisphere.

Northern sea-ice varies greatly in extent from season to season and throughout the year. Where the warm waters of the north Atlantic drift reach the northeast Atlantic, the sea-ice limit is furthest north. Northwest of Svalbard it reaches latitude 81°N, and north of Jan Mayen it is in latitude 72 to 75° N. Sea-ice forms on the whole on the wide continental shelf off Siberia in winter, assisted by the low salinity.

The extent of sea-ice influences the air pressure and cyclone tracks, the latter tending to swing equatorwards in years of heavy ice. Pack-ice makes it difficult to approach the coast of east Greenland north of 77°N and from Cape Farewell to 70°N. Between these latitudes, coastal lanes and more open pack-ice make approach possible in most summer seasons. The total amount of ice leaving the polar basin has been estimated at 12,700 km³ between Greenland and Svalbard, 5000 by Baffin Bay, and 2000 between Bear Island and Franz Jose Land, each year. This amount of ice must play an important part in the formation of the deep water-mass of the north Atlantic. It has been calculated that 32 million km³ in the southern hemisphere and 7 million km² in the northern hemisphere melt each year, while the total range between the winter and summer ice-fields has been suggested to be about one-eighth of the whole earth's surface. This greatly changing area of ice cover has a considerable effect on the climate, and is in turn influenced by it. There is thus a close correspondence between the processes affecting the atmosphere and the ocean surface.

4.2 Icebergs

There are two distinct types of icebergs, which are typical respectively of the northern and southern hemispheres. The northern bergs are derived largely from calving glaciers, while the flat, tabular icebergs of the south originate by calving of large blocks from the ice-shelves around parts of the Antarctic continent. A typical northern iceberg is 200 m across, 25 m maximum elevation, with 90 per cent of its mass below water, and its volume would be 5×10^9 kg. The small number of true glacier icebergs in the Antarctic is due partly to the slow movement of Antarctic glaciers compared with those in Greenland, where iceberg formation is most active. The rapidly moving Greenland glaciers produce the maximum number of icebergs. The Jacobshavn glacier, which moves at 20–25 m/day, produces over 1000 icebergs each year. The icebergs characteristic of the northern hemisphere are generally much smaller and more irregular than those of the Antarctic, being calved from crevassed, fast-flowing glaciers, which break up into fairly small pieces. The Antarctic bergs are usually composed more of firn than ice. They may be very large, and are flat-topped, and being less dense, float higher in the water. Bergs up to 100 km

in length have been observed, but they are mostly much smaller, many being less than 6 km. About 400 m length and 30–40 m height is common. Large tabular bergs are sometimes 500 m thick. They move right through the westerly wind belt, and penetrate far into the oceans, where they may take up to 10 years to melt. Although they rarely move north of 40°s, some have been sighted in the tropics.

Calving of Antarctic icebergs is very irregular, and some years produce many more than others. Abundant bergs were produced in 1832, 1854, 1893, 1897, and 1922. In the last decade of the nineteenth century so many icebergs calved that shipping between South America and Australia had to use a more northerly route. The distribution of icebergs in the ocean is of considerable importance and reflects in part the source of ice and also the current pattern in which they move. Icebergs of the southern hemisphere are all produced around Antarctica, but they are scarce in the longitude from 140°E to 170°w, off south Victoria Land. Icebergs drift much further north than the pack-ice, reaching a limit of 35°s in the Atlantic, 45°s in the Indian Ocean, and 50°s in the Pacific; the cold Falkland current helps to account for their more northerly extent in the Atlantic.

Where the icebergs float in the pack-ice they sometimes move at considerable speed through it, as their greater draught makes them respond to the ocean currents, while the pack-ice is influenced more by the wind. This can be dangerous to ships held in the pack-ice.

In the northern hemisphere the sources of icebergs are numerous. Greenland, Franz Josef Land, and Novaya Zemlya are the main sources, but the origin of icebergs is localized along these coasts. In Greenland, bergs are most numerous off the northwest and southeast coasts. From both sources the icebergs drift south in the Labrador and east Greenland Currents respectively, where they merge in the 'Gateway of ice bergs', east of the Grand Banks between 43 and 47°N. It is in this zone that the International Ice-Patrol has operated since the loss of SS *Titanic* in 1912. The southerly extent of the icebergs varies greatly from year to year. The ice patrol sighted over 1300 icebergs in 1929, but only 11 in 1924. Between 1900 and 1953 an average of 407 icebergs reached 48°N, although the number was very variable. In years of many icebergs they penetrate at times as far south as 30°N, when they drift east towards the Azores or the coast of Britain. Most of the Greenland icebergs last less than 2 years, a berg decreasing from 50 million ft³ (1,415,000 m³) in the Davis Strait to only 6–8 million ft³ (170,000–226,000 m³) at the Grand Banks. The main iceberg season lasts in the Grand Banks area from mid-March to mid-July, with the maximum number in mid-May. For the rest of the year there are very few icebergs in this area. Sonar provides the best method of detecting icebergs, using underwater sound waves. Icebergs move dominantly by current action because of their great depth penetration. There are virtually none in the Arctic Ocean. Prediction of iceberg numbers off Newfoundland, based on the strength of winds and temperatures along the Labrador and Newfoundland coast the preceding December to March, gives reasonable results. The trend in the number of icebergs between 1927 and 1964 suggests an erratic decline.

Plate 2 Ice in the sea (*Institute of Oceanographic Sciences*) :
a (*above opposite*) An iceberg between South Orkney and South Shetland. The tabular berg is 23 m high, and shows firn stratification very clearly.
b (*middle opposite*) A weathered iceberg in loose pack-ice, showing old water levels and stratification.
c (*below opposite*) A very weathered iceberg, showing cave formation.

Numbers vary from only about 12 to more than 1000. There were few in 1924, 1940, 1941, 1951, 1958, and many in 1909, 1912, 1929, and 1945.

The total mass of icebergs in the world is estimated at $7 \cdot 65 \times 10^{18}$ g, of which 93 per cent are in the southern hemisphere. There is 18·7 per cent of the world ocean in which icebergs may occur, of which 56 million km^2 is in the south.

4.3 Ice islands

Another type of sea-ice is made up of ice islands which have been discovered fairly recently in the Arctic Ocean. The movement of one of the largest of these, T-3 or Fletcher's Ice Island, has been traced since 1947, when it was first sighted. It is 50–60 m thick and has a surface dimension of 14 by 8 km. It has been used as a floating research station for some time since its discovery. It moves at about one-fiftieth of the windspeed. Because of their greater thickness than the pack-ice, ice islands drift rather differently from the polar pack-ice. Ice islands probably originated from the ice-shelves around the northern edge of Ellesmere Island. These ice islands are similar to the large tabular icebergs of the Antarctic, which also come from ice-shelves. As they are formed partly of névé, rather than dense glacier ice, their density is less and they float higher out of the sea. The surface of the ice islands is smoother than the pack-ice in which they occur. As they drift slowly round with the pack-ice they melt from below and are built up from above by snow precipitation, because of their greater elevation above the surrounding pack-ice.

Further reading

COCHRANE, J. D. 1958: The frequency distribution of water characteristics in the Pacific Ocean. *Deep Sea Res.* **5,** 111–27.

COX, R. A. 1959: The chemistry of sea water. *New Scientist* **6,** 518–21.

GOLDBERG, E. D. 1954: Marine geochemistry. *J. Geol.* **62,** 249–65.

HILL, M. N. (editor). 1963: *The sea.* Vol. II. New York: Wiley. Chapters 1: E. D. Goldberg, The oceans as a chemical system, 2: A. C. Redfield, B. H. Ketchum, and F. A. Richards. The influence of organisms of the composition of sea-water, 4: W. Broecker, Radioisotopes and large-scale oceanic mixing, 5: D. E. Carritt, Chemical instrumentation, 6: H. F. P. Herdman, Water sampling and thermometers.

HORN, M. K. and ADAMS, J. A. S. 1966: Computer-derived geochemical balances and element abundances. *Geochem. Cosmochim. Acta* **30,** 279–97.

LEE, A. and ELLETT, D. 1967: On the water masses of the northwest Atlantic Ocean. *Deep Sea Res.* **14,** 183–90.

MACINTYRE, F. 1970: Why the sea is salt. In J. R. Moore (editor), *Oceanography.* Readings from *Sci. Amer.* San Francisco: Freeman, 110–21.

MONTGOMERY, R. B. 1958: Water characteristics of the Atlantic Ocean and of the world ocean. *Deep Sea Res.* **5,** 134–48.

POLLAK, M. J. 1958: Frequency distribution of potential temperature and salinities in the Indian Ocean. *Deep Sea Res.* **5,** 128–33.

SVERDRUP, H. U., JOHNSON, M. W., and FLEMING, R. H. 1946: *The oceans.* Englewood Cliffs: Prentice-Hall.

3 The circulation of the ocean

1 Causes and character of the ocean currents: *1.1* Types of currents, *1.1a Geostrophic currents 1.1b Inertial currents, 1.1c Wind drift currents, 1.1d Slope currents;* 1.2 Westward intensification; 1.3 Eastern boundary currents; 1.4 Equatorial undercurrent

2 The surface currents: 2.1 North Atlantic Ocean; *2.1a North equatorial current, 2.1b The equatorial counter-current and south equatorial current, 2.1c The Gulf Stream system;* 2.2 The currents of the south Atlantic Ocean, *2.2a Benguela current, 2.2b Brazil current;* 2.3 Currents of the southern or Antarctic Ocean; 2.4 Indian Ocean currents; 2.5 The surface currents of the Pacific Ocean, *2.5a South Pacific Ocean, 2.5b South equatorial current, 2.5c Currents of the north Pacific*

3 Deep water circulation: 3.1 Theoretical analysis; 3.2 Observations

4 Circulation of the oceans: 4.1 The Atlantic Ocean; 4.2 Southern ocean; 4.3 The Pacific Ocean; 4.4 The Indian Ocean

5 General conclusions

The main pattern of currents closely follows the average wind systems over the oceans. There are, however, variations between the different oceans, and as data become more detailed and plentiful, local anomalies are being increasingly revealed. The circulation of the oceans is an important aspect of oceanography from the geographical point of view. It helps to distribute heat received in low latitudes to certain areas in higher ones. The distribution of the fertile areas of the ocean depends to a considerable extent on the oceanic circulation, which brings the nutrients to the surface. In the days of sailing ships the major ocean currents were important in establishing trading routes.

Oceanic circulation depends fundamentally on two factors. The wind stress is very important in its influence on the surface water, while the distribution of density is both partly caused by, and partly causes, the movement of the water. When a steady state is achieved, the pressure distribution in the water is so arranged that it is balanced by the Coriolis effect. The wind-driven current velocity is adjusted to the pressure gradient force. Strong pressure gradients lead to a strong wind stress and a strong Coriolis effect, and fast currents result. In some areas, however, the thermohaline density effects cause surface gradients, which in turn may influence the wind field to some extent. Wind stress and water density distribution are, therefore, very closely linked and the circulation of the oceans represents the balance between these two forces, both on the surface and at depth.

1   Causes and character of the ocean currents

No part of the water of the oceans is completely stationary. The major elements of the oceanic circulation can be explained by the drag of wind on the sea surface and the distribution of density in the ocean. Major directions of flow can be calculated from the observed temperatures and salinities. Upper layers are strongly influenced by the frictional and inertial forces due to the stress of the wind, while one of the fundamental causes of deep water movements is the distribution of mass within the ocean. The flow is related to the distribution of mass, while it in turn affects this distribution, leading to a feedback situation.

Surface winds can only affect the uppermost layers of water down to a depth of about 100 m directly. Phillips (1957) considers that there is no minimum wind velocity for disturbing the sea surface. The stability of the air exerts a strong effect on the disturbance of the sea surface, and must be taken into account. Wind stress causes an upslope parallel to the wind flow in shallow constricted areas, where transverse currents cannot operate. Observations in shallow seas, such as the Baltic, suggest that the slope depends on the depth and the square of the wind force. The wind in the open ocean piles up the light surface water—for example, on the western side of the ocean in the trade-wind belts. The effect extends to 150 m in the Atlantic and 300 m in the Pacific. The trade winds, therefore, actually blow uphill, as the isobaric surfaces slope at 3.8×10^{-8} (radians) in the Atlantic and 4.5×10^{-8} in the Pacific. The equatorial counter-currents flow downslope between them to compensate. The lighter water is on the right in the northern hemisphere. In this instance the distribution of mass is maintained by the current. Types of ocean current include: geostrophic flow, inertia currents, wind drift currents, and thermohaline currents.

One of the difficulties of calculating ocean movements from the data giving the form of the isobaric surfaces is the lack of sufficiently detailed observations of temperature and salinity at all depths in enough places to draw accurate isobars. Conditions change rapidly near the surface, necessitating observations taken over as short a time as possible, but in the deep ocean conditions are stable enough to allow observations taken at long time intervals to be used, even covering different years. It has been shown that the isotherms and isohalines (lines of equal salinity) are often parallel at different levels. Their direction, therefore, coincides with that of the relative isobars or contour lines on an isobaric surface. This suggests strongly that the main oceanic circulation is related intimately to the pressure field or the internal distribution of mass.

The surface of zero velocity must be known to calculate absolute current velocities. Various methods of arriving at a value for the level of no motion have been adopted by different oceanographers. Such a surface need not coincide with an isobaric surface. Stommel (1957) has shown that the layer of no horizontal flow may well be that in which vertical movement is at a maximum. Where detailed observations are available, it has been shown that computed currents often conform closely to the observed ones, for example in the Straits of Florida. That such calculations are sufficiently accurate to be of practical application is shown by the use made of these methods of computation in the preparation of forecasts of iceberg movement off the Grand Banks of Newfoundland. Ocean-current velocities here are estimated from the observed temperatures and salinities above the 1000 decibar surface at a depth of nearly 1000 m. It is assumed that the flow follows the contour lines, and from this the drift of the icebergs can be predicted with results that justify the continuation of the method (Carruthers, 1956).

It is also possible to calculate the mass transport under these assumptions, from the temperature and salinity observations. Again the depth of motion must be known, and where questions of continuity arise this may not be at the bottom. It can often be assumed, for example, that in one cross section the amount of water flowing in one direction must be compensated at depth by an equal amount flowing in the opposite direction. It is frequently found that the surface of no motion follows the isothermal and isohaline surfaces. The surface slopes from about 1450 m off South America to about 1200 m off South Africa, for example.

Current mechanisms have been studied by means of dye by Assaf, Gerard, and Gordon (1971). They studied the mixing of ocean water near Bermuda. The first few metres were subject to helical flow except under calm conditions. The cells were spaced at 3–6 m apart as the wind increased in strength. The irregularity of currents has been demonstrated by Gould (1971) from observations made in the Bay of Biscay. The water moved mainly east and then turned northwards. The disturbance, called an 'event', moved westwards according to direct observations at a speed of 0·56 ± 1 km/day at 400 and 1400 m depth. The disturbance was caused by the passage of a depression moving at 70° to north. The speed of movement of the event agreed with the theoretical analysis of Longuet-Higgins (1965).

The general pattern of surface ocean currents is well known and is shown in figure 3.1. In the northern hemisphere there is a clockwise circulation in the Atlantic and Pacific, where it is in two separate gyres. In the southern hemisphere the currents in general move anti-clockwise, while in high southern latitudes there is a continuous west to east movement of water all around the southern ocean.

1.1 Types of currents

Currents depend on the nature of the distribution of mass in the ocean. There are two possible states. A baratropic field is one in which the isosteric surface coincides with the isobaric surface. The isosteric surface is one of equal specific volume. When these surfaces intersect the field is called baroclinic, which is a field that cannot remain without motion, whereas the baratropic field can be motionless. The baroclinic state is normal in the ocean. The equations of motion can be expressed either as Eulerian equations or Lagrangian ones. The former describe the changes of fluid motion with time at fixed points in a Cartesian coordinate system. The Lagrangian system follows a specific element of fluid over time as it moves. In most ocean currents it is necessary to take both the Coriolis effect and the frictional effects into account. The simplest ocean current is one in which the sum of all the forces is equal to zero, when the current is non-accelerated.

1.1a Geostrophic currents

These are frictionless currents, and they flow horizontally without change of velocity. Gravity is the only external force. The Coriolis effect and pressure gradient force balance each other. The horizontal current vector is parallel to the isobars, with high pressure to

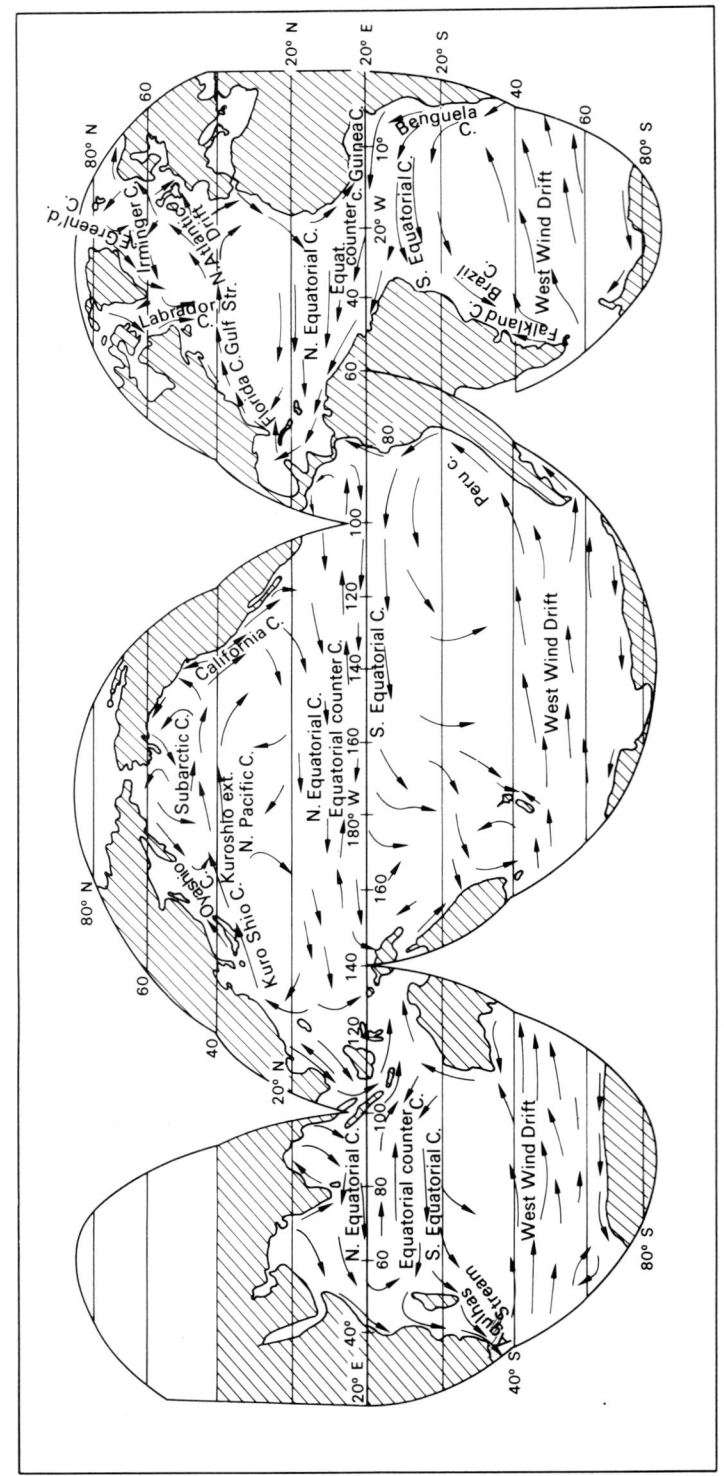

Figure 3.1 The main surface currents of the oceans. (*Modified after Sverdrup et al., 1946.*)

the right in the northern hemisphere and vice versa in the southern. The isobaric surfaces slope across the current, as do surfaces of equal temperature, salinity, and density, the latter being called isopycnal surfaces. The slopes are proportional to the current speed. Across the Gulf Stream at about 35°N, 71 to 74°w the isopycnals slope at 700 m/100 km, but the isobaric surface slopes less steeply at 1·5 m/100 km.

Where currents are flowing strongly, as in the Gulf Stream and Kuroshio, a nearly geostrophic flow is closely approximated. Classical methods of analysis are successful if the depth of no motion can be adequately established. Both theory and observation suggest the presence of a deep south-flowing current under the Gulf Stream. The depth of the Gulf Stream itself increases from 500 m on the continental side to 2000 m on the ocean side. The geostrophic method is not successful where water movement is accelerated, and frictional or vertical forces are significant. At the sea surface, layers are often not in geostrophic balance, due to acceleration and friction. At times bottom friction also causes complications. Wind stress on the surface can cause sudden accelerations that disturb the geostrophic balance.

1.1b   Inertial currents

These are due to sudden changes in the forces that cause currents. They result from a lack of balance between the pressure gradient, friction, and the resultant geostrophic acceleration. Inertia currents in the northern hemisphere have a clockwise circular movement, the circle of inertia, if the latitude is considered constant. The inertial period is given by $T_p = 2\pi/f$, where $f = \omega \sin \phi$, or by $2\pi r/c$, where r is the radius of the inertial circle and c is the speed of the water particle. T_p depends only on the latitude ϕ; it is 12 hours at the poles and 24 hours in latitude 30° and infinite at the equator. Inertial currents deep in the Atlantic have been reported from observations of neutrally buoyant floats at 28°N 55°w at a depth of 2340 ± 20 m. The average speed of movement was 8·2 cm/sec and the inertial period observed was 25·2 hours, which is close to the calculated value of 25 hours. The observed diameter of the inertial circle was 2·4 km, which agrees with the theoretical value. Currents cannot be in geostrophic equilibrium if they are in circular or meandering motion, with a curvature approaching the radius of the circle of inertia, as this requires acceleration.

The centrifugal force as well as the Coriolis effect must be considered. A current that is affected by centrifugal force is called a gradient current and operates when the isobars are curved. If the curvature is small, the gradient and geostrophic currents are essentially the same, and this is common in ocean currents. In a stratified ocean stream lines of geostrophic flow and isopycnals must be parallel.

1.1c   Wind drift currents

These do not flow in the wind direction owing to the friction between the wind and the sea surface, and friction within the water. This observed fact was studied by Ekman (1905), who developed the idea of the spiral called after him the Ekman spiral. This indicates that the current gradually changes further to the right of the wind direction and becomes slower as the depth increases. This change with depth is due to the action of the Coriolis effect, which must be balanced by the resultant of the frictional forces with stationary and non-accelerating motion. This condition of balance can only be fulfilled in the northern hemisphere if the current turns to the right as it decreases in velocity with increasing depth. Deflection of the current to the right in the northern hemisphere and to the left in the southern is the most important aspect of the wind-driven current. Ekman's theoretical value indicates that a current should flow at 45° to the wind direction at the surface to the right in the northern hemisphere, and vice versa in the southern. At a depth D the current flows in the opposite direction to the surface wind and the speed is reduced to $e^{-\pi}$ of the surface speed. The depth D increases as the current velocity and wind velocity increase. D also depends on the latitude, increasing as the latitude decreases. D is about 52 m at 45° and 126 m at 5°. The variation with latitude cannot be expressed exactly owing to variations due to wind stress. The total mass transport by wind-driven currents is theoretically directed exactly to the right of the wind direction, and is proportional to wind stress. As the water becomes shallower the current depends on the ratio d/D, where d is the depth. When the water is very shallow (d/D is 0·1) the current does not change direction with depth, but the speed is reduced due to greater friction.

Pure drift currents cannot develop in the oceans as currents cannot continue indefinitely without piling up water, and winds are not constant. Slopes therefore develop, and horizontal pressure gradients build up.

1.1d Slope currents

Those that result also decrease and change direction with depth, a horizontal plot of the current vectors forming a logarithmic spiral in infinitely deep water. Near the surface the current flows parallel to the isobars, and only at depth can it flow at right angles to them to equalize the pressure gradient.

The direction of current flow in relation to the wind requires more refinement. In shallow water observations show the same angle between the current and surface wind. Observations made by recording the drift of plastic envelopes on the surface of the Atlantic west of Britain do not confirm the theoretical results. The observations showed that the surface water drifts parallel to the gradient wind, determined by isobar spacing, at 2·2 per cent of its velocity. There is reason to suppose that surface wind was nearly parallel to gradient wind. Current velocity would be about 3·3 per cent of the surface wind speed, assuming this to be two-thirds of the gradient wind. Deviation of the current from the wind direction might occur in deeper water than that in which the envelopes floated (Hughes, 1956).

Close to the coast a wind blowing nearly parallel to a shore on its right in the northern hemisphere will cause transport of water towards the coast and the surface will slope up in this direction, with lighter water converging in this region (figure 3.2). The convergence is compensated by a relative current in the direction of the wind parallel to the shore. If the coast is on the left of the wind, upwelling will take place as the transport will tend to be offshore, to the right of the wind direction. This upwelling brings water from depths of about 200 to 300 m. If the water is stratified into two different densities, with the denser layer to the left of the other, the two waters will move at different velocities and the boundary between them slopes. The water-masses are so arranged in this instance that the boundary between them slopes down to the right, while the lighter water lies on the right-hand side of the current in the northern hemisphere, with its surface sloping up in this direction.

Another cause of ocean currents is the thermohaline condition, which affects density

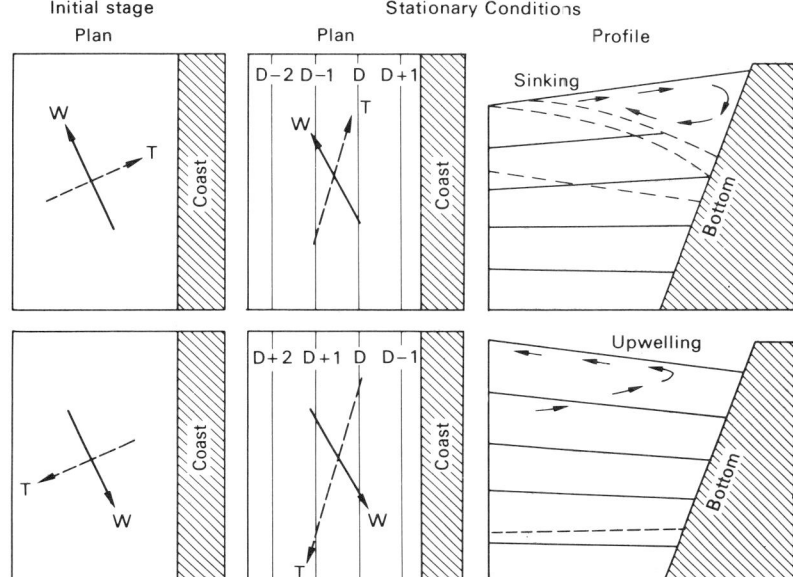

Figure 3.2 Diagram to show the effect of a wind blowing along a coast in the northern hemisphere. W refers to the direction of the wind, and T to the direction of water transport. D, D + 1, etc. are contours on the sea surface. (*After Sverdrup* et al., *1946*.)

and thus the vertical stability of the ocean. Vertical movements in turn lead to horizontal differences in density, and thus horizontal currents are set up; these are related to the wind-stress currents. On the whole thermal currents are weak, although a model by Bryan and Cox (1967) showed that the thermohaline component of a combined system was strong, indicating a strong poleward current along the western side of a model ocean in the northern hemisphere.

Ocean currents feel the bottom relief in areas where they extend to the bottom. As the current flows over increasing depths a positive curl, or anti-clockwise deflection will occur; while the reverse occurs, with a negative curl or clockwise deflection when the water becomes shallower in the direction of flow. The Gulf Stream, for example, is deflected south when it flows over decreasing depths of the mid-Atlantic Ridge, turning back north where the depths increase again. This deflection results in a tongue of cold water pushing south in 50°w.

Bryan and Cox (1967) have used a computer to calculate the circulation and temperature characteristics of a model ocean of uniform depth and bounded by two meridians. Input included temperatures and wind stress at the upper surface. The system was shown to depend on five basic parameters, which were tested in 8 numerical experiments. The five parameters were thermal diffusion coefficient, wind stress, momentum diffusion, depth, and rotation. Each was varied in turn in the model. The computations were made with and without wind stress to indicate the interaction of wind-stress circulation and thermohaline circulation. In the absence of wind stress there was one large anticyclonic gyre covering the whole basin, with a shallow western boundary current extending to high latitudes. There was a vertically uniform southerly drift in the interior extending down to the thermocline. Below the thermocline there was a sluggish cyclonic gyre.

The wind stress was tested in three cases: 1) with no wind, 2) with an intermediate wind stress, and 3) with a high wind stress. The wind stress added a cyclonic gyre in subarctic latitudes when the wind stress was at a maximum in the westerly zone at 45°N. Details of the subtropical gyre compared favourably with reality. The boundary between the two gyres that occur when the wind stress was added lay in the zone of maximum westerlies. The western boundary current became progressively more effective as the wind strength increased and the thermal gradient became stronger. The southward-moving current on the east side now had a maximum velocity at the surface, as in reality.

The results in general showed that the total poleward transfer of heat and turnover rate of the thermohaline circulation are not markedly affected by other variables. There was, however, a discrepancy in the renewal time of the deep water in the Atlantic, which has been estimated as 8–10 centuries by carbon 14, but was calculated to be less than 2 centuries. The study could be made still more realistic by adding the effect of bottom relief, bottom friction, and non-linear terms. The addition of transient disturbances may cause concentration of the Gulf Stream into a jet, instead of causing diffusion.

1.2 Westward intensification

There is one very striking feature of the pattern of surface currents, and it also probably applies to deep water circulation. This is the strong asymmetry of the systems. In nearly all the ocean basins the currents are so arranged that there is a very strong, concentrated flow on the western side, while the currents on the eastern side are more diffuse. According to the theoretical analysis of Stommel (1958b) this applies to both surface and deep circulation, and observations have on the whole confirmed his views.

Various possible causes for the asymmetry have been suggested. The necessity for the boundary current is due to the imbalance between the pressure gradient and the geostrophic influence. The main cause of the imbalance is the fact that the Coriolis effect varies with latitude. Reference to the character of the forces involved shows that factors other than the geostrophic balance between the Coriolis effect and the pressure gradient must also be involved.

The wind stress in a symmetrical system is directed to the north on the west of the ocean in the northern hemisphere, and to the south on the east. This stress develops a vorticity to the right of the flow in each instance. If the deflection is clockwise, the vorticity is said to be negative, thus the wind-stress vorticity is negative on both sides of the ocean. The Coriolis effect causes a deflection to the east on both sides. This produces a negative vorticity on the west, but a positive one on the east. Assuming these forces to be 1, and the frictional force to be 0·1, the north-flowing current on the west side would have a vorticity tendency of −1·9, while on the east it would cancel partially to give +0·1. In a very asymmetrical system the frictional force would be increased on the west, as a result of faster flow, to 10 times the wind-stress vorticity, for instance. The Coriolis effect would also be greater, being −9·0, with a negative wind-stress vorticity of −1·0; the vorticity tendency would cancel to zero (wind-stress vorticity = −1·0, friction = +10·0, Coriolis effect = −9·0, total = 0). It would also cancel to zero on the east, where the smaller frictional force of +0·1 and the Coriolis effect of +0·9 balances the negative wind stress of −1·0. The argument only applies to the viscous theories of ocean currents, but it does illustrate how a strong boundary current could preserve a state of zero vorticity tendency, which would result in a steady state.

Stewart (1969) has explained the western intensification in a rather different way. Water converges at the subtropical convergences owing to the deflection to the right of the wind direction in the northern hemisphere. This results in a downward force on the water, causing lateral displacement at depth, as the water is virtually incompressible. As the water is pushed out laterally, its radius of gyration and its moment of inertia increase, and its rate of rotation must, therefore, decrease. If the rotation slows, it no longer fits with the rotation of the earth in that latitude. It can achieve the correct rate of rotation by moving towards the equator, where the vertical component of the earth's rotation is smaller, or it can rotate faster than the earth. The former is more commonly done. A body of water whose rotation is speeded up by being expanded upwards to replace water at a divergence will tend to move towards the pole as its speed of rotation will be increased. Water thus moves towards the equator as a result of the subtropical convergence in latitude about 30°. This movement must be compensated for by a return flow. The return flow must also have the correct speed of rotation to fit that of the earth. If the water is flowing north, as in the Gulf Stream, it must gain anti-clockwise rotation or lose clockwise rotation. This can be achieved by a strong western boundary current. Friction at its left side against the sea floor changes its rotation appropriately. The Gulf Stream thus represents the return flow of water squeezed equatorwards as a result of subtropical convergence. Such boundary currents must always be on the western side of the ocean in both hemispheres whether the flow is north or south. This is because the earth's rotational angular velocity is greatest in an anti-clockwise direction to an observer looking down at the north pole and a maximum clockwise when looking down at the south pole. A south-flowing current must gain clockwise rotation in the northern hemisphere, and it obtains this by friction on its right side, which is the west side of the ocean. North-flowing currents must keep their left side against the edge; again this is on the west of the ocean.

If deep waters mix upwards, at a rate of a few metres/year, lateral inflow must occur to compensate. The inflow must result in an increase in the speed of rotation, leading to a poleward movement of water. In order to reach low latitudes to accomplish this, a deep western boundary current must form; there is some evidence for such currents.

The western intensification can be considered in terms of the relative and planetary vorticity, as described by Longuet-Higgins (1965). The mean horizontal velocity of a current eastwards is u, and northwards it is v, and the horizontal velocity is given as u, v, = u. The fluid vorticity of the movement is important, and it is equivalent to the angular velocity. If the fluid velocity is called w, and it were instantaneously stopped, its angular velocity would be 0.5 w. The vertical component is the important one in this connection, and it is called the curl$_z$ of u. The curl$_z$ is equal to w, where u is the velocity relative to the rotating earth, and $u = dy/dx - du/dy$. The total vorticity is the sum of the relative vorticity, w, and the planetary vorticity, which is f, and $f = 2W \cos \phi$. The sum $w + f = $ constant. This is a fundamental relationship and is known as the conservation of total vorticity.

If a particle moves towards the equator in the northern hemisphere its planetary vorticity, f, decreases, with the result that its relative vorticity, w, must increase. The fluid must, therefore, spin faster in an anti-clockwise sense relative to the earth. In slow currents the relative vorticity is negligible compared with the planetary vorticity, f, owing to the slow flow and small internal friction, compared with the wind stress.

Western intensification takes place in all oceans both north and south of the equator. Strong currents are, therefore, found on the west side of all oceans. In the northwest Pacific the Kuroshio is the western boundary current. It carries about two-thirds of the volume carried by the Gulf Stream. The southern western boundary currents are weaker than the northern ones. The Brazil current carries only 17 million m³/sec. The Somali current off southeast Africa is strong, flowing at 3·5 m/sec, which is about the mean velocity of the Gulf Stream, at its maximum. The east Australian current is a complex western boundary current (Hamon, 1965). The volume carried above 1300 m varies between 12 and 43 million m³/sec, being strongest between December and April. In some ways it resembles the Gulf Stream. The narrow, swift current flows south, following the continental shelf between 27 and 32°s, in a way similar to the Gulf Stream. The equally narrow north or northeast return current, however, is not present in the northern western boundary currents. The east Australian current only transports about half the volume carried by the Gulf Stream, whereas, in view of the width of the Pacific compared with the Atlantic, it would be expected to carry more.

One of the main features of the Gulf Stream and Kuroshio, the best developed western boundary currents, is their intense baroclinicity. This characteristic causes a rapid decline of density lines across the current. The course of the Gulf Stream is guided by the continental slope along which it flows. Bottom relief probably initiates its meandering characteristic. The pattern depends on the angle at which the stream crosses the isobaths. It feels the bottom as far as the tail of the Grand Banks. Its lower boundary is 600 m on the west side and 2000 m on the Sargasso Sea side. Swallow and Worthington (1957) have shown by float observations that the lower boundary was between 1500 and 2000 m at 33°N. A deep south flow under the western boundary currents has been predicted theoretically by Stommel for the Gulf Stream and it has been observed below the Kuroshio. Eddies can form in the Gulf Stream; they shift laterally at 5–6 cm/sec with the upper and lower water moving in opposite directions. Observations made by Knauss (1966) in 37°N suggest that the current flowing south below the Gulf Stream may not always be present.

1.3 Eastern boundary currents

Eastern boundary currents have been less intensively studied than the strong western ones. They are, nevertheless, of importance for several reasons, one of which is their upwelling characteristic. On the eastern side of the oceans the wind-driven currents indicate that the vorticity balance is achieved mainly by the wind-stress curl and the planetary vorticity. The lateral stress and inertial terms are of relatively little importance, while these play a dominant role in the strong western boundary currents, which are swift, narrow and extend to great depth, in contrast with the opposite characteristics of the eastern boundary currents.

The eastern boundary currents have a number of features in common, including their source waters, surface characteristics, nature of the flow, as well as features related to the presence of neighbouring land-masses. The source of the water is that moving in the west wind drift. This flow is completely blocked in the north Pacific and partially in the south. It is not blocked in the Indian Ocean, so that no well-developed eastern boundary current exists in this ocean. In the north Atlantic the Canary current is the eastern boundary one, while the Benguela current, flowing north along the west coast of south Africa, is the equivalent in the southern hemisphere.

These currents, which flow towards lower latitudes, are generally colder than the water in the centres of the oceans. Upwelling produces even colder water. The salinity is sometimes low, owing to the origin of the water in high latitudes. Upwelling increases the salinity in the Californian current, but decreases it in the Peru and Benguela currents. The circulation depends on the vorticity of the meridional winds, the balance being between wind-stress vorticity and planetary vorticity, resulting in slow, broad, and shallow currents. Lateral stress vorticity plays only a minor part. Average speeds are 25 cm/sec or less, most of the flow taking place above 500 m, in contrast to depths in excess of 2000 m in the Gulf Stream. Average transport volumes are about 15 million m/sec. Isotherms and isopycnals rise towards the coast. There is sometimes a polewards undercurrent. Isopleths of dissolved oxygen also rise towards the coast due to coastal upwelling, which at times is so intense that nearshore waters are undersaturated, particularly off Peru. Isopleths of phosphate also ascend towards the coast, being especially high in the coastal waters of the Peru and Benguela currents. This property gives these waters their very high productivity.

The eastern boundary currents are exceptionally fertile. Records of carbon fixation rates include the Canary current at 0·11–0·67 g cal/m²/day, the Benguela current at 0·46 to 2·5, rising to 3·8 in Walvis Bay, the Californian current at 0·24 to 0·9, the Peru current at 1·02 g cal/m²/day. The average value in the open ocean is less than 0·2 in middle to low latitudes. The high fertility leads to high values of zooplankton in these areas, and very abundant fish. A strip 1300 km long and 48 km wide off Peru, covering only 0·02 per cent of the ocean surface, provided one-seventh of the world fish catch in 1959, which if the fish caught by birds is included, brings the total to 4 million tons.

Upwelling takes place from a few hundred metres depth. In most areas the index of upwelling is highest in spring and summer, but off Peru north of 30°S there is a winter maximum. There is a double circulation in the Benguela current at 28·5°S, with one upwelling cell at the edge of the continental shelf and another over the inner shelf, and offshore flow in the surface layer. These eastern boundary currents often have south-flowing coastal currents between them and the coast in the southern hemisphere. There is also a small clockwise vertical cell over the outer part of the shelf, set between the two anti-clockwise ones. Normally, upwelling is greater in the Atlantic and Indian Oceans, but the Peru current is again an exception. Negative values occur in the north Atlantic south of 10°N and north of 50°N, the Indian Ocean north of 20°S, the north Pacific north of 40°N and the south Pacific south of 45°S. In the Canary, Benguela, and Californian currents, the maximum index of upwelling migrates seasonally, moving south in spring. Estimates of speeds of upwelling range from 10–20 m/month to 80 m/month, while theoretical calculations suggest 50 m/month. Poleward counter-currents sometimes develop close to the shore in eastern boundary currents. One of these occurs from the tip of Baja California to 45°N, and similar currents occur in the Benguela system and the El Niño off Peru. The distribution of density suggests the presence of undercurrents, and one has been observed in the California current throughout the year. A similar poleward flow is apparent on the dynamic charts of the 200-db surface off southwest Africa.

Upwelling is a process of great importance in oceanic circulation from the biological point of view and forms one of the major links between the physical aspects of oceanography and the biological ones. Smith, Mooers, and Enfield (1971) report a study of coastal upwelling in two regions where it is particularly active, off Oregon and Peru. The physical records include recording current meters on the continental shelf off Oregon, operating during the upwelling season in the northern summer. Hydrographic data and airborne infra-red thermometry were also obtained. In the Oregon area the general flow pattern consists of 1) an equatorward flow of 20 cm/sec in the upper 40 m, with an offshore component in the surface Ekman layer, about 20 m thick. 2) An undercurrent flows poleward at 10 cm/sec beneath an inclined permanent pycnocline with an onshore component in the upper pycnocline and below it. 3) The cold saline water upwelled from beneath the permanent pycnocline. It rises and absorbs heat near the surface and flows offshore along the base of the pycnocline. This water is warmer and causes a temperature inversion.

Similar measurements were made off Peru in 1969 at 15°s. The density stratification was small. The current field below the shallow Ekman layer was quasi-baratropic, with a mean flow velocity of 20 cm/sec. The mean flow below the surface layer was poleward, to the southeast against the prevailing winds. A large-scale baratropic 'event' caused the currents to reverse and flow equatorward for four days, but the upwelling was not affected. The mean poleward flow is the Peru–Chile undercurrent over the continental shelf. The actual currents were found to be more variable than theory suggested. Off Peru the most intense upwelling takes place in the southern winter. The nitrates and silicates were similar in their distribution to the temperature, with lower values near the coast. The most active upwelling is very close to the coast within 20 km of the shore, but the effects extend to the edge of the continental shelf. The undercurrent may be tied to the continental slope. In order to obtain reliable data an area 110 km long by 50 km wide is required for detailed observations and good meteorological data are essential as wind stress is the prime cause of upwelling. A vertical current meter has been developed to measure upwelling directly, and a numerical model could also provide valuable information.

Off California upwelling is seasonal, taking place from spring to early autumn, when north and northwest winds cause upwelling. The process is most effective on the equator-side of capes, such as Cape Blanco in 43°N. The temperature changed 2·5°c in 10 km and the Ekman layer was 10 m thick. There is a net onshore motion below the surface to compensate the offshore surface flow in the Ekman layer in the upper part of the pycnocline, which rises in level shoreward. When the wind is strong the pycnocline can break the surface to form a surface front or discontinuity.

1.4 Equatorial undercurrent

Another type of current is the equatorial undercurrent, which was discovered in 1951 but not explored until 1955 (Longuet-Higgins, 1965). It is caused by a downward sea slope along the equator from west to east. In the Pacific the current carries about 40 million m^3/sec. At the equator the wind blows water westwards and the Coriolis effect is zero so the currents act as on a stationary earth. The flow is, however, blocked at the western side of the ocean. A bottom is provided by the change of density downwards and friction at the boundary is zero. To the north of the equator the current direction rotates clockwise as in the Ekman spiral; at depth, therefore, the current turns towards the equator. The same occurs south of the equator where currents are also directed equatorwards below the surface. On the surface currents tend to diverge and westerly momentum will be lost, but at lower levels eastwards momentum will tend to strengthen the easterly flow. An eastward current naturally tends to become stabilized at the equator, while a westerly one does not. The east-flowing equatorial undercurrent has been observed in the Atlantic and in the western Pacific between 142 and 172°E, although it is more complicated here than in the east, due to annual variation in the winds. It is particularly well developed in the eastern Pacific and it occurs at all seasons in the eastern Indian Ocean also, but its strength is half that in the Pacific; in the western Indian Ocean currents are much more variable. It was observed between March and June from 58 to 68°E in 1964 when the slope was reversed in the northern winter. The undercurrent was missing in the Indian Ocean from June to September 1962 during the southwest monsoon when the surface sloped down to the west at the equator. In the Indian and Atlantic Oceans the current is associated with a core of high salinity water centred on the equator. The thermocline weakens in the current and the isotherms spread out. The source of the high salinity water is probably to the north. Salinities can reach 36·2‰. The current sometimes moves a little north or south of the equator. Its course is stabilized by the Coriolis effect, which acts with the downslope of the sea surface from west to east. Winds can cause deviations from the stable course along the equator. Free oscillations of the order of three days have been predicted theoretically, and observations suggest a possible 2·5-day oscillation.

Gill (1971) has described a three-dimensional model in a homogeneous ocean of the equatorial current system. The currents appear to be controlled by lateral friction rather than vertical friction. There is a westward current on the surface with a westward stress, but below the flux is eastwards on the equator. The wind stress at the surface produces an Ekman flux, which is at right angles to the wind stress at some distance from the equator, where the Coriolis effect is associated with the flux balancing the wind stress on the surface. At the equator, where the Coriolis effect vanishes, the wind stress tends to accelerate the surface water in the down-wind direction until a balance with the friction forces is achieved. A westward flux results in strong upwelling at the equator. Equatorward flow is required below the equator by continuity. Where the Coriolis effect is zero, at the equator, the pressure gradient tends to accelerate the fluid in the direction of decreasing pressure. This down-gradient current is the undercurrent and it is limited by lateral friction where this is important. The transport of the current, which is independent of the friction parameter, is equal to 14×10^6 m^3/sec for each dyne cm^{-2} (10^{-1} newton m^{-2}) of wind stress.

2   The surface currents

Surface currents fluctuate with time, and these fluctuations are such that mean values may be one or two orders of magnitude less than observed velocities. The fluctuations may drive the mean motion by providing changes of vorticity and momentum transport. Some of the variations in the speed of ocean currents may be due to wave fluctuations. A numerical example illustrates the possible effect. A cyclone with a wind speed of 60 knots 500 km from the storm centre over an ocean 5 km deep in 45° latitude moving northwest at 500 km/day could generate a current of 10 cm/sec 3 days after the storm passed. The motion would be baratropic and extend all the way to the bottom. Variables that complicate the current pattern include boundary effects, variation of depth, internal friction, and non-linear interactions.

2.1 North Atlantic Ocean

2.1a North equatorial current

The north equatorial current is primarily dependent on the trade winds. It is a shallow current moving westwards slowly within the uppermost 200 m of water. The current covers a wide belt extending from 10°N to nearly 30°N. Its course is not always straight, as it is influenced by the mid-Atlantic Ridge. North of 20°N it is more variable, but south of this the average speed is 28–31 km/day. The water in the current is derived from the southeasterly current flowing off the coast of northwest Africa. This Canary current is one of the eastern boundary currents, and is characterized by relatively cold, dense water near the coast, and upwelling of cooler water from moderate depths occurs. On reaching longitude 60°w the current divides into two branches. One flowing into the Caribbean Sea eventually re-enters the Atlantic from the Gulf of Mexico via the Straits of Florida. The other branch joins the first again by a much more direct route north of the West Indies at 8°N.

2.1b The equatorial counter-current and the south equatorial current

The trade winds of the southern hemisphere also generate a westerly-flowing current, which is stronger and more uniform than its northern counterpart. It has two streaks of maximum velocity at 2°N and 3–5°s and the current pattern is complex. The velocity in June and July exceeds 37 km/day. Between the two equatorial currents, in the zone of equatorial calms, where little stress is exerted by the wind, the equatorial counter-current flows to the east. It is strongest in the summer months. The equatorial counter-current is a downslope current, between two upslope ones. It is confined to the shallow upper layer above the main thermocline. The thickness of surface light water increases from 40 m in the east to 140 m in the west. As the counter-current flows fairly fast in the opposite direction to the currents on either side, the system is influenced by friction. The trade winds, in maintaining the slope, provide the energy to compensate for the loss by friction. Lateral friction is developed, which acts at right angles to the wind stress. This results in some mixing of the waters of the different currents and sets up a transverse circulation as well.

The south equatorial current is north of the geographical equator because the thermal equator is in the northern hemisphere. The equatorial counter-current also lies always in the northern hemisphere. This means that the Coriolis effect, which acts symmetrically about the equator, does not influence the currents in a symmetrical fashion. As a result of the lateral mixing and frictional forces, a secondary flow at right angles to the main east–west direction is set up. The fairly rapid flow of water causes the discontinuity between the surface water and denser subsurface layer, the thermocline, to slope. This slope is down to the right of the direction of flow in the northern hemisphere, and to the left in the southern. The secondary flows are directed in the same sense. The sea surface slopes in the opposite direction to the thermocline, but at a smaller gradient.

The result of superimposing these forces on the current pattern is shown diagrammatically in figure 3.3. Where the currents are flowing towards each other, convergences are set up, and where they separate divergences occur. The north equatorial current, lying entirely in the northern hemisphere, has a northerly-directed subsidiary current. Its surface slopes up to the north and the thermocline below it slopes down to the north. The counter-current flows east, south of the north equatorial current, its subsidiary flow is, therefore, to the south, as it lies entirely within the northern hemisphere. Between these two currents there is a zone of divergence at about 10°N. The part of the south equatorial current lying in the northern hemisphere is affected in the same way as the north equatorial current, as it flows west. There is a zone of convergence between the equatorial counter-current and the south equatorial current at about 4°N, and the discontinuity surface sinks down to this point from either side. The south equatorial current

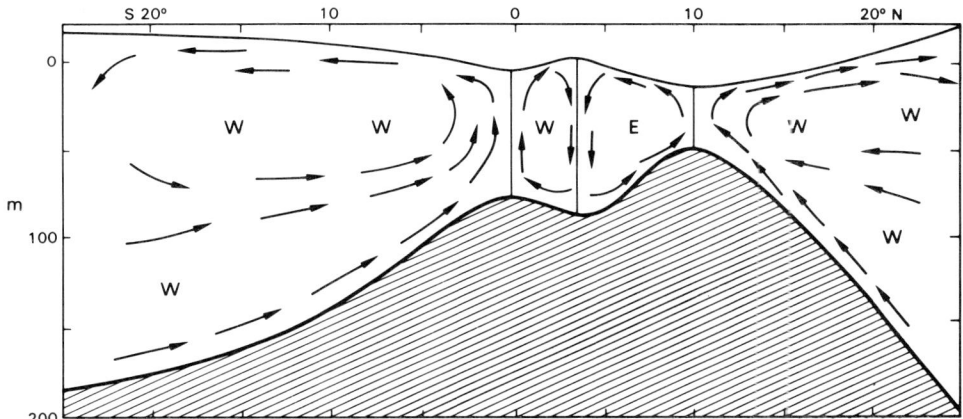

Figure 3.3 Vertical circulation giving convergences and divergences in the equatorial Atlantic. W and E refer to the direction of the main currents. The discontinuity above the Intermediate Water is shown, which is supposed at rest. (*After Sverdrup* et al., *1946.*)

lying to the south of the equator has a secondary flow to the south. There is as a result a zone of divergence at the equator, and this is especially noticeable in summer, when it is indicated by a tongue of colder water, upwelling from shallow depths below. The whole system is shallow, varying between 50 and 100 m in depth. Stalcup and Parke (1965) have recorded the equatorial undercurrent in the Atlantic in February and March 1963 at 25–35°w, where it covered a depth range of 75–200 m and had an eastward velocity of up to 75 cm/sec. The surface current varied between 60 cm/sec to the southwest and 60 cm/sec to the east. The variation in surface flow may be due to the decrease in wind stress in the area as the intertropical convergence moves through the area.

About 6 million m³/sec of water crosses the equator from the southern hemisphere to the northern on the surface in the south equatorial current. This amount of warm surface water moving to the north may have helped to displace the thermal equator northwards. The south equatorial current, which covers approximately the zone between 20°s and about 4°N, on reaching the east coast of Brazil divides at Cape Sao Roque, part flowing northwest towards the Caribbean, where it joins the north equatorial current, and part flowing south as the Brazil current, which flows south fairly fast along the South American coast. The equatorial counter-current, at the eastern side of the ocean, merges into the east-flowing Guinea current along the east–west African coast just north of the equator.

2.1c The Gulf Stream system

The waters driven westwards by the trade winds, forming the equatorial currents, provide the greater elevation on the western side of the Atlantic. Sea level in the Gulf of Mexico is raised and this initiates the Gulf Stream system. The system can be divided into three parts. The southerly part, the Florida current, extends from the Straits of Florida to Cape Hatteras; the Gulf Stream proper extends from there northwards to the east of the Grand Banks in longitude 45°w. Thereafter the current is more diffuse and branches, and in this area it is called the north Atlantic drift.

The character of the Gulf Stream has sometimes been misinterpreted. It is not a stream of warm water flowing through the ocean, but as Stommel (1958a) has pointed out, it is a narrow ribbon of fast-flowing water that acts as the boundary between two very different types of water. It prevents the warm water of the Sargasso Sea, on its right, from over-flowing the colder, denser water on its left and inshore side. At the surface there is a very rapid fall of temperature between the zone of maximum flow, which runs along the edge of the continental slope, and the coastal water. The difference is at a maximum in winter, when the Gulf Stream water, in its southern section, is nearly 20°c, while the coastal water is only 14°c. This sharp change in surface temperature was noted by early travellers. It was commented upon in 1609 by Lescarbot (Stommel, 1958a). In a vertical section the Gulf Stream can be divided into three layers. The surface layer is a few metres thick and varies greatly in temperature with the seasons. Below this is a layer in which the temperature falls off rapidly with depth, called the main thermocline. Under this is a thick layer of cold water below 1500 m. Rapid movement is restricted to the surface layers. There is probably a slow sinking from the surface to intermediate levels in the north, with a slow upward mixing with surface water in lower latitudes from 20 to 45°N. In winter at least a surface transference of water to the north must exist to keep the circulation going.

The total transport of the Gulf Stream system off Chesapeake Bay (estimated at 82 million m³/sec in April 1932) also includes some water which circulates to the southeast of the stream, but which is not properly the north equatorial current. This volume may be too high. Munk's theoretical analysis gives half the amount calculated from the density observations. The reason for the high value could be due to taking the depth of no motion at the bottom, instead of at 1500–2000 m, below which observations show southerly flow (Charnock, 1960).

The width of the Gulf Stream is narrow compared with the currents that feed it. It forms a band stretching from the continental slope off Cape Hatteras to 50°w longitude, south of the Grand Banks. On either side of the main stream there are distinctive water types—the slope water on the west and the Sargasso water on the east. The submarine relief influences the position of the current. It flows along the Blake Plateau at about 800 m depth to 33°N, but beyond this it flows through an area where the depth is 4000–5000 m, and where its position is less determined by the relief of the sea floor, because the movement below 1500 m is not great. Seamounts in this area may, however, have some influence on its course. Observations suggest that the water carried northwards increases in volume in this direction. Wüst (1933) calculated that 26 million m³/sec passes east through the Straits of Florida as the Florida current, an amount much less than that moving northeast further north. Additional water is partly derived from the Antilles current, which supplies about 12 million m³/sec. The Sargasso water on its right side supplies most of the remainder, making the total given already. The pattern is shown in figure 3.4.

Schmitz and Richardson (1968) have estimated that the transport of the Florida current

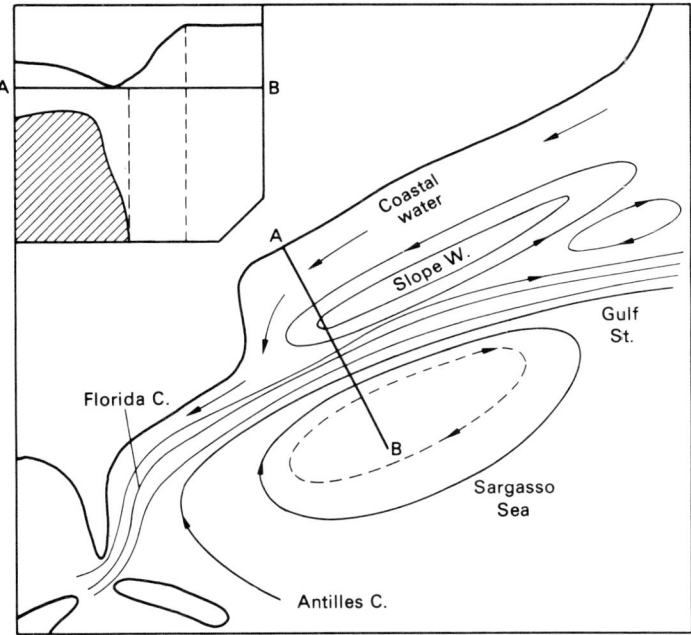

Figure 3.4 The approximate transport of the Gulf Stream system is indicated. Each full line represents a transport of 10 million m³/sec. The slope of the water surface across the stream is shown. (*After Sverdrup et al., 1946.*)

is 32 ± 3 million m³/sec in the steady state, based on 50–60 transects across the current. The total fluctuation bound is ±12 million m³/sec. The tidal components are 3·5 ± 1 million m³/sec for the lunar semi-diurnal and both diurnal constituents, and 1·5 ± 1 million m³/sec for the solar semi-diurnal part. There is no evidence for total transport fluctuations at non-tidal periods (for example, seasonal periods) of amplitudes larger than 10 per cent of the mean. Reproducibility of the transport in different years indicates that the system is stable.

The limits of the Gulf Stream can be defined in different ways. There are the obvious surface features, such as the colour change and the lines of Sargassum weed, but these may not coincide with the inner edge or western side of the current as defined by the pressure gradient between the warm, saline water to the southeast and the colder, fresher water to the northwest. On this basis the limits of the Gulf Stream can be defined as the points where the pressure gradient is zero. This may not coincide with the surface change in temperature and can only be located if there are detailed observations of temperature and salinity at depth. The surface phenomena are associated with shear zones at the edge of the 'warm core'. This is defined as that part of the Gulf Stream in which the water is warmer than that at the same depth to the east. It is usually at 300–400 m depth, within the maximum temperature anomalies at about 100 m. There can be abrupt changes in surface temperature, colour, and other features at the left edge of the warm core.

The actual speed and position of the main current varies, as shown by the detailed

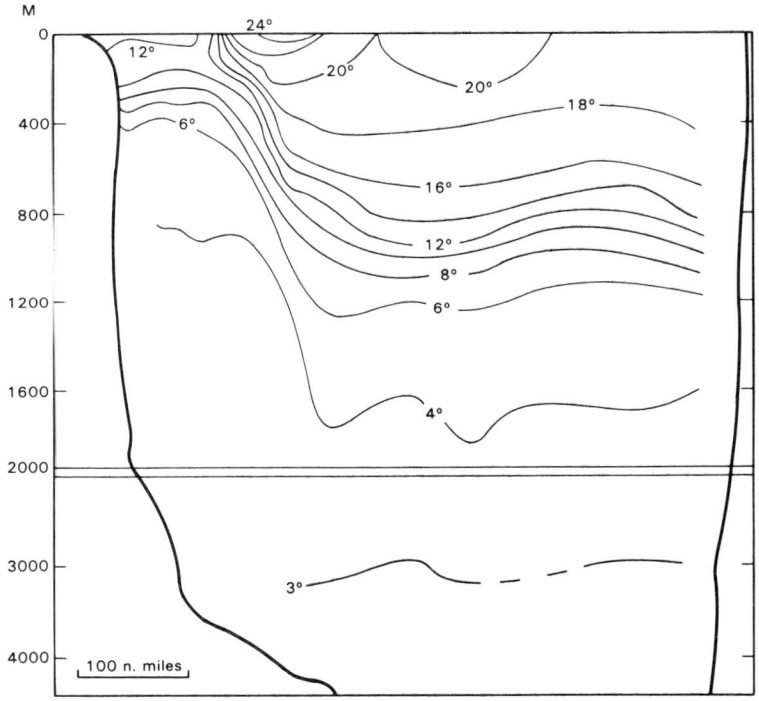

Figure 3.5 Temperature section across the Gulf Stream from Chesapeake Bay to Bermuda, 11–18 Feb. 1932. (*After Stommel 1957.*)

observations of Worthington (1954b). He has indicated that the current is very narrow and has a high velocity. Where the flow is fastest it reaches over 160 km/day. The high velocity is confined to a narrow band about 74 km wide, which reaches a speed of 250 cm/sec. A counter-current flows at 50 cm/day approximately in the opposite direction. The positions of the high temperature and high velocity do not always coincide. In the main thermocline the 10°C isotherm drops from 200 to 900 m across the stream in less than 110 km (figure 3.5). The position of this zone of rapid temperature change can be recognized fairly readily. Many observations of its position at varying times have shown that the Gulf Stream develops large eddies or meanders. Fugilister and Worthington (1951) have shown that two of them moved eastwards at about 20 km/day, while their amplitude doubled in a fortnight. Their wave length is about 200 km. The movement of the eddies can be traced by an airborne radiation thermometer. As the Gulf Stream merges into the north Atlantic drift in about 40°N the meanders intensify and break up into detached eddies at times.

Another point of interest in connection with the Gulf Stream is the variation of sea level along the east coast of the United States, deduced by precise levelling. The sea level slopes up to the north from zero in Florida to 6 cm at Norfolk, Virginia, 1000 km along the coast, and to 20 cm at Atlantic City, New Jersey, 1400 km along the coast. It is 28 cm at Boston, Massachusetts, 2000 km up the coast, and 35 cm at Halifax, Nova Scotia, 2600 km to the north. A south-flowing downslope current over the continental shelf results. The slope may be related to the piling up of water in the northwestern north Atlantic, due to the southwest wind and oceanographical distribution of mass. According to Iselin (1940) the variation in level noted on the tide gauges of the east coast of the United States is associated with the changing volume of water carried by the Gulf Stream. His evidence suggests a weakening of the Gulf Stream since 1934.

The energy of the current is derived from the difference in level of 19 cm which exists between the Gulf of Mexico and the Atlantic, as measured by precise levelling across

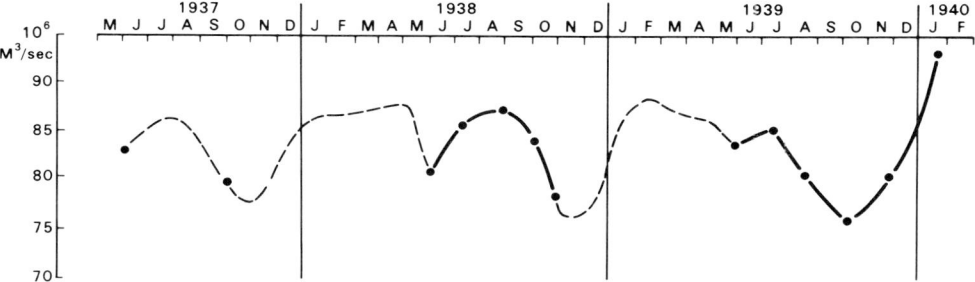

Figure 3.6 Gulf Stream transport determined by hydrographic stations. (*After Stommel, 1957.*)

the Florida peninsula. The average amount of water passing through the Straits of Florida has already been given. The speed of flow is considerable and the current is confined within the straits between Florida and Cuba. The water surface, therefore, slopes up to the south, while the lower interface slopes down in this direction. The maximum slope across the stream occurs in July (when the low is greatest), while at the same time there is a maximum downslope from west to east on the coast of Florida. The highest velocity, which can exceed 150 cm/sec, is found nearer the coast of Florida, the speed falling off with depth. Fluctuations in volume can be linked with changes in the major atmospheric circulation in the north Atlantic, particularly in the breakdown of the subtropical high-pressure cell, although the relationship is not direct.

Large fluctuations in the transport of water by the Gulf Stream further north have also been reported. The well-defined July maximum and November minimum in the speed of the surface current in the Florida area continues to a position south of Cape Hatteras. Northeast of the Cape the maximum flow occurs in May and the minimum in October, the variation amounting to about 15 million m³/sec (figure 3.6). Southwest of the Grand Banks the maximum is in March and the minimum in November. It might be supposed that the transport of the Gulf Stream would increase in winter due to the increase in wind strength, which would reduce the diameter of the current curve, and thereby interrupt or decrease the northeast discharge of water. The southerly movement of the wind system should cause the Gulf Stream to be further offshore in winter, north of Cape Hatteras. The Caribbean and Straits of Florida, however, influence the circulation in such a way that this does not occur. In summer the flow is increased by the water moving directly into the Gulf Stream, as a result of the northerly movement of the wind system.

The relation between the kinetic wind energy stress currents and the potential energy of the warm mass of Sargasso water affect the constancy of the Gulf Stream system. Stommel has calculated that the potential energy stored in the warm core of the Sargasso water, over and above that which would exist if the thermocline were level, is very great. It would be sufficient to keep the current system going for 1700 days in the absence of the driving force of the wind. Thus the current system is to a considerable extent independent of the fluctuations of the weather. The density field in mid-latitude ocean areas does respond to variations in the wind stress, but for variations within the period of a week and a year the response does not reach equilibrium. The main thermocline in the north Atlantic is influenced by the movement of the main high-pressure centre, but is not entirely adjusted to it. Fluctuations of less than one week cannot produce any changes. The Gulf Stream carries an enormous amount of water northwards, amounting to 33 times the water flowing in all rivers and glaciers on land, while the amount of salt carried north is 1,210,000 tons/sec (Defant, 1961, p. 641). A considerable amount of heat is carried north by the northerly extension of the Gulf Stream, the north Atlantic current.

Detailed studies of the temperature, salinity, oxygen, nutrients, runoff, and wind data on the shelf off North Carolina are reported by Stefansson, Atkinson, and Bumpus (1971). Four processes are considered by which the shelf water can be renewed: 1) horizontal advection near the coast from the north, 2) meanders of the Gulf Stream, 3) subsurface intrusion, and 4) cascading and runoff. In winter runoff is high, resulting in fresher shelf water in winter and spring. The pattern of freshwater distribution is affected by the wind. With strong northeasterly winds fresh water from Virginia intrudes into Raleigh and Onslow Bays, but with southwesterly winds the shelf water becomes much warmer and more saline, changing from 4–5°C and 29.9‰ with northeast winds to 12.5–13.6°C and 34–35‰ with a southwest wind. Wind and runoff both affect the process.

In Onslow Bay the Gulf Stream meanders through 10–30 nautical miles across the shelf, causing considerable changes in the water type. Subsurface water from the Caribbean sometimes intrudes over the outer shelf as a bottom layer. When this takes place the supply of nutrients is enriched. On the other hand, nutrients are lessened by the fourth process, which is the cascading of high-density water beyond the shelf break. The density increase is due to cooling in the winter. Northeasterly winds will accelerate the cascading process. The effect on the nutrient supply of these different processes will lead to variations in the primary production on the shelf area.

The volume transport of the Gulf Stream was measured by Richardson and Knauss (1971) using transport floats, which integrate the horizontal velocity over the whole depth. The three values obtained were 58, 67, and 64 × 10⁶ m³/sec, with a mean value of 63 × 10⁶ m³. The measurements indicated that the Gulf Stream extends to the bottom beneath and offshore of the high-speed surface layer, where the observations were made off Cape Hatteras. A deep southward flow was found on both sides of the northward flow, but temperature and salinity observations did not reveal any discontinuities in the water type between the north- and south-flowing water. The southward-flowing water is the western boundary undercurrent. Beyond Cape Hatteras the Gulf Stream starts to meander and its position is not constrained by the edge of the shelf as it is to the south of the Cape, where it flows over the Blake Plateau and has a minimum temperature of 7°c. To the north it entrains cold water as its minimum temperature falls to 2·5°c in depths in excess of 4000 m. The geostrophic flow obtained from hydrographic observations agrees with at least some of the direct observations. Figure 3.7 shows that although the Gulf Stream becomes narrower with depth it does extend to the bottom in 3000 m. Reversals of flow occur at either side and within the gap between two south-flowing zones. Comparison of the counter-current data for direct observations and geostrophic flow are compared in the following figures:

	Width		Transport	
	Direct	Geostrophic	Direct	Geostrophic
Left counter-current	30 km	40 km	$1 \times 10^6 m^3/sec$	$2 \times 10^6 m^3/sec$
Right counter-current	Variable	80 km	Variable	$10 \times 10^6 m^3/sec$

The counter-current appears variable, but bottom photographs by Rowe and Menzies (1968) suggest a width of over 300 km extending from 1100 to 5100 m. Streaks of high velocity appear to be separated by zones of little or no motion. The counter-current is the western boundary undercurrent, which is continuous along the continental slope and rise from southeast of Cape Cod to 200 km south of Cape Hatteras. This statement is based on much evidence including direct measurement, water-mass analysis, bottom photography, and distribution of sediments. The water characteristics cannot be used to separate the two currents, which can, however, be separated by their velocity fields. The Western boundary undercurrent may pass through the Gulf Stream according to Knauss.

The water-masses and their circulation in the Blake–Bahama Outer Ridge area in the western Atlantic have been studied by means of salinity, temperature, and depth observations by Amos, Gordon, and Schneider (1971). Their results were converted into a dynamic system that revealed the western boundary undercurrent flowing with a velocity up to 26 cm/sec at the bottom on the eastern flank of the outer ridge. The volume transport was computed to be $22 \times 10^6 m^3/sec$. This water contains up to 20 per cent of Antarctic bottom water, suggesting that the Antarctic bottom water flows north into the western Atlantic and is deflected by abyssal hills around 35°N to merge with north Atlantic deep water before returning south along the eastern continental margin. The western boundary undercurrent follows the contours in this area. The relief and current pattern (figure 3.8) shows the geostrophic velocities and deep transport values along the line of observations, and relates these to evidence derived from bottom photographs. The calculated volume transports vary from previous estimates for a number of possible reasons, including variations in the horizontal extent of sections and their depth and in times and seasons of observations. There are, however, considerable similarities in the general pattern and that

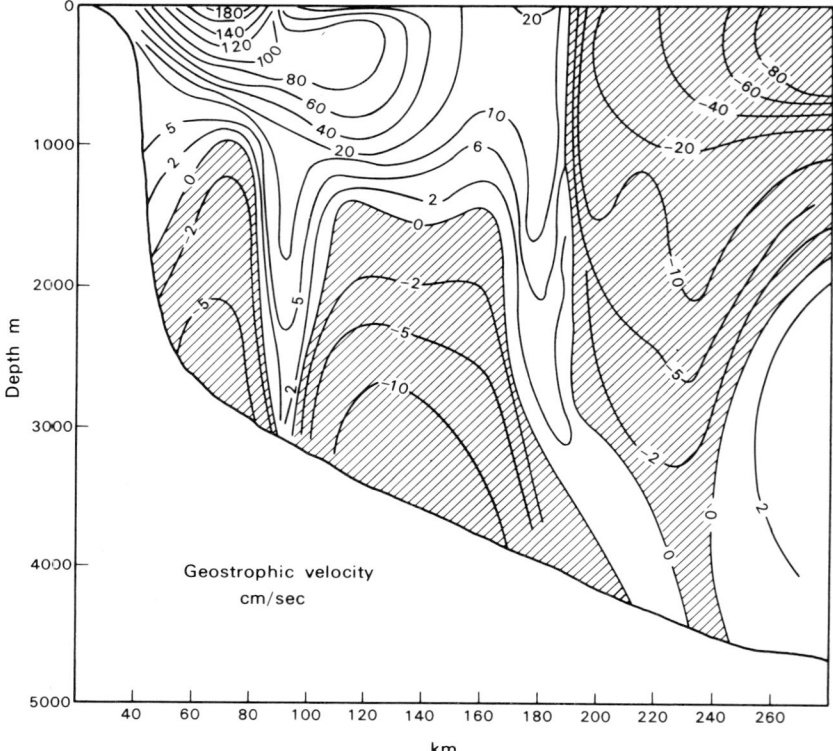

Figure 3.7 Geostrophic velocity pattern across the Gulf Stream. The shaded area represents south-flowing water. (*After Richardson and Knauss, 1971.*)

described by Richardson and Knauss (1971) off Cape Hatteras a little further north. In the Blake Plateau–Bahama Ridge area there is a cold isothermal layer on the bottom 20–170 m thick, in depths from 3800 to 5400 m. This layer, which is sometimes capped by a high-temperature gradient layer, contains a large amount of Antarctic bottom water. It may result from the confluence of Antarctic bottom water and north Atlantic deep water. This bottom layer may be a lower turbulent layer owing to bottom roughness, on analogy with similar characteristics in the lower atmosphere.

The north Atlantic current is not a narrow band of fast-flowing water, but consists of several broader bands or variable filaments of current. In this form it reaches as far as 65°w, the zone where the westerly concentration decreases and is no longer strongly marked. Further east the water transport branches, part being carried to the east and southeast to form the Canaries current, while the rest passes northeastwards and northwards into the northern North Atlantic and Polar Ocean. Surface water is influenced by the west wind drift, but beneath this there are eddies and fragmentary remnants of the Gulf Stream. In the zone of decay of the Gulf Stream, the surface conditions of the now slowly moving water are considerably altered by the climate. In the subpolar regions of the northern part of the system thermohaline processes become important.

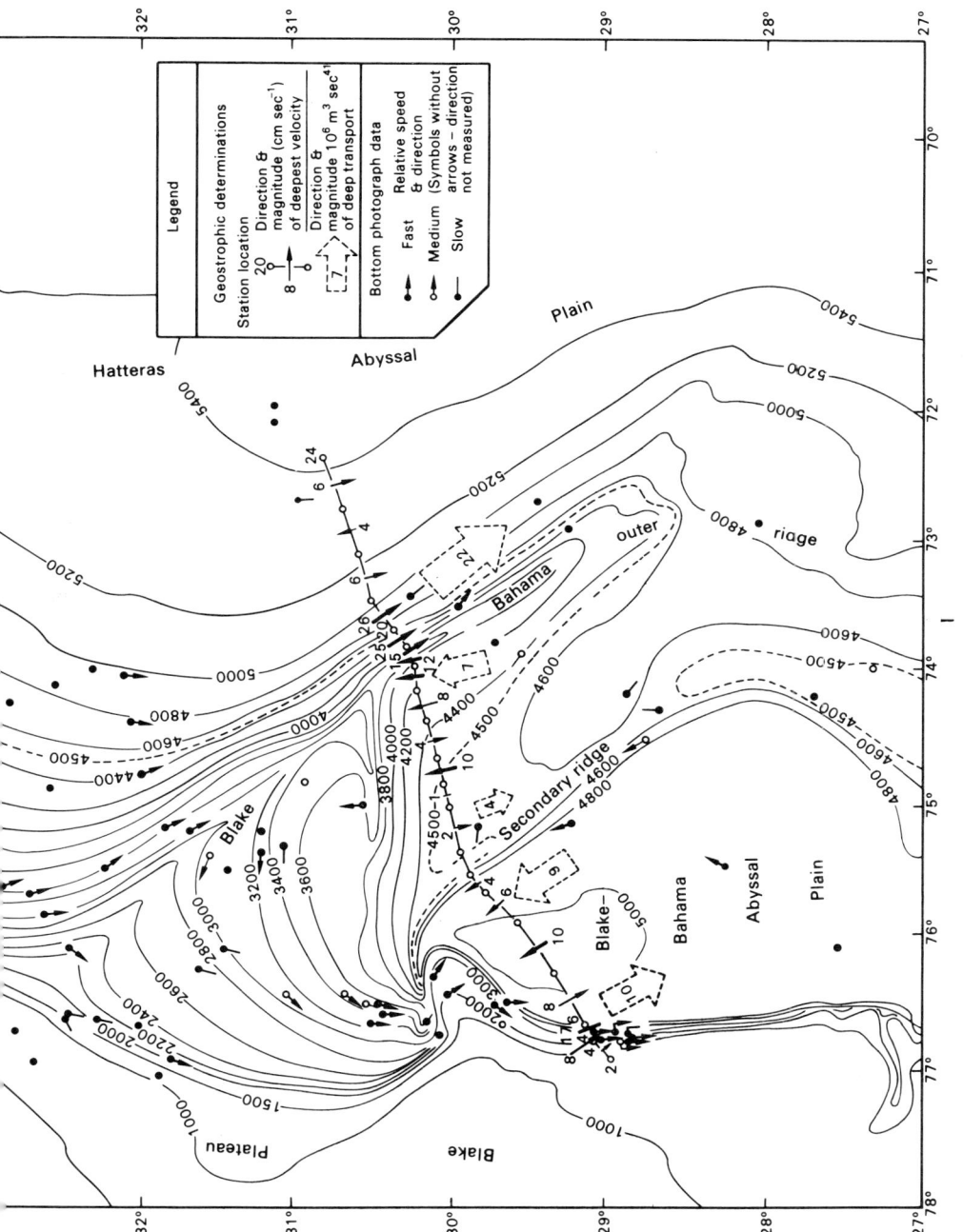

Figure 3.8 Comparison between deepest geostrophic velocities and data obtained from bottom photographs in the Blake Plateau–Bahama region. (*After Amos, Gordon and Schneider, 1971.*)

The central part of the north Atlantic Ocean contains a mass of warm water covering the uppermost 700 m. This water drifts slowly to the southwest as a result of the trade winds and the rotation of the earth. Carruthers has reported interesting evidence of the currents in the Sargasso Sea area, shown by the drift of the *Fanny Wolston*, which was abandoned in October 1891, and drifted for three years. The wreck zigzagged in the Sargasso Sea area until February 1893, when it got into the Gulf Stream, in which it drifted northeast until October 1894, finally sinking in 38°N after travelling 14,000 km.

The Gulf Stream system forms the edge of the mass of slow-moving warm water and prevents it flowing over the colder, denser water of the northern part of the ocean. The Gulf Stream is not, therefore, a river of warm water flowing through the ocean, but a limit to the northern spread of warm water. Its essential climatic role is not carrying warm water towards northwest Europe, but preventing the northwards extension of this warm water. Iselin (1940) has suggested that increased transport by the Gulf Stream may be associated with cooler conditions in Europe. This is based on the argument that a deepening of the thermocline in the Sargasso Sea might be associated with a radial shrinkage of the current system, when the transport is increased. This would lead to a southern retreat of the warm surface water in the north Atlantic. There is little evidence that this does in fact happen, although the tide gauge data for the east coast of America supports it. There is a suggestion that the warming of the Norwegian Sea by about 2°C and the decrease in depth of about 50 m in the 10°C isotherm throughout the Sargasso Sea would cause a slowing down of the north Atlantic circulation. This would decrease the transport of the Gulf Stream at the same time as the climate of northern Europe is ameliorating. It is significant that the annual variations in flow are greater than the year-to-year mean variation in the transport of the Gulf Stream.

Mann (1967) has considered the connection between the Gulf Stream and the north Atlantic drift, or the north Atlantic current as it is now called. The north Atlantic current is a mixture of slope water and Gulf Stream water in about the proportions of 15 million m³/sec to 20 million m³/sec respectively. The currents are shown in figure 3.9, which indicates that the slope water crosses 50°W north of 41°N and the Gulf Stream flows south of 40°N. At times the Gulf Stream moves north of 41°N, but at others a counter-current exists between the two streams, which may be part of a Gulf Stream meander. The Gulf Stream migrates between 39 and 41°N, according to the meander pattern. The total transport across 50°W is estimated at 65 million m³/sec above 2000 m. Thus although the major part of the warm water of the Gulf Stream passes south, 20 million m³/sec pass north into the Atlantic current. It is this warm water that ameliorates the climate of western Europe. The north Atlantic current is, therefore, a mixture of slope water and Gulf Stream water. It has a higher oxygen content than the Gulf Stream water. The oxygen is derived from the anticyclonic gyre that exists on the right of the system after the Gulf Stream has split. This eddy consists of water similar to that of the Gulf Stream.

One of the characteristics of the Gulf Stream system that has been studied recently is the meander system associated with it. Ruddiman (1968) has discussed the stability of the Gulf Stream meander system in the past by reference to evidence from foraminifera. The meanders are large off Nova Scotia, extending through a zone of 2° although the actual current remains only 32 km wide. The position of the late Holocene equatorial foraminifera in the bottom deposits allows the position of the current to be reconstructed, as these organisms are carried by the current. Four species are found that identify the stream. They are *Globigerinoides sacculifer*, *Pulleniatina obliquiloculata*, *Globorotalia menardii*, and *G. tumida*. The zone in which they are especially abundant is only slightly wider than that in which the meander belt has been observed to lie during the last two decades. Both terminate in the same area, where the streams weaken markedly. The results indicate that the meander system is stable, while its past pattern could be identified by studying the deeper layers of sediment cores.

Warren (1963) has examined the effect of relief on the Gulf Stream east of Cape

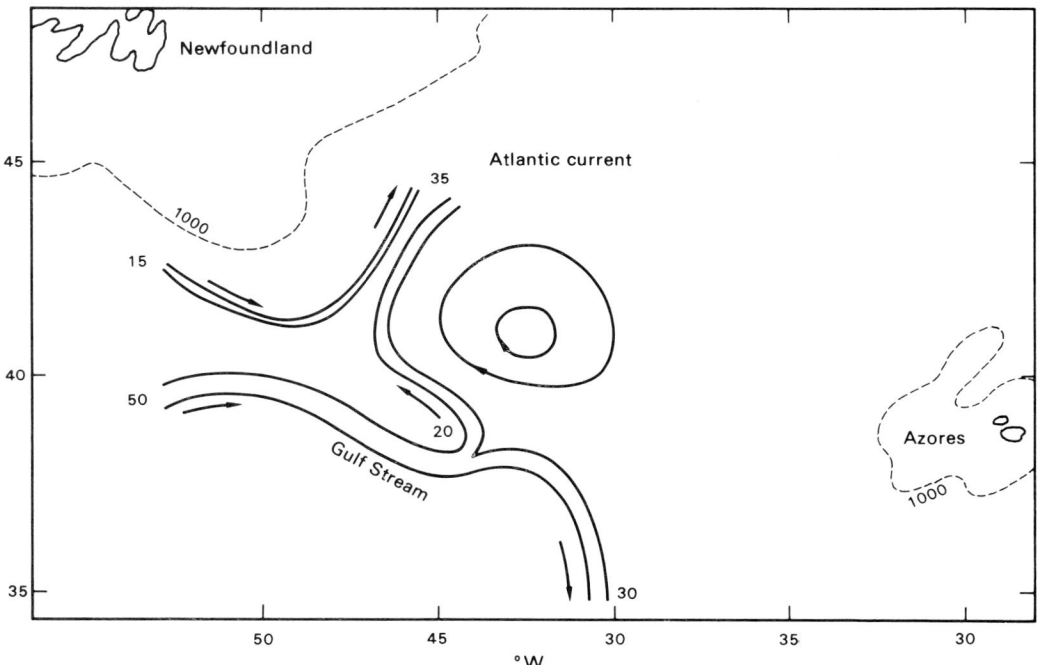

Figure 3.9 Transport volumes in the northern part of the Gulf Stream and the north Atlantic Current (*After Mann, 1967.*)

Hatteras. He suggests that its path is controlled by the continental rise where it extends to the bottom. Some observations by Fugilister indicate that it does in fact extend to the bottom. The Coriolis effect seems to exert less control than the relief on the path of the Gulf Stream. There are marked correlations between the isobaths and the pattern of curvature. The current can be explained by the constraints exerted by the sloping bottom on a quasi-geostrophic current extending to the ocean floor. Some features of the current can only be explained by this mechanism in quantitative terms. The meanders are topographically controlled stable waves. This is supported by the sediment foraminiferal data of Ruddiman. The stream flows along the Continental Rise for more than 1000 km after it leaves the slope. The meanders increase suddenly in size between 65 and 62°w, bending north and forming anti-clockwise flowing curves. The isobaths also bend sharply north in the same vicinity. Bottom velocities of 8 to 10 cm/sec have been recorded in the same direction as the surface stream, but observations are very few.

Holland (1967) has studied the same problem theoretically. He takes into account inertial and frictional effects and various bottom reliefs, but he neglects density stratification. The wind-stress curl is assumed steady. The results show that the separation, meandering, and variable transport of the Gulf Stream could be due to relief. The theoretical model of Bryan and Cox (1967) does not lead to separation of the boundary current from the coast. The density field influences the way in which the ocean feels the bottom, but the analysis is very complex, so simple models must be used. The reaction of the water depends on the nature of the assumptions. In the highly frictional case, in the zone of the western boundary current formation a southward-directed counter-current forms. In the highly inertial case, on the other hand, the water turns directly north. Where the current feels the shoaling bottom the separation region begins. The current leaves the western boundary in this area and follows the relief contours to the northeast. The effect of friction is dependent on the inertial parameter, W_i, the width of the inertial zone. In the topographic region now entered, divergence and convergence are of major importance. Friction and wind stress are negligible. Meandering becomes important in the highly inertial situation, and Warren's theory is successful. Friction does, however, allow the particle to complete the circuit with a rotation proper to the situation, and no energy or net vorticity accumulates in the circuit. The effect of slope steepness was calculated; as the slope increased in steepness so the streamlines were deflected south in the topographic region. When the slope is very steep the streamlines follow the slope and no western boundary current forms on the shelf as it does with a gentle slope. The separation of the current from the shore is a very complex process, but topographic effects are important in both frictional and inertial situations. The application of the simple model results to the real ocean is complicated by the little-known baroclinic and time-dependent processes, and the lack of knowledge of turbulent diffusion processes.

The more diffuse currents of the north Atlantic Ocean do carry warmer water into Arctic latitudes, some of this water flowing west, south of Iceland, as the Irminger current. The water eventually comes into contact with the cold-water current flowing southwest along the east coast of Greenland. The zone in which the two waters meet is the place of origin of part of the north Atlantic deep water. Some of the Gulf Stream water penetrates west and north of Iceland towards the Norwegian Sea, while the more easterly branch also flows into this sea along its eastern boundary, off the west coast of Norway. Some of this water penetrates as far as the Arctic Ocean, in which its character is changed by cooling and excessive runoff and precipitation, to become relatively fresh and cold. In this form it flows out of the Arctic Ocean as the east Greenland current.

The greater effectiveness of the north Atlantic current in warming the land of northwest Europe, compared with the Kuroshio, its Pacific counterpart, is partly due to its shorter course, of 5000 km compared with 8000 km in the Pacific, and the favourable arrangement of the land-masses, which allow greater penetration. The addition of the warm Mediterranean water at depth also helps to reduce downward conduction of heat.

Part of the southern branch of the north Atlantic drift also changes its character. Some water enters the Mediterranean as relatively cool, fresh water, to emerge below the surface as the very distinctive warm, saline Mediterranean water. The rest continues to the south to join again into the circulation as the north equatorial current, completing the gyre. The water, flowing south along the coast of North Africa, is the Canaries current. It is associated with upwelling particularly from January to May, when temperatures along the coast may show an anomaly of -7 to $-10°c$ between the Canaries and Cape Verde. Figure 3.10 illustrates some aspects of the oceanography of the area off northwest Africa.

2.2   The currents of the south Atlantic Ocean

2.2a   Benguela current

The Benguela current flows north close to the coast of Africa and is particularly strongly developed between the Cape of Good Hope and latitude 17 to 18°s. The denser water lies on the right, close inshore because it is in the southern hemisphere. The coldness of the offshore water is also increased as a result of the upwelling of water from moderate depths, caused by offshore winds. A belt of cool, relatively fresh water spreads to a distance of 200 km offshore. This upwelling is important as it brings nutrients to the surface.

The Benguela current is estimated to carry about 16 million m^3/sec to the north. As the water moves north beyond 20°s it gradually flows away from the coast to form the northern part of the south equatorial current. It helps to produce the tongue of cooler water that extends along the equator, reinforced as a result of the divergence. The Benguela current carries more water to the north than the southwards-flowing Brazil current on the west side of the ocean.

The Benguela current is the cool coastal current that flows north along the coast of southwest Africa between 15 and 34°s, within 100 nautical miles of the coast. Intensive up-welling takes place at different places in different seasons, being further north in the winter. A detailed study of the current was made by recording the temperature, salinity, oxygen, nitrate, phosphate, and silicate content of the water (Calvert and Price, 1971). The upward inflection of the isopycnals was most intense in the south, where upwelling was active. The water was derived from 150 to 250 m depth. The highest concentration of nutrients occurred off Sylvia Hill at 25°s in the area of active upwelling. The water is warmed as it moves up onto the shelf, but the salinity is little altered. The upwelled waters are very rich in nutrients, being more concentrated than at the level from which upwelling takes place, suggesting nutrient regeneration on the shelf. This process is probably due to entrainment of decomposing plankton in the shoreward-moving mass of upwelled water, on the shelf bed, while water moves offshore on the surface.

Bang (1971) uses detailed bathythermograph data to study coastal upwelling in the Benguela current area between 24° and 31°s. Data in figure 3.11 show the inclination of the isotherms to the shore indicating rising cool water in an area of active upwelling. Zones of upwelling are shown in figure 3.12 between 29 and 32°s. A southerly gale caused substantial changes in the temperature pattern and intensified coastal upwelling, with a 2°c drop of temperature. The isotherms became much steeper.

Four distinct meridional belts are shown in figure 3.12. The inshore zone (a) is one of discontinuous upwelling, driven directly by short-term wind stress. The intermediate zone (b) consists of water derived from earlier upwelling that is being warmed. This zone is often separated from (a) by a sharp discontinuity, marked as (2) and indicated by a change of colour. There are minor inversions, labelled (1) suggesting over-turning as

D

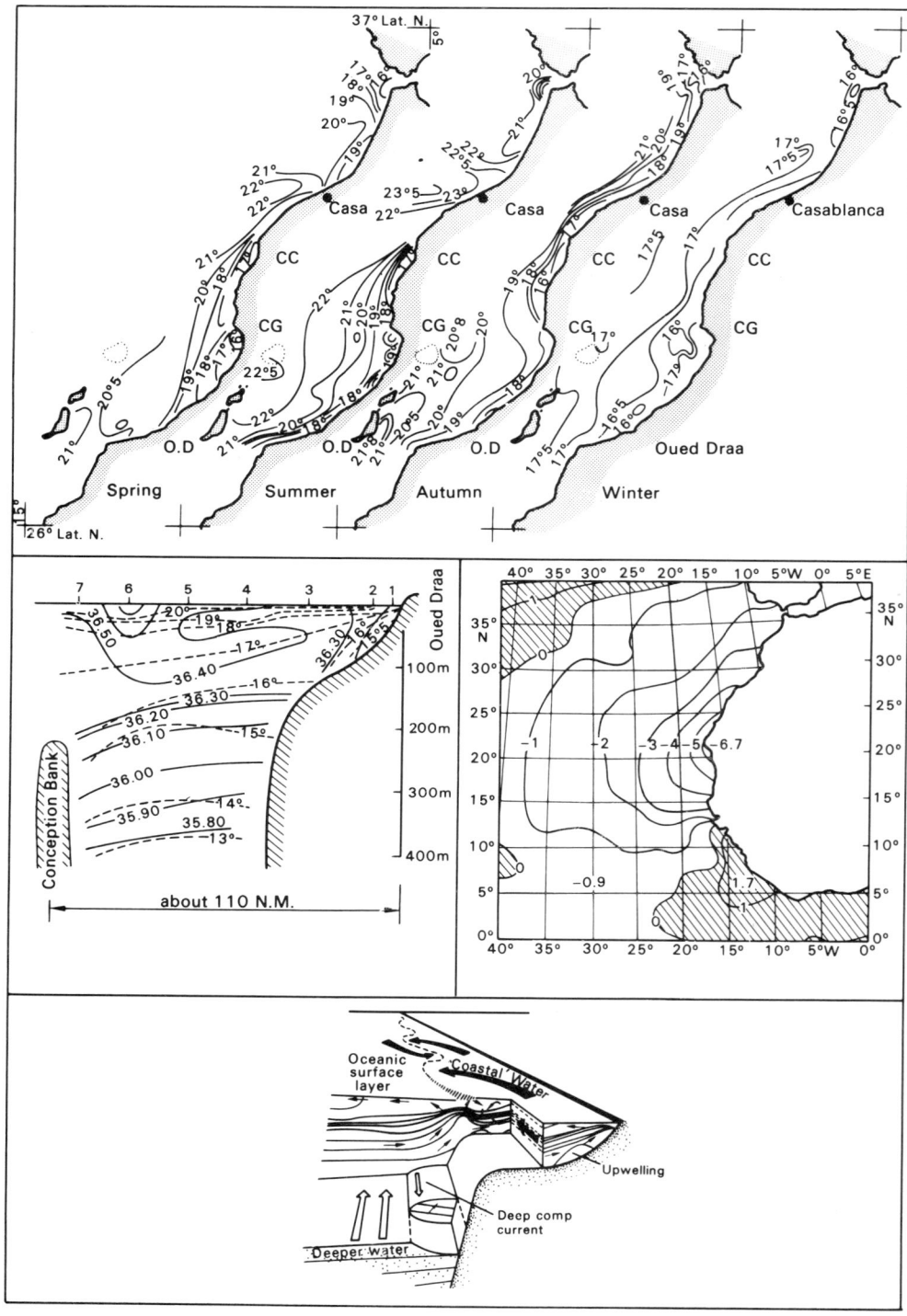

Figure 3.10 Hydrographic data off northwest Africa and the pattern of coastal upwelling and currents (*After Lacombe, 1970.*)

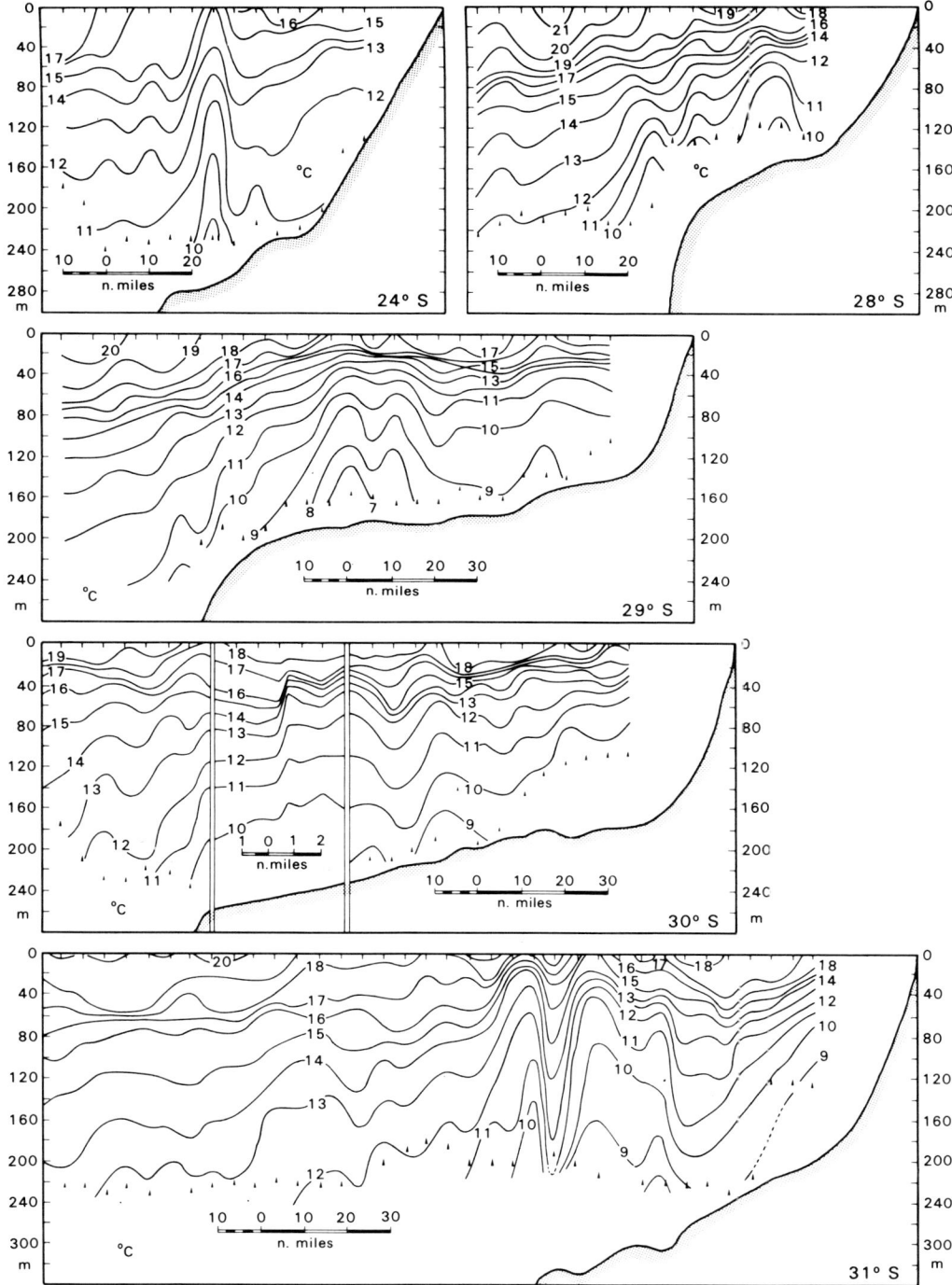

Figure 3.11 Vertical temperature sections along east–west lines south of the equator at the stated latitudes in the Benguela current region in February 1966. (*After Bang, 1971.*)

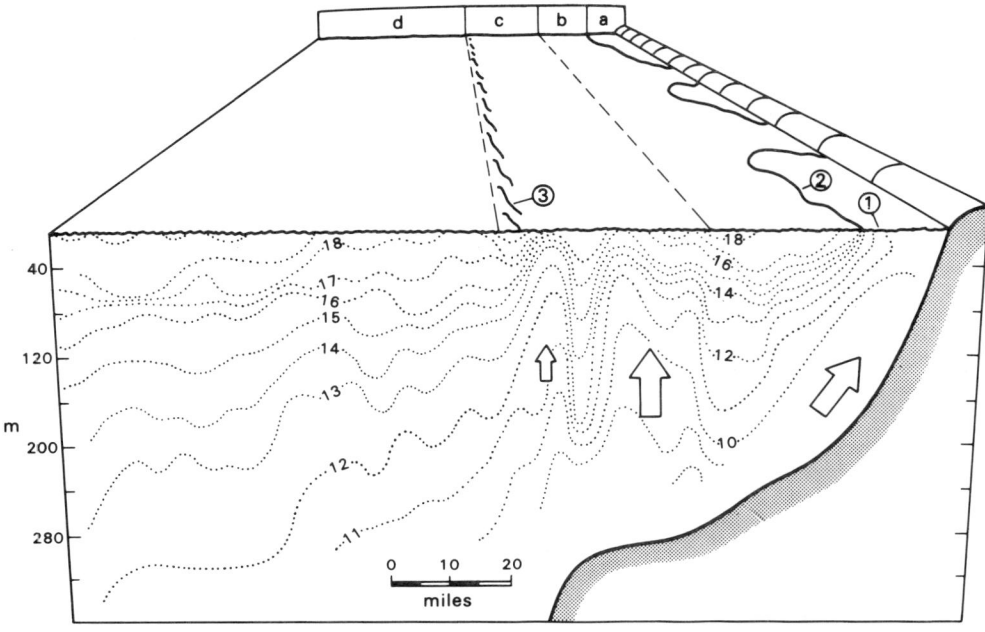

Figure 3.12 Diagrammatic representation of the main structural features of the Benguela region between 29° and 32°S, showing upwelling. See text for further details. (*After Bang, 1971.*)

upwelled water moves seawards. The zone (c) is one of offshore divergence where upwelling water may be exposed although this was not directly observed; however, slicks (3) were seen. The fourth (d) zone is the oceanic zone of simpler structure, consisting of warmer, northward-moving water in the Benguela system. The term Benguela current is restricted to the area east of the offshore divergence, which is characterized by short-term atmospheric interactions, while offshore of this zone is the trade-wind drift. The Benguela system is also related geomorphologically to the continental slope break. The Benguela current system is weather dominated, while the trade-wind drift is climatically controlled and is part of the south Atlantic general circulation.

2.2b Brazil current

The Brazil current is fed by the southernmost part of the south equatorial current. Ten million m³/sec flow south along the coast of Brazil. This current is warmer than the surrounding water as it is flowing away from the equator. It flows south to about 30°s, where it meets the cold Falkland current, which flows north along the coast of South America from the southern ocean. The Brazil current carries only one-tenth of the amount of water carried by the Gulf Stream. The difference is due to the greater stress of the wind in the north Atlantic, resulting from the shorter distance between the trade winds and the westerlies, which is 60 per cent less in the northern than the southern hemisphere. The northerly position of the thermal equator (relative to the geographical one) resulting from the distribution of land sea is the cause. According to Munk (1950) the permanent currents are related to the rotational component of the wind-stress field over the ocean. Hence the vector difference between the trades and westerlies is of great importance.

The Falkland current brings water of lower temperature and salinity into contact with the warmer water coming from the equator. Together these waters turn east to flow across the south Atlantic as the west wind drift. This completes the anti-clockwise circulation of the south Atlantic Ocean.

2.3 Currents of the southern or Antarctic Ocean

The flow of water in the southern ocean, which is continuous round the Antarctic continent, results from both the wind drift and the distribution of mass. The water of greatest density is found near the Antarctic land. The isobaric surfaces, therefore, slope up to the north. This results in a current directed from west to east, because the denser water lies to the right of the direction of flow in the southern hemisphere. The slopes are not great and the velocities of the currents are low, being about 15 cm/sec around 50°s and 4·4 cm/sec near 60°s on the surface. Deacon (1960a) has shown that drifting matter travels eastwards round the Antarctic at about 13 km/day, taking 3–4 years to complete the circuit. The flow falls off to low values at depth. There is a relatively sharp bend in the isobaric surfaces in the position of the Antarctic convergence, where a wedge of lighter water is found above a denser layer. The general west to east flow of water around the Antarctic continues all round the continent, although in places its direction is affected by submarine ridges. It has been estimated that between 80 and 100 million m³/sec is transported by this current (Stommel, 1957). The effect of the submarine ridges is to bend the current to the north as it approaches the ridge and to the south again as it passes across it. Five such deflections of the currents have been located. The southern extension of South America constrains the drift into an abnormally narrow zone, but on passing into the Atlantic the current bends north to form the Falkland current. It is probably only in this region that there is much flow uphill across the isobaric contours. Near South America the current is not purely zonal, but has a considerable meridional component. The southern part of the west wind drift may be called the circumpolar current.

Close to the Antarctic shore the direction of the current is reversed, owing to the effect of the prevailing winds, which are easterly in this zone. Between 40 and 60°s the strong winds are westerly and therefore reinforce the currents due to the distribution of mass. Between the two wind systems there must be a zone of divergence. The transport of water due to the prevailing westerly wind will have a component towards the north, which must be compensated by a southwards flow at a greater depth, resulting in the climb of water towards the surface. This has an important effect on the fertility of this part of the ocean. A large amount of nutrient material is brought to the surface in this zone, which accounts for the great productivity of the southern ocean.

The southern ocean is the only continuous ring of water around any latitude (Deacon, 1959). There is a low-pressure belt in about 65°s to the north of which the westerly winds blow. The current follows them and has a northerly component. Its speed is about 13 km/day, or about 2–3 per cent of the wind speed. South of the low-pressure belt the winds are more variable, but the currents mainly set towards the west. There is a southward diversion in this area, leading to a zone of divergence between the two systems. It is in this zone that upwelling occurs, while deep water probably moves east in this area despite the surface winds, apart from the Atlantic section and the eastern approaches to the Ross Sea.

Gill and Bryan (1971) discuss the influence of the Drake Passage on the southern ocean circulation in a series of numerical models. The models have 1) a complete barrier, 2) an opening 0·5 the total ocean depth, and 3) an opening to the full depth of the ocean. The circulation is imposed by wind and temperature stress. Five cases were tested as follows: 1) closed gap—linear surface temperature distribution, 2) deep gap—linear surface temperature distribution, 3) deep gap—curved surface temperature distribution, 4) deep gap—uniform temperature (barotropic case), and 5) shallow gap—linear surface temperature distribution. The effect of the gap is shown strongly on the pattern of meridional circulation. When the gap is deep it tends to prevent meridional transport across it so that two zones of transport result on either side of the gap. The zonal transport in the closed model must be accompanied by up- and downwelling, but when the gap is present transport takes place through it. The models illustrate some points concerning the formation of the Antarctic intermediate water. When there is a gap, water is forced to sink as it approaches the latitude of Cape Horn, and this sinking helps in the formation of the Antarctic intermediate water. The downflow regions north of the gap can be identified with the downflow near the Antarctic convergence. The circumpolar current is reproduced in the model, showing it to be an eastward-moving one, decreasing in strength with depth and moving north after passing through the gap. There is a westward counter-current near the bottom close to the southern boundary, whether the gap is there or not. This current has also been identified in reality. There is a dramatic increase in transport when the bottom half of the gap is closed, the increase being three-fold. The increase in transport is due to pressure differences on either side of the gap and these act in the same direction as the wind stress, thus enhancing the wind effect. This driving mechanism is therefore thermal, as the water is colder on the western boundary than the eastern. In reality the water is denser on the Atlantic side of the Drake Passage than on the Pacific side, so that a similar pressure difference may exist in reality. Data to calculate the pressure difference are not yet available, however. The predominant driving force is the wind when the gap is deep, but when it is shallow the predominant driving mechanism is thermal, and transport is three times greater than with the deep gap.

2.4 Indian Ocean currents

The west to east current of the southern Atlantic continues into the southern Indian Ocean, forming part of the anti-clockwise circulation of the southern hemisphere, similar to that in the south Atlantic. The seasonal wind variations in the northern Indian Ocean cause a complete reversal of the currents with the seasons in the northern hemisphere. The west to east current in the south in the southern summer bends northwards before reaching Australia, where it is reinforced by a current flowing to the west along the south coast of Australia. In the southern winter this current flows east along the coast of Australia. The south equatorial current flows to the west, north of 20°s, flowing faster in the southern winter, when it is augmented by water flowing west along the north coast of Australia. In the southern summer this flow is reversed. On reaching the coast of Africa the water turns south to form the strong Agulhas stream.

South of 30°s the Agulhas current forms a well-defined narrow flow extending less than

100 km from the coast. The coldest water lies nearest to the shore and the surface slopes up away from the coast at a gradient of 29 cm in about 110 km. This current mainly turns east on reaching the southern tip of Africa, although some of the water may flow west into the Atlantic where strong vortices develop. The current is caused by the piling up of water on the coast of Africa and is a typical gradient current.

In the northern part of the ocean, the currents reflect closely the seasonal changes in the wind direction as the monsoon develops. The north equatorial current is well developed in February and March, when the northwest monsoon is blowing. An equatorial counter-current is also strongly developed, with its axis at 7°s. When the southwest monsoon develops in August and September the north equatorial current is replaced by the monsoon current, which flows towards the east. At this season the water flowing along the coast of Africa moves northwards from latitude 10°s and some of this crosses the equator. At this season the equatorial counter-current is also lacking. These movements take place above the thermocline.

Dietrich (1963) has suggested that the transport by the Agulhas stream amounts to rather more than 20 million m³/sec. The surface currents of the Indian Ocean show that in the upper layers of this ocean the transport by wind is important, as the direction of flow changes with the variations in direction of the monsoon.

One of the strong currents of the Indian Ocean is the Somali current, which flows northeast with increasing volume (Swallow and Bruce, 1966). Deep measurements were variable, but they tended to indicate a southwards motion below 1000 m at between 1·4 to 28 cm/sec. The surface speeds, on the other hand, exceeded 200 cm/sec near the coast in the Somali current. The current turns east near 9°N.

Swallow (1964) has observed the Indian Ocean equatorial undercurrent at 58°E and at 67·5°E in May. Some subsurface movement was always found at a depth of 50–100 m The current at 58°E was weaker in June and shallower, but it was still present, flowing at 84 cm/sec at 50 m depth. Speeds up to 120 cm/sec were recorded in March. The equatorial undercurrent is more variable in the Indian Ocean, but it does exist.

Swallow's results may be compared with those of Knauss and Toft (1964), who also recorded the Indian Ocean equatorial undercurrent. The core of the current was in the middle of the thermocline at 100 m depth, with a maximum speed at the equator. The current was stable over the period of a week with a velocity of 75 cm/sec at 92°E. Knauss and Toft, unlike Swallow, found that the current was variable in the west of the ocean and was not always present. In most places the thermocline was spread where the current velocity was large, but not where it was absent. The strongest current recorded occurred during the northeast monsoon.

The circulation in the Indian Ocean illustrates well the influence of the wind system on the ocean currents because of the monsoon reversal in wind direction. Taft (1971) has shown that the situation along the Arabian coast is more complex than in the centre of the ocean, where the near surface of the northern Indian Ocean is primarily wind-driven with a one-month lag in the centre. A zonal band of vortices may extend east from the Somali coast during both monsoon phases and continuous zonal currents do not exist. Surface temperatures also change with the seasons. Coastal upwelling occurs in the south-west monsoon and is greater on the west than the east side of the ocean. The upwelling north of the Somali current and off southwest India is not wind-driven, but off the Arabian coast cool water rises to the surface through the wind-drive. At the equator there is no upwelling with either monsoon, due to the east wind stress. The Indian monsoon winds are symmetrical, blowing northeast in January and southwest in July, the latter being the stronger and generating the stronger currents that move east north of 5°s. The Somali current forms an intense western boundary current as far as 8°N, where it leaves the coast and turns east. Off the Arabian coast weak currents flow northwest. The averaged currents hide vortices, which show up on the 800-db surface as highs and lows 5° of latitude apart,

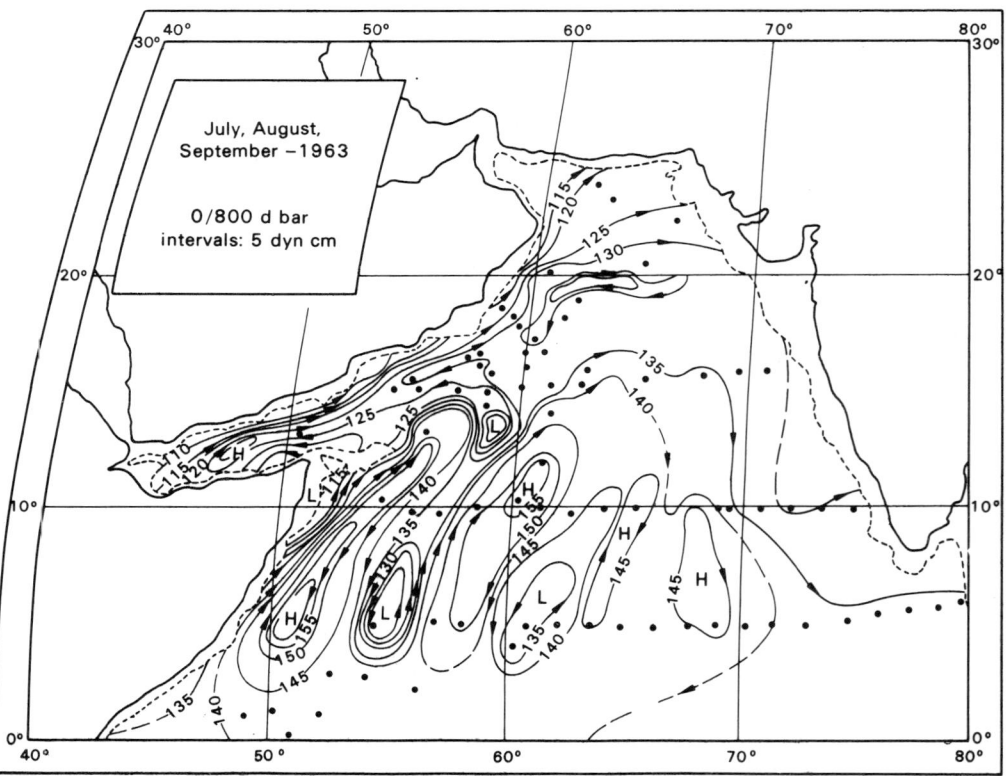

Figure 3.13 Dynamic height of the sea surface relative to the 800 db surface for summer 1963. (*After Taft, 1971.*)

forming south-southwest to north-northeast ridges and troughs, with highs at 50°E and 60°E, and troughs at 55°E and 63°E, as shown in figure 3.13. These vortices would cause mixing and increase the nutrients off Arabia.

In January the western side of the Indian Ocean is cooler than the Indian coast, the isotherms running southwest to northeast. The only upwelling at this season that is suggested takes place off Pakistan in a small area; there is no upwelling off the Indian coast because of the slack northeast monsoon winds. The southwest monsoon causes large changes of temperature, cooler water occurring off the Somali and Arabian coasts. The temperature falls to less than 22°c in places, and 14°c was recorded in August–September

1964. The water is low in oxygen where the current turns offshore to the east. Parallel to the coast the winds should cause upwelling, but there is little evidence of this along the coast south of 8°N. This may be due to mixing resulting from the strong horizontal gradient across the current. North of 8°N the thermocline appears to break the surface, because the Coriolis effect causes the isotherms across the current to slope more steeply northwards. The outcrop of the current is probably due to dynamic constraint. The cold water off the Arabian coast is due to upwelling with strong winds blowing parallel to the coast. The surface currents are not so strong in this area, and the southwest winds drive water offshore so that upwelling brings up nutrient rich water. No temperature minimum has been recorded along the equator in the Indian Ocean where the wind stress is eastward, which would lead to convergence and not divergence so that wind-driven upwelling would not be expected in the Indian Ocean.

2.5 The surface currents of the Pacific Ocean
2.5a South Pacific

A double anti-clockwise circulation probably exists in the south Pacific. The eastern gyre is better developed than the western one. The latter appears to be very variable with the seasons.

The Peru or Humboldt current flows north along the coast of Chile and Peru. Its vagaries have an important influence on certain aspects of the life of the area. The water of this current is derived from the subantarctic water, which flows eastwards across the Pacific. The volume transported is about 10–15 million m³/sec. The current extends to about 900 km from the shore and it turns west just south of the equator. Upwelling water close to the shore is thought by Gunther (1936) to come from depths between 40 and 360 m, the mean being 133 m. It is restricted to certain regions, four of which occur between 3 and 33°s. This leads to zones of alternating cold and warm water offshore. Beneath the northward-flowing surface current a counter-current flows to the south above about 400 m. This is formed of Pacific equatorial water, which is less saline than the surface water off Peru, but more saline than off Chile to the south.

The southward extension of warm, relatively fresh water is associated with the Peru current. The amount and distance it reaches varies from season to season, but it can have very serious consequences. During periods when the warm water extends south along the coast there is liable to be heavy rain on the usually dry coast, as the winds change from southerly to northerly. The incursion of warmer water is known as 'El Niño' and it occurs roughly every 7 years, for example in 1911, 1918, 1925, 1932, 1939, and 1941. Particularly severe periods occurred in 1891, 1925, and 1953. The anchovies, which are the basis of the massive fishing industry, die or migrate during the periods of 'El Nino'. It is thought that the anchovies either move south into cooler waters, or further offshore, where the temperature is also lower. One problem is to establish whether these disastrous consequences are due to warming of the water only, or to a decrease in the nutrient supply, warming only being a secondary feature.

There are various possible ways in which the water can be warmed. Firstly, warm water may extend southwards along the coast. Secondly, warm water may come close inshore from out to sea when upwelling is not active. Thirdly, solar radiation, in the absence of upwelling, may cause local warming. The warming of the coastal waters extends at times as far south as Pisco Bay in latitude 13°45's, although this may be due more to warming as a result of lack of upwelling, than to the extension of El Niño this far south. The observed warming of Pisco Bay can be accounted for by solar radiation, particularly as the salinity tends to increase rather than decrease, which it would do if the warming were due to an incursion of warmer, less saline water from the north. The actual surface temperature may be about 20 to 21°c in both normal and abnormal years, but the downward gradient varies. In abnormal years the 14°c isotherm is about 50 m below its level in normal conditions. In a normal year repeated upwelling prevents any warming at depth and the average temperature down to about 50 m is about 14·5°c as against an average of 18°c in 1941, an abnormal year.

The biological effects of abnormal conditions are of two types. The first leads to the disappearance of the anchovies. This could be due either to excess warming or to the absence of upwelling water. The second possibility is the appearance of 'red water'. This leads to large-scale mortality of fish and has usually been associated with the incursion of El Niño. The term 'red water' refers to the colouring of the water as a result of the excessive numbers of small planktonic organisms. These small phytoplankton grow when the water is extra warm, but other factors on which their abnormal production depends are not clearly known, although it has been suggested that fresher water is sometimes present. The development of red water does not necessarily depend on the incursion of water from elsewhere; the usual upwelling of the Peru current, if warmed on the surface, may give rise to red water (Sears, 1954).

2.5b   South equatorial current

This current flows astride the equator and is separated from the west-flowing north equatorial current by an equatorial counter-current, flowing east. This latter current is much more strongly developed than in the Atlantic. Defant (1961) shows that the current, unlike that in the Atlantic, is due entirely to the wind-stress system. The Cromwell current, called the Pacific equatorial undercurrent by Cromwell, who first discovered it, also flows east a short distance below the surface along the equator. It was first found in 1951, when fishing for tuna. Long-line fishing gear drifted to the east below the surface, where the surface current was flowing west.

The equatorial counter-current flows at up to 100 cm/sec on the surface during the northern summer, when it lies further north of the equator. It has been estimated to carry about 25 million m³/sec eastwards, an amount that probably increases towards the east. A similar system of transverse circulation is set up as a result of the rotation of the earth as that described in the Atlantic. Similar zones of divergence and convergence also develop.

Some aspects of the equatorial undercurrents have already been mentioned. The Cromwell current in the Pacific is one of the best developed of these currents. It is a thin ribbon of fast-flowing water, moving east beneath the south equatorial current at the equator. At 140°w the current velocity was 125–150 cm/sec at a depth of 100 m. The rate fell off with depth to 10 cm/sec at a depth of 350 m. The change of direction took place at a depth of 20 m. The current extended laterally 2° of latitude either side of the equator. The velocity gradient above the maximum was greater than that below it, changes of velocity of 150 cm/sec taking place over a depth range of 70 m. The current is nearly symmetrical about the equator. It is 0·2 km thick and 300 km wide within the speed of 25 cm/sec (figure 3.14). The amount of water transported is large, averaging 39 million m³/sec. The current is steady, especially compared with the changes observed in the counter-current, which flows east 800 km north of it. The Cromwell current has been traced eastwards to 92°w, near the Galapagos Islands, but it has not been found east of 89°w. The current weakens as it is traced eastwards.

At 140°w the temperature fell from between 25 and 28°c at a depth of 100 m to between 10 and 12°c at a depth of 300 m, at the base of the thermocline. This was better developed on either side of the equator than at the equator. The oxygen content was also greater

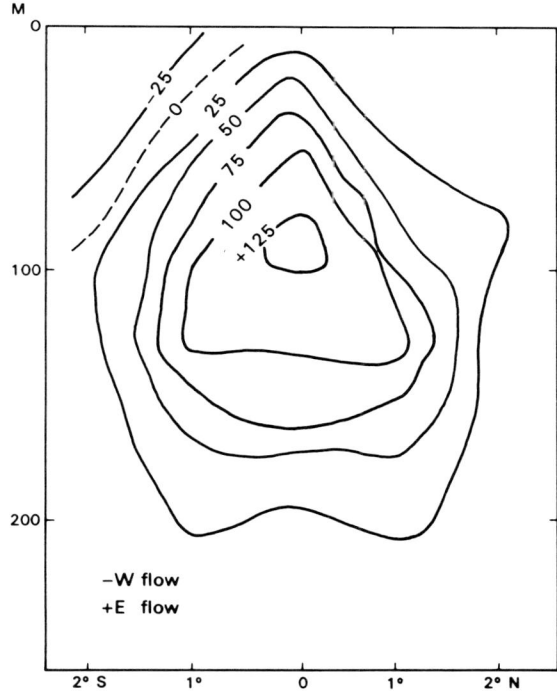

Figure 3.14 A veolcity cross-section in cm/ sec of the Cromwell current at 140°W on 20– 22 April 1958. The total transport was 42 million m³/sec. (*After Knauss, 1960.*)

beneath the equator than the low values found at similar depths of 100–300 m on either side. This is indicative of considerable vertical mixing of water at the equator, probably accompanied by upwelling. The marginal limits of the Cromwell current to north and south coincide with the limit of the zone of mixing.

Knauss (1960) suggests that the Cromwell current can be accounted for as a result of the mixing that takes place at the equator. If there were a simple two-layer structure extending across the equator, there would be no north–south horizontal pressure gradient and no resulting geostrophic current. If, however, the water at the equator is mixed, the density structure is altered with respect to that to the north and south. This will result in the formation of a geostrophic current. Both the pressure gradient and the deflection due to latitude change sign at the equator, so the resulting current is symmetrical about the equator. The Cromwell current, therefore, seems to be in geostrophic balance and results from mixing across the thermocline at the equator. The current becomes shallower to the east as the thermocline rises in this direction. In order to account for the Cromwell current in this way it is necessary to explain the vertical mixing at the equator to a depth of 300 m. This could be the result of wind-induced divergence.

The equatorial undercurrent in the Pacific Ocean near the Galapagos Islands has been studied by Christensen (1971). The core of the current is not always centred on the equator and it varies in depth, and in width by a factor of 2; the transport volume varies over an order of magnitude compared with the value in the central Pacific. The current ends in the area studied by the Equator Expedition, which made detailed current and temperature observations. The vertical temperature pattern was shown to be variable, by means of chains 270 m long with 34 equally spaced thermistor beads and 4 current meters. The variability was greatest away from the current, where the pattern was very smooth. The vertical temperature gradient weakens in the current. The north–south current components were small compared with the east–west ones. The undercurrent is narrower between 92 and 97°w than in the central Pacific, and at 97°w the greatest easterly speed was 0·45 m/sec at a depth of 75 and 150 m on the equator.

The distribution of nitrite in the equatorial Pacific has been discussed by Hattori and Wada (1971), who show that concentrations are very high in the subsurface waters on both sides of the equator. South of the equator, between 2·5 and 7·5°s the nitrite maximum was centred at a depth of 150 m, where the salinity was a maximum. The production was about 0·06–0·08 μg at N/l/day, with a residence time of about 30 days. Nitrate-reducing bacteria were abundant in the area, particularly south of the equator, while to the north reduction of nitrate is accomplished by phytoplankton.

At the western side of the Pacific Ocean the south equatorial current in part turns south to form the east Australian current. It is often U-shaped with a southward current near the continental shelf and a northwards component offshore, according to Hamon (1965). Anticyclonic eddies 250 km in diameter occur near the continental shelf, moving southwards away from the coast at 130 km/month. The width of the current is about 150 km and the horizontal shear is greater on the right-hand side of the current. The velocity falls to half the surface value at a depth of 250 m and is zero at 1300 m. The current causes a sea-level change of $\pm$30 cm on Lord Howe Island (160°e, 32°s) as it changes, but the change is not seasonal. The volume transported according to Hamon is 12–43 million m³/sec, although Rotschi and Lemasson (1967) give a range of 10–25 million m³/sec southwards flow. They consider that the current starts at about 20°s between 158 and 153°e, as the westerly-flowing water is diverted south by the land-mass of Australia. The current does not extend as far east as its equivalents, the Gulf Stream and Kuroshio in the northern hemisphere. The current does have some features of the western boundary currents, however. There is a narrow swift current near the edge of the continental shelf between 27 and 32°s according to Hamon, but there are also equally narrow north or northeast return currents, which contrast to the pattern in the northern hemisphere. The surface temperature structure is also poorly developed. The fall is only 1·5–3·5°c across the current, compared with 12°c across the Gulf Stream.

Eddies are shed more frequently and in a definite area; they move south. The current is probably wind-driven. Its volume is half that of the Gulf Stream, a fact that may be due to the west wind current absorbing some of the energy of the wind, and therefore the gyre set up in the south Pacific is not as strong as that in the north Atlantic. The east Australian current could be expected to have double the volume of the Gulf Stream in view of the greater width of the Pacific, if other factors did not intervene.

2.5c   Currents of the north Pacific

The north equatorial current in the Pacific resembles that of the Atlantic in many ways. It runs from east to west, increasing in volume as it flows west. It starts from the west coast of America, where some of the water of the counter-current flows north to meet some flowing south in the Californian current. As the current flows west its volume is augmented by some of the central water-masses. It is broad and deep, but not very fast, its velocity mostly being below 20 cm/sec. As it approaches the western part of the ocean it divides, some feeding the counter-current and some flowing to the north to follow the northern Philippine Islands and the coast of Formosa.

The Kuroshio current is formed of the north equatorial current water as it turns north, and it is the Pacific equivalent of the Gulf Stream. It is also divisible into three sections. The Kuroshio current forms the southern portion, and flows from Formosa to the Ryu Kyu and as far as 35°N, close to the coast of Japan. The Kuroshio extension branches there to form two streams flowing to about longitude 160°E. Finally the north Pacific current flows eastwards to around 150°W, sending various branches off to the south as it progresses.

The Kuroshio current is in many respects similar to the Florida current. According to Japanese oceanographers, the current extends to a depth of 700 m, and attains a velocity of 90 cm/sec on the surface in summer, while in winter the maximum is 61 cm/sec at a depth of 150 m. The amount of water carried in this zone reaches 21 million m³/sec in winter and 23 million m³/sec in summer, the transport increasing a little towards the northeast. The temperatures of this water are similar to those in the Atlantic, but the salinities are considerably lower, owing to the generally lower values in the Pacific. The current is variable and has shown considerable fluctuation during the period 1919–1950. A large cold eddy evolved during 1935–1942 off Cape Shiono Misaki, diverting the current 200 km offshore, but it has since returned to its former position (Stommel, 1958a).

In latitude 35°N the current branches to form the Kuroshio extension. It forms a fairly well-defined narrow current. The northern branch mixes with the southerly-flowing cold water of the Oyashio current. This mixture forms the subarctic water of the north Pacific. Most of the north Pacific current turns south again before it reaches the longitude of the Hawaiian Islands, turning eventually back to the west. In the northern part of the ocean the Aleutian current carries different type water, the subarctic water mentioned above. Some of this water flows anti-clockwise around the Bering Sea, but 15 million m³/sec flows east on the southern side of the Aleutian Peninsula and Islands. It does not by this time resemble in any way the water of the Kuroshio, from which it was originally derived, owing to mixing and to external changes. The Kuroshio current has less effect on the climate of the north Pacific than the Gulf Stream in the north Atlantic, owing partly to the greater distance it must cover before reaching the eastern shores of the ocean. The Aleutian current also divides, one branch flowing into the Gulf of Alaska and another flowing south along the west coast of North America.

The branch flowing south along North America is the Californian current, where it flows south from 48°N to 23°N. Its offshore boundary is about 700 km away from the coast, and its total transport is only about 10 million m³/sec. It is, therefore, only a sluggish flow to the south, very different from the fast-flowing western boundary currents. Like the Peru current in the southern hemisphere, particularly during spring and early summer, upwelling occurs. In areas of active upwelling this results in lower temperatures in spring than in winter. The regions of upwelling are localized and result in the formation of swirls. The upwelling water does not come from depths greater than 200 m and ceases towards the autumn.

The total transport of water by the surface currents of the north Pacific is less than that in the Atlantic. The maximum transport of the Kuroshio system is about 65 million m³/sec compared with 82 million m³/sec in the Atlantic. However, both patterns illustrate the strong asymmetry of transport, with the strong currents and transport concentrated on the western side.

The pattern of surface currents depends partly on the distribution of land and sea, so that the pattern in the past may not have been the same as the present one. Fell (1967) has considered the surface current pattern during the Cretaceous and Tertiary periods. He bases his results on permanency of the oceans, but he does make allowance for the movement of the pole through 70° in the last 500 million years. He considers that a rotation of the major axes of the current gyres best explains the palaeontological distributions. He reaches the conclusion that the west wind drift, the Benguela current, and the south Atlantic equatorial currents all existed through the Tertiary, and probably in the late Cretaceous. Extrapolation of time-dependent variables can be carried back to the late Cretaceous, including ocean mass, salinity, movement of the earth, moon, and sun systems. In the late Cretaceous reconstruction, the equator is placed through Panama and just north of Australia. Fell shows the Gulf Stream flowing through the Tethys Sea to the south of Great Britain, during the period of extensive marine transgressions. The pattern he considers has remained similar through much of the early Tertiary. The Panama seaway remained open until the Pliocene in his view. He states that the Atlantic is poorer in fauna than the Pacific and Indian Oceans because the fauna is younger, although the ocean is not. Both Atlantic and Pacific faunal realms can be recognized in the Cambrian–Ordovician period. The faunal paucity of the Atlantic is due to faunal losses consequent upon the middle Tertiary changes in the circulation pattern, following the separation of the Tethys into two portions, and the diversion of the Gulf Stream to the north.

3   Deep water circulation

Deep water circulation of the ocean can be studied in three main ways. Methods of measuring the flow directly are now giving much more reliable results. Deductions can also be made on the basis of the temperature and salinity observations. These provide a picture of the distribution of water-masses and of the density and mass balance of the water. Theoretical studies can also be made.

3.1 Theoretical analysis

The work of Stommel and others has given a reasonable picture of the factors on which the deep water circulation of the oceans depends. Some of the features predicted by Stommel's theory have subsequently been verified by direct observations. His analysis is based on steady geostrophic flow, except in narrow bands where other forces such as friction or inertia play an important part. Circulation is driven by zones of sinking water, themselves related to the surface circulation. Vertical movements are essential to an explanation of the deep water circulation (Stommel and Arons, 1960; Stommel, 1958b).

One of the major elements of Stommel's theoretical circulation is the narrow western boundary current, similar to that on the surface. The circulation at depth is based on a small source region of deep water in high latitudes, compensated by a slow upward movement at mid-depths in other areas. There are only two important sources of deep cold water on which the oceanic circulation depends. These are situated in the north Atlantic, where the north Atlantic deep water forms, and in the Weddell Sea, where most of the Antarctic bottom water originates.

The theoretical model of Stommel is idealized, but it does, nevertheless, bear a striking resemblance to conditions as they are known. Assuming an idealized earth, which corresponds approximately to the actual conditions, it can be shown that there must be a strongly localized current flowing southwards at depth on the western side of the Atlantic Ocean. There is also a general upward movement at mid-depths. The necessity for this strong localized boundary current results from the incomplete geostrophic balance between the pressure gradient forces and the Coriolis effect. In order to compensate for this imbalance, a strong current in which frictional and inertial forces are important, is invoked. Such a current would compensate for the net transfer of water in the opposite direction at other depths.

Stommel's analysis is based on a model of the ocean in two layers. The upward movement at mid-depths would lead to divergence, which would be associated with poleward flow. This must be compensated by equatorward flow in a concentrated current. A strong boundary current, therefore, would be expected to flow south on the western edge, while a more diffuse northerly flow would be expected towards the east. The northerly flow is fed by an eastern flow from the southerly boundary current. Part of the southward boundary current continues across the equator to about latitude 35°s, where it meets a northerly flow from the other source of deep water in the Weddell Sea. The water then turns eastwards, to produce a great zonal flow, which continues all round the Antarctic. This flow has a slight southerly component that brings it eventually to the region south of the southern tip of South America. From this position the current again becomes concentrated at the western side of the ocean, flowing northeast to join the northerly flow from the Weddell Sea source in about 35°s as shown in figure 3.15.

Stommel's theory suggests that in the Pacific Ocean there is also a western boundary current, but it is directed north to about 30°N, where it meets a southerly-directed boundary current. The difference between the oceans is due to the lack of a deep water source in the Pacific. The main source for the deep water circulation in this area is the west to east moving water, which reaches the Pacific to the south of New Zealand. The theoretical circulation of the Indian Ocean is similar to that in the south Pacific. A

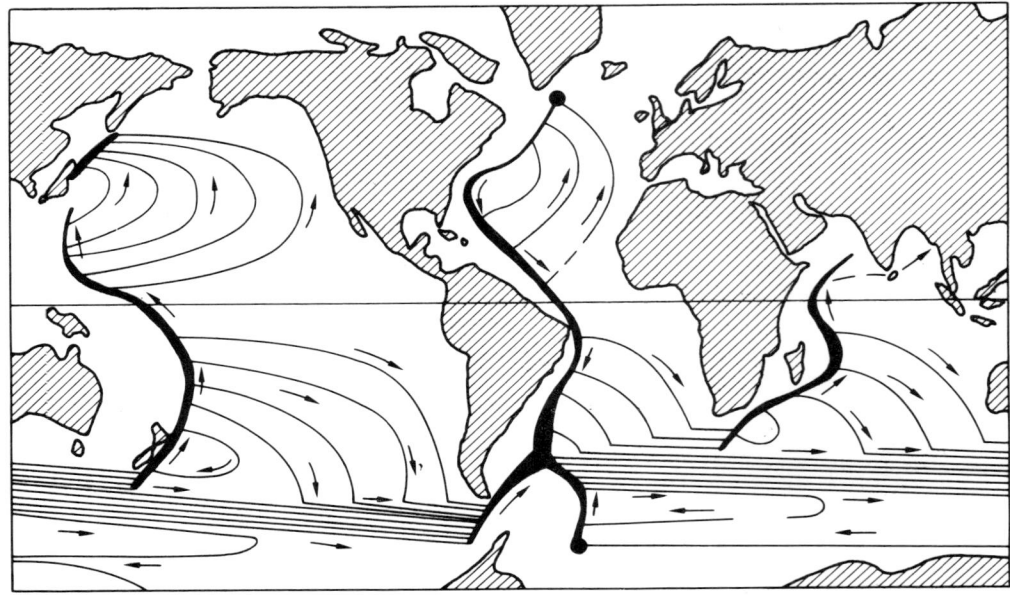

Figure 3.15 Deep-water circulation, showing two source regions in the north and south Atlantic (*After Stommel, 1957.*)

concentrated northward movement on the west side is balanced by a more diffuse circulation to the east and south, in the central low latitudes and east respectively, thus completing a clockwise gyre. In the north Pacific the direction of movement, apart from the northerly-directed boundary current between 0° and 35°N, on the west side of the ocean, is in the form of an anti-clockwise circulation as in the north Atlantic.

Stommel and Arons (1960) have also attempted to estimate the possible volume of water taking part in these movements. They suggest that 20 million m³/sec move south from the northern source in the western boundary current of the north Atlantic. This assumes a level of no motion at 1500 m. The Weddell Sea source is estimated to produce another 20 million m³/sec. The 40 million m³/sec that sink are compensated for by a similar upward flux, which is spread throughout the ocean at mid-depths. It is estimated to amount to about 1 million m³/sec for each area of 7·5 million km² rising across the 3000 m level. The amount of water assumed to be moving south across the equator in the Atlantic Ocean is 16 million m³/sec, while 2 million move north in the Indian Ocean and 10 million in the Pacific cross the equator. Other volumes of flow are shown diagrammatically in figure 3.16.

The essential cause of this suggested abyssal circulation lies in the character and mechanism of the main thermocline, which divides the two layers of water. This is the result of heating at the surface and the effects of wind stress. It demands the upward flux of water from mid-depths. The position of the main sources of deep water is a climatological and geographical accident, which could change, given different circumstances, without affecting the basic theory of the circulation.

In the more traditional concept of oceanic circulation, winter cooling is fundamental

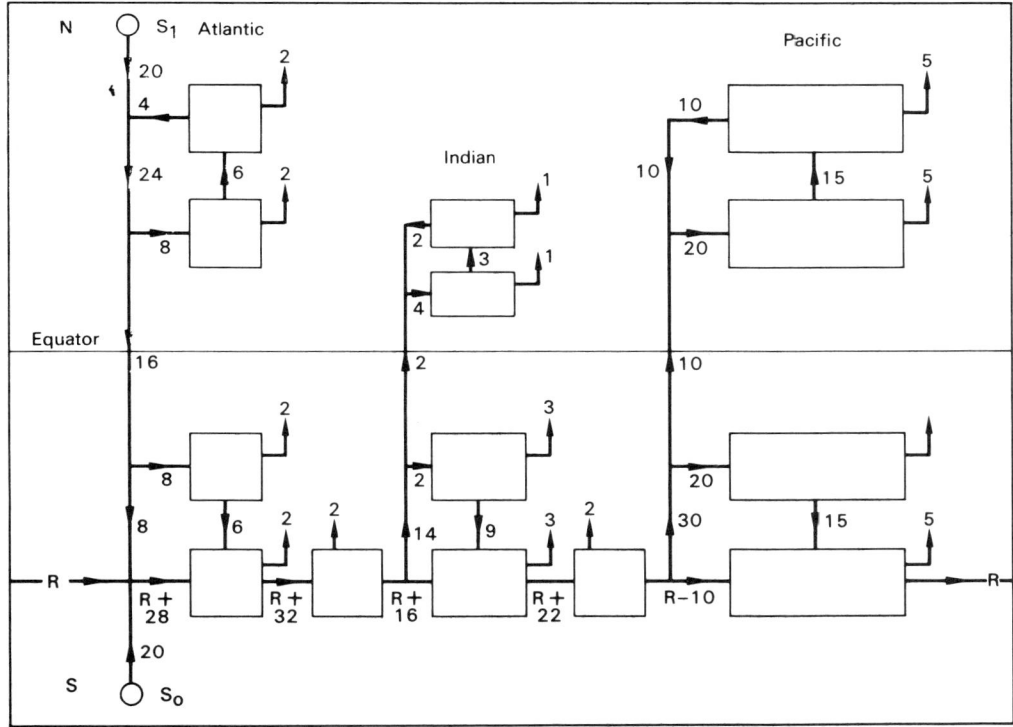

Figure 3.16 Schematic budget of transport of water in different parts of the oceans. R refers to the volume of water that is continuously moving around the southern ocean. The figures refer to million m³/sec. (*After Stommel and Arons, 1960.*)

to the formation of deep water and a reduction in this, due to increased warmth in high latitudes, would lead to a reduction in the intensity of the deep oceanic circulation. Stommel, however, maintains that warming of the polar regions by 1–2°c would not slow down the abyssal circulation, although it might alter the source regions and change the boundary currents. The thermocline affects the vertical component of velocity at mid-depth. An estimate of the speed of this upward movement gives an average of 4×10^{-5} cm/sec, reaching a maximum at a depth of about 1000 m.

The renewal of deep water would take about 200 years, given an average depth of 3000 m below the thermocline. This value does not take local variations into account, and it is based on assumptions that may not be justified. The result is uncertain and other methods of calculation of the deep water replacement time give very different values, varying from 300 to 1800 years. Estimates of the total flux taking part in the abyssal circulation vary between 15 and 90 million m³/sec respectively. The total flux of the abyssal circulation is assumed to be 2–3 times that of the western boundary current of the Atlantic Ocean.

Arons and Stommel (1967) suggest a theoretical value of about 6 million m³/sec for transport into the western boundary current of the north Atlantic basin below 2000 m.

This value compares with Swallow and Worthington's estimate of 6·7 million m³/sec below 1400 m on the Sargasso Sea side of the Florida current off Cape Romain, Dietrich's value of 5·6 million m³/sec passing into the Labrador Sea through the Denmark Strait, and Worthington and Volkmann's value of 4·6 million m³/sec flowing across the Iceland–Faroes Ridge and a gap in the mid-Atlantic Ridge into the Labrador basin. Arons and Stommel obtain their estimate by analysing data on tracers, using the oxygen content and carbon 14 observations of the water. Their model forms a consistent picture that includes the deep southward-flowing western boundary current, an interior geostrophic circulation, and a dynamical thermocline mechanism which returns abyssal water to the upper layers. It explains the observed west to east decrease in dissolved oxygen concentration. The eddy diffusion coefficient is found to be low, and this agrees with the long life times of the Gulf Stream eddies that persist for months.

Nowlin (1967) has calculated a theoretical model of a steady, wind-driven frictional circulation consisting of two moving layers in a rectangular ocean basin with a horizontal bed. It is assumed that the wind stress varies latitudinally as a cosine. The results show that the western intensification is greatest in the lower layer. It is also assumed that the model ocean is situated on a β-plane, which implies that the Coriolis effect is reduced to a linear function with respect to latitude on a plane, instead of varying with the cosine of the latitude on a sphere, $\beta = df/dy = $ constant, where $f = \omega \sin \phi$. Both upper and lower circulations are anticyclonic, and fit lower and middle latitudes. The result of the intensified western flow in the lower layer is to cause a deep recirculation in the western boundary region. The omission of many important factors, such as bottom relief and lateral effects, detract from the applicability of the model results to reality.

3.2 Observations

A technique of observing the actual speed of flow at different depths has been developed by Swallow at the National Institute of Oceanography. The movement of a float, the density of which is such that it will remain at a predetermined depth, is followed from a ship above. The free-moving float is followed by pulses emitted from it and received in the ship, and its movement relative to the ship can be followed. The position of the ship must be accurately known. Measurements of the currents can be carried out over several days or weeks.

The first measurements were made as a joint project between the National Institute of Oceanography and the Woods Hole Oceanographic Institute in the area around 35°N and between longitude 75 and 76°W (Swallow and Worthington, 1957). Nine floats were followed, three of them at 2500 m depth and four at 2800 m. The former moved south and southwest at 2·6–9·5 cm/sec, while the deeper ones moved at speeds of 9·7–17·4 cm/sec.

Another method of measuring near-bottom current velocities was used by Laughton, who showed a southward movement of 5 cm/sec at a depth of 3200 m only 50 cm from the bottom. This result was obtained by underwater photography of the deflection of a ball suspended on a string. Hydrographic data suggested that the level of no motion in this area must lie around a depth of 1500–2000 m. The observations confirm the presence of deep south-flowing water beneath the Gulf Stream in the western Atlantic Ocean.

Since the early experiments the technique has been improved and further observations have been carried out. In the eastern Atlantic (Swallow, 1957) results off the Straits of Gibraltar agreed fairly well with the values deduced from hydrographic observations. Depth accuracy improved to ± 30 m, and observations were made north of Scotland and in the area between 30 and 40°N and 10 to 20°W. The currents in the Faroe–Shetland Channel were calculated from hydrographic data and compared with direct observations, and although the velocities differed the direction of the currents agreed.

A longer series of observations has been carried out to the west of Spain near the edge of the Iberian abyssal plain in latitude 41°N (Swallow and Hamon, 1960). The floats could be followed for periods up to 12 weeks. Their position was fixed by reference to small seamounts on the bottom, as other navigational aids were not available in the area. An attempt was made to locate the level of no movement by observing the variation of the current with depth. Five floats were used between a depth of 1500 and 4500 m. The floats all moved in nearly the same direction of about 143° True, but they varied in speed. There was no uniform decrease of velocity with depth, nor was there a depth of no horizontal motion. The movement of the floats was variable. Two of them only 25 km apart moved with speeds differing by a factor of 5. One float at a depth of nearly 3000 m was followed for a total period of 48 days, and it moved at 1·2 cm/sec for the first 6 days in a direction 156°. It then remained stationary for a period of 22 days, after which it moved at 1·0 cm/sec in a direction at right angles to its previous movement. There was little agreement in this instance between the observed currents and the geostrophic currents calculated from hydrographic data. Short-term fluctuations in deep water movements, when the velocities are between 1 and 5 cm/sec, are such that it is difficult to compare them with steady-state models. The details of the recorded movement are shown in figure 3.17.

Pochapsky (1966) recorded 5 neutrally buoyant floats in 27°56′N, 55°22′W, at depths around 2300 m. The centre of the cluster moved southeast at a mean speed of 2 cm/sec. The velocity of the whole group agreed with the dynamic height measurements. One float, however, 140 m deeper than the other, moved in an almost circular path relative to the other. The radius was 2 km and the period was 25·5 hours. The deep circulation in this position was inertial in type, showing the correct period for the latitude. It is thought that internal waves could be due to the relative movement of different layers of water. The temperature varied through $\pm 0·02$°C and the temperature gradient was not monotonic, even near the bottom.

Swallow and Worthington (1969) have extended their observations into the Labrador Sea. They followed floats for periods between 12 and 71 hours at depths from 1664 m to 2396 m. Mean speeds varied between 0·8 and 10 cm/sec. The observations support the idea of an anti-clockwise flow in the deep layers of the Labrador basin, and deep water transport below 1200 m is estimated as about 10 million m³/sec. Deep transports were very variable. There was a weak zone of flow around 1000 m depth, with northwest flow above and below. There was an inflow of deep water southwest of Cape Farewell and an outflow on the Labrador side. The boundary at 1000 m separates the deep, higher salinity water below from the Atlantic or Labrador Sea water above.

A new version of an instrument designed to measure deep sea currents is described by Sasaki, Watanabe, and Oshiba (1967). The instrument is a ping-pong ball pendulum, the movement of which is photographed against a compass. The instrument has been

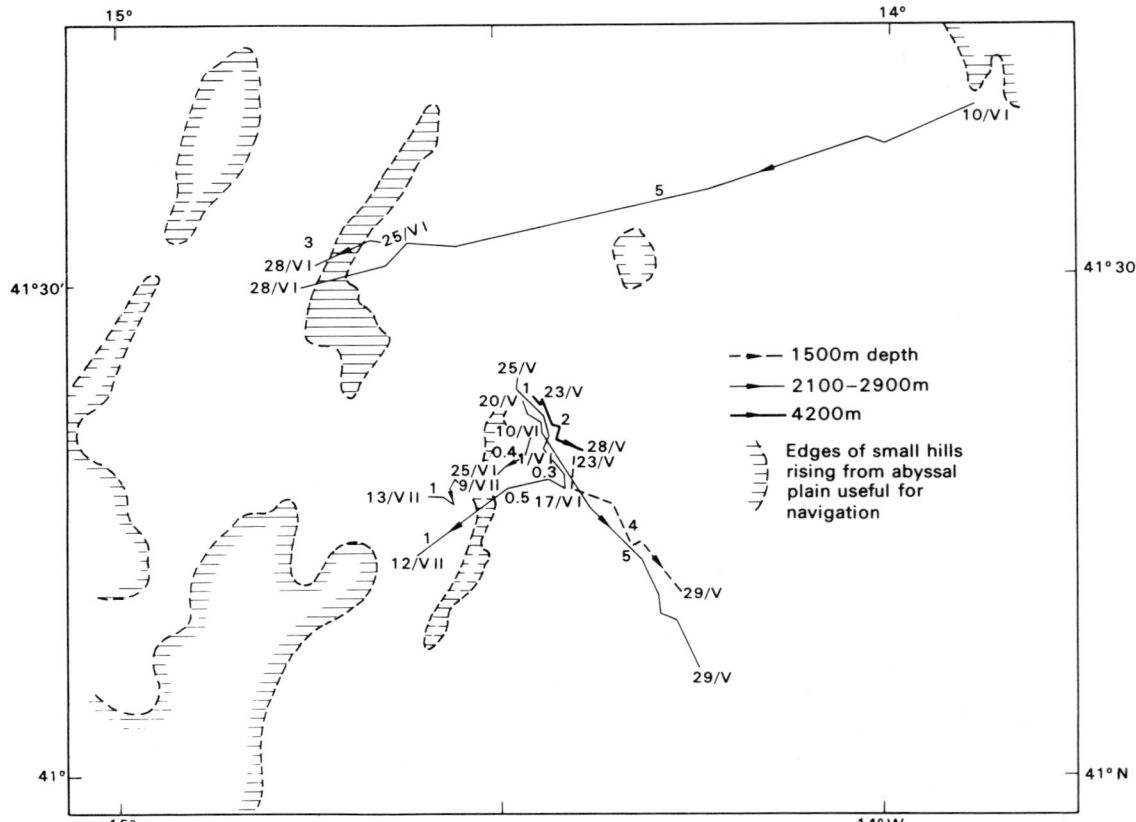

Figure 3.17 Movement of floats in the eastern Atlantic Ocean. Observed depth of floats in m: 1 2760 ± 110, 2 2120 ± 50, 3 2460 ± 50, 5 2590 ± 70, 6 2430 ± 130, 7 1560 ± 50, 11 4240 ± 160, D 2940 ± 70 (*After Swallow, 1957.*)

tested off Japan in depths up to 5270 m. The deepest successful test in 4255 m indicated bottom currents of 1·2–4·3 cm/sec over 13 minutes. At a depth of 1280 m a speed of 10·4 ± 2·9 cm/sec was recorded.

The direct measurement of deep sea currents raises the problem of their representativeness, a matter that has been considered by Webster (1969). He has analysed data collected at a single location 39°20′N, 70°W by means of a modified Richardson-type current metre. The depth at the point selected was 2600 m, about 50 km from the continental shelf, which may complicate the values. It was 150 km from the axis of the Gulf Stream. The tidal motion introduces a velocity of about 8 cm/sec, and a slow drift to the south-southwest was also recorded over a period of 10 days. Longer-period observations show considerable variations in direction and speed; for example, in one day a net flow of 27 km was recorded, while in another period of 9 days only 33 km net flow was observed. These observations covered a 2-month period in June–August. Another set of observations in October and November showed considerably greater velocities, with a mean flow of 62 km on one day.

Depths of observations were 120 m in the summer and 98 m in the autumn; a similar movement was recorded at 87 m and 7 m depth, although the inertial period movements were better developed near the surface.

The data were subjected to spectral analysis, which revealed peaks of activity at 12·42 hours, the semi-diurnal tide; 18·9 hours, the inertial period; the diurnal period, a 36·4 hour period, representing the interference of inertial and tidal movements; and a period of 60 hours, the cause of which was unknown. A comparison of the spectra with those for other periods showed changes of energy density by as much as a factor of 10. The study suggests that caution is needed in interpreting short-term deep current data; they cannot be considered representative of longer periods. Averages can also be misleading, and the variability consists of a considerable number of elements in the spectrum. The variability seems to be greater in the vertical sense than the horizontal, comparing points 3 km apart horizontally.

Deep current observations on the continental rise off New England have been made by Zimmerman (1971) within 2 m of the bottom in depths of 2900–5000 m. The western boundary undercurrent had a range of 5 to 25 cm/sec between depths of 2000 and 3000 m according to the field observations. The bottom currents, which were measured by current meters, were capable of carrying sediment. The greatest recorded velocity was 265 cm/sec in a southwesterly direction parallel to the regional contours. These observations confirm the importance of deep currents in sediment transport.

Duing and Johnson (1971) have measured currents in the Florida Straits. The current profiles show that the lower part of the water flows south, at times with velocities up to 30 cm/sec. Large temporal variations occurred. There appeared to be a deep counter-current while the surface current appears to meander horizontally. In June 1971 a flow of 80 cm/sec was recorded at times in the deepest part of the Florida Strait off Miami.

In the north Pacific a narrow eastward bottom current flows along the southern flanks of the mid-Pacific Mountains through a deep passage 10 km wide south of Horizon Guyot. The flow continues around the southeast end of the Hawaiian Islands from observations of temperature profiles and hydrographic data according to Edmond, Chung, and Sclater (1971).

Reid and Nowlin (1971) have examined the transport of water through the Drake Passage, and they concluded that a geostrophic relative transport of 113 million m³/sec takes place. The 'absolute' transport adjusted to current meter results gives a value of 237 million m³/sec, which is higher than earlier estimates and is also more evenly spread across the passage. The current meters were released to fall to the bottom; when they have recorded the current their anchor is released and they rise to the surface and transmit a radio signal to aid recovery. The mean daily speeds measured ranged from 0·5 to 14·7 cm/sec in the southern half and 8 cm/sec in the northeast. The pressure field was stable and agreed with earlier results.

Moored current meters and Swallow floats have been used by Webster (1971) to record the intensity of horizontal currents in relatively quiet parts of the Atlantic Ocean. All the values with the exception of one site north-northeast of Bermuda, which may have been in a meander of the Gulf Stream, showed very similar values. The kinetic energy of the currents was also very similar at all the localities tested. The values are compared with those obtained at a position 39°20′N, 70°W. The measurements used in the analysis

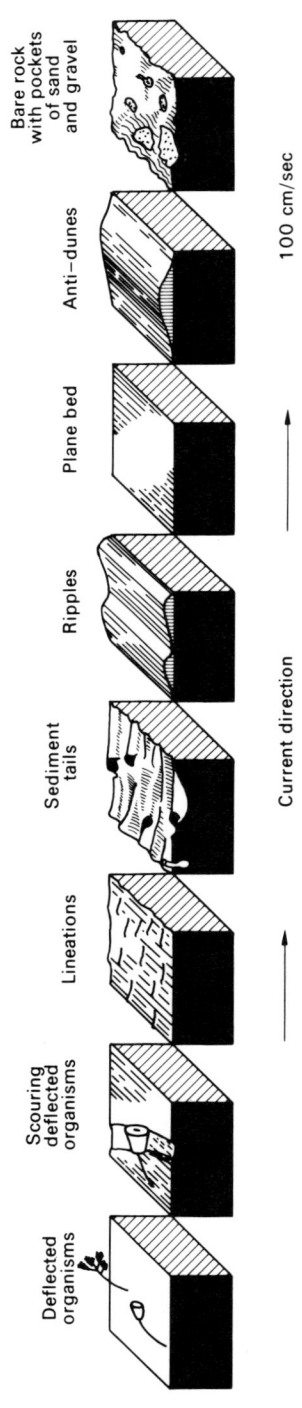

Figure 3.18 Bottom morphology in relation to increasing current velocity. (*After Heezen and Hollister, 1971.*)

were made west-southwest of Bermuda, north-northeast of Bermuda, in the Bay of Biscay, and in the Mediterranean. Values were recorded at depths of 10, 100, 500, 1200, 2000, 3000, and 4000 m, among others. The speeds of the currents fell off with depth, but were remarkably constant in all the sites sampled at any one depth, and agreed with those previously obtained at the control site. The values fell close to a line falling from a velocity of nearly 20 cm/sec at a depth of 100 m to about 5 cm/sec at a depth of about 3000 m. Total kinetic energy also fell off with depth at a rather greater rate, from 100 ergs/cm³ at 100 m to about 10 ergs/cm³ at about 4000 m. There was a reasonably good agreement between the current meter and float observations, the values being closer for the speed than the kinetic energy, which is easier to calculate from the current meter data. The areas in which the measurements were made differ, ranging from an enclosed sea, the Mediterranean, to the Bay of Biscay and the open Atlantic Ocean, yet they all appear to be characterized by similar current speeds and energy. The observations also cover a considerable range of time and were made at different seasons. The narrow range of current speed and energy may indicate that there is a saturation mechanism that prevents energy exceeding a certain natural limit. Another possibility is that the energy may be kept above some natural level by a long time-constant related to energy dispersion and dissipation. Global effects as opposed to local effects may also be important in the generation of oceanic energy and its uniform distribution. The absence of severe fluctuations of energy suggest support for the global energy idea. Figure 3.13 illustrates diagrammatically the effect of currents at different intensity up to 100 cm/sec on the bottom morphology.

4 Circulation of the oceans

4.1 The Atlantic Ocean

Direct observations have not yet been sufficiently numerous or prolonged to give a clear picture of the oceanic circulation in depth. The circulation can best be deduced indirectly from observations of temperature and salinity at depth in relation to theoretical concepts.

The Atlantic Ocean is easier to consider from this point of view as it contains very distinctive water-masses, which can be followed by means of their characteristic temperature and salinity. This method of tracing water movement is called the core method. The Atlantic contains the only sources of deep water, and these distinctive water-masses can be traced as they move away from their sources. Another significant point is that in the Atlantic Ocean there is a transference of water across the equator on the surface, estimated at 6 million m³/sec. This must be compensated by a return flow at depth as the north Atlantic is not in free communication with the rest of the world ocean. Stommel and Arons (1960) have, however, suggested that the compensation for the sinking of water at either end of the Atlantic is widely distributed throughout the whole oceanic area.

The general pattern of oceanic movement at depth in the Atlantic Ocean is based on the interdigitation of distinctive water-masses, formed in high latitudes, each lying above the other in order of their density. A vertical section, running north–south the length of the ocean, reveals the variations of temperature and salinity. Figure 3.19 shows that a tongue of low-temperature, low-salinity water extends from the Antarctic coast northwards,

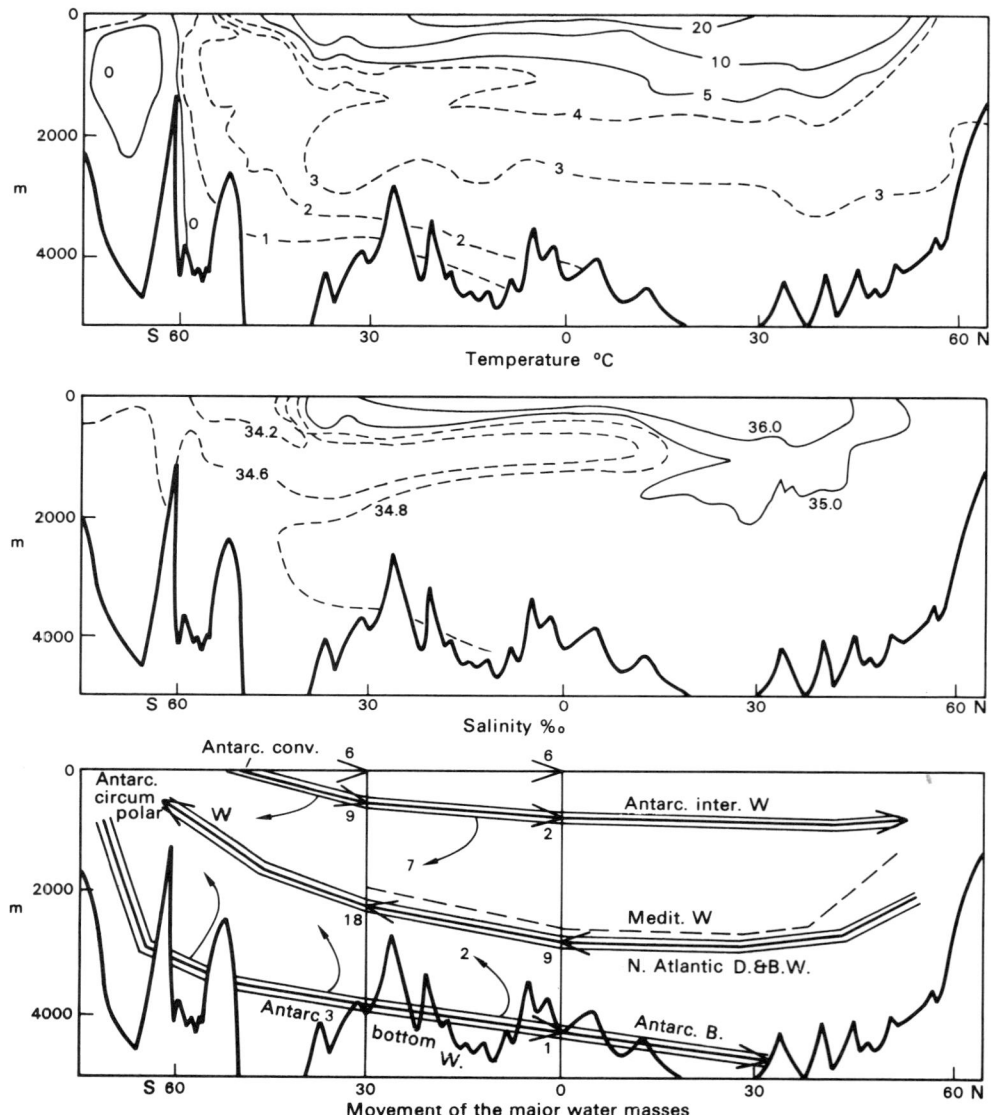

Figure 3.19 North–south sections through the Atlantic Ocean to show temperature, salinity and the movement of the major water masses. The figures give the volume in million m³/sec. (*Partly after Sverdrup et al., 1946.*)

reaching across the equator as far as 35°N. The salinity of this water is below 34·9‰, and temperatures below 2°C extend to 20°N on the bottom. Above this is a tongue, stretching south, characterized by its high temperature and high salinity. Salinities above 34·9‰ extend in a broad wedge as far as 40°S. The high salinity of this tongue of water is partly due to the very saline Mediterranean water, which mixes with the north Atlantic deep water. This water is also warmer than that below it. In the southern hemisphere it forms a sandwich between two layers of fresher water.

The uppermost of these layers consists of the fresh, cool Antarctic intermediate water, which has a salinity minimum of less than 34·4‰, extending to nearly 10°s. Mixing reduces this minimum northwards. The water-mass can be traced far into the northern hemisphere. The shallow surface water-masses lie above this water-mass.

The distribution of oxygen also gives useful information concerning the water movement of the Atlantic Ocean. The oxygen content of the Atlantic waters is greater than that of the other oceans. Within the Atlantic there is in general a diminution of oxygen from

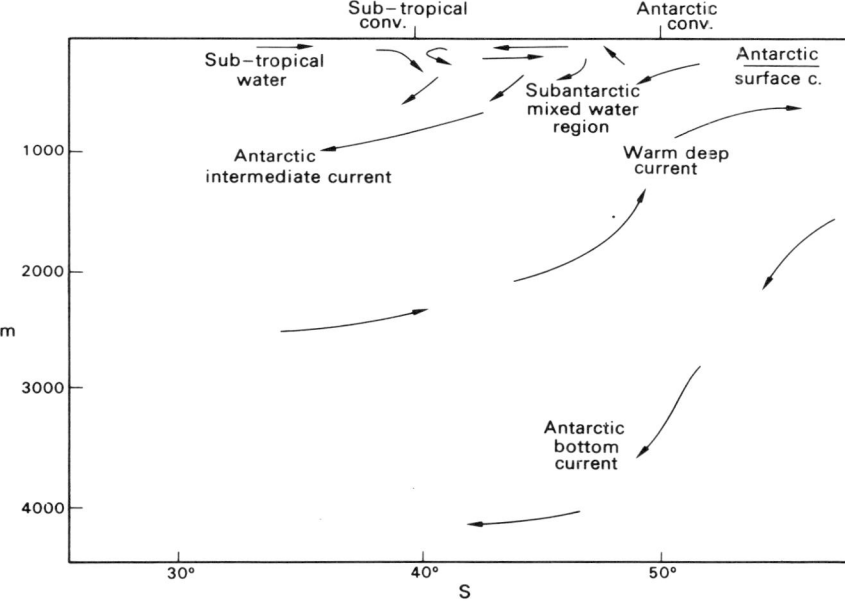

Figure 3.20 Vertical circulation in the southern ocean showing meridional flow, which is superimposed on a general west to east zonal flow. (*After Deacon, 1960.*)

north to south, particularly in the north Atlantic deep water and the Mediterranean water. Another zone of relatively high oxygen content is found in the deeper layers of the south Atlantic, within the Antarctic bottom water. The minimum is found above 1000 m just north of the equator.

The oxygen values suggest a circulation consisting of two elements. Firstly, there is a movement in the southern hemisphere to the north at the bottom and relatively near the top, with a compensating southerly movement between, shown in figure 3.20. Secondly, there is a movement across the equator at the surface and a reverse movement at depth. The amount of these transfers can be generalized from considerations of continuity, shown in table 3.1 (Sverdrup, Johnson, and Fleming, 1942). These figures may be compared with the theoretical estimate of Stommel and Arons (1960). They suggest that 16 million m³/sec move south at depth, in concentrated currents on the western side of the ocean.

Table 3.1 Water-mass transport in the south Atlantic Ocean

| Latitude | Water-mass | Current | Transport in million m³/sec | |
			North	South
	Upper water	Benguela	16	
		Brazil		10
		Central part of gyre	7	7
30°S	Intermediate water		9	
	Deep water			18
	Bottom water		3	
	Upper water		6	
0°	Intermediate water		2	
	Deep water			9
	Bottom water		1	

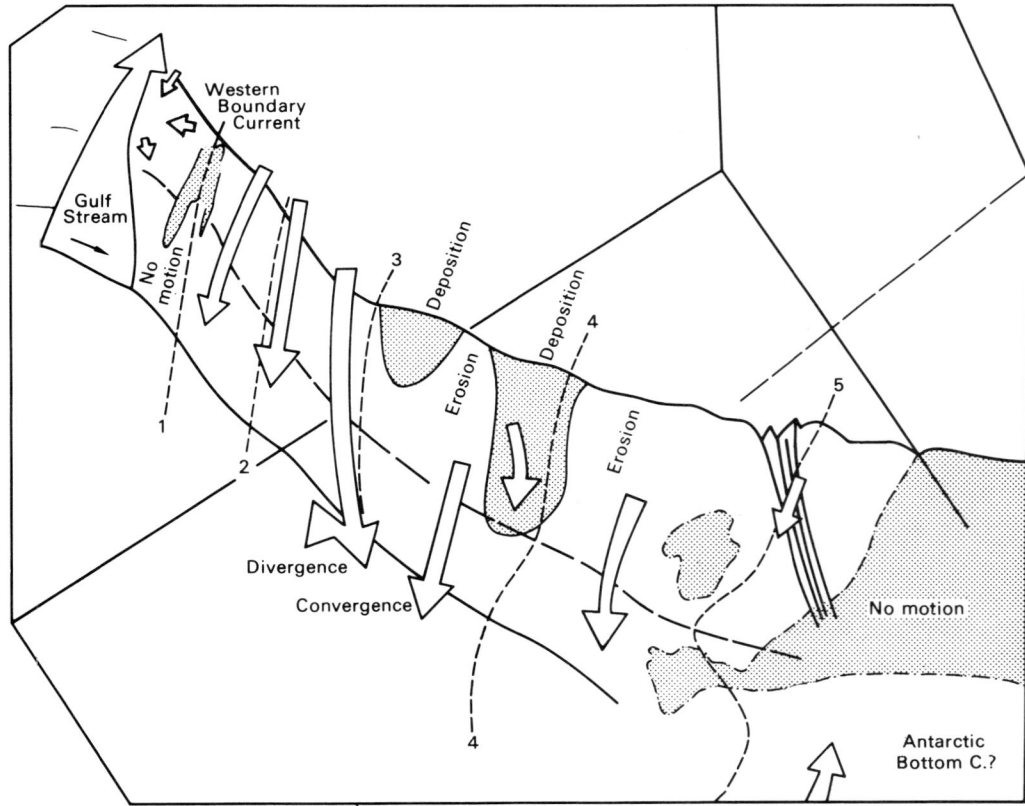

Figure 3.21 Block diagram to illustrate the three-dimensional flow pattern off the eastern coast of the USA. The relative current velocities are indicated by the size of the arrows. Stippled areas are apparent zones of no motion. Figures indicate depth in 1000 m (*Rowe and Menzies, 1968.*)

It is presumed that the compensating flow takes place at higher levels, partly on the surface and partly in the much shallower intermediate water. These figures suggest a more vigorous circulation across the equator than that derived from hydrographic data and considerations of constancy of salt content in both hemispheres and similar continuity features. The former figures are derived from the work of Wüst.

Lynn and Reid (1965) trace the movement of abyssal waters by means of their potential density. The potential density is calculated from the *in situ* temperature and salinity moved adiabatically to sea surface pressures. The west Atlantic water is unusual in that its greatest potential density is above the bottom between 40°N and 35°S. The high values are found in the lower north Atlantic deep water, which originates in the Greenland–Norwegian Sea area. The maximum value of the potential density allows the water to be traced to 35°S at least. A deeper reference level of 4000 dbars can be used to trace the water further. The potential density, calculated in this way, increases monotonically to the bottom in very deep water. Despite the inversion of potential density, the water is stable because the warmer water is less compressible than the colder underlying water.

Figure 3.21 shows a diagrammatic representation of the three-dimensional pattern of water movement along the western boundary of the Atlantic Ocean off the southeast United States. Three major currents are indicated in contact with the sea floor (Rowe and Menzies, 1968). The Gulf Stream flows strongly north along the outer shelf and upper continental rise. It is separated by a zone of no motion from the south-flowing western boundary current beneath it. This water flows between depths of 1000 and 5000 m along the edge of the continental slope and rise. The south-flowing water is separated by another plane of no motion from the northward-moving Antarctic bottom water at depths greater than 5000 m. This water moves over the Hatteras Abyssal Plain, which extends to 5335 m off North Carolina.

4.2 Southern ocean

The water moving south in the Atlantic becomes increasingly diluted by mixing with surrounding waters. It gradually loses its high salinity and, eventually rises to the surface to form the Antarctic circumpolar water-mass (figure 3.22). In this form it continues to move eastwards into the Indian and Pacific Oceans. The flow, according to Stommel and Arons, continues at greater depth, where they suggest that the Antarctic bottom water, moving north, is partly deflected to the east. The water character and oxygen content also indicates an eastward flow of deep water around the southern continent, derived from both northern and southern sources of deep water. If the estimate of Stommel and Aron of 20 million m^3/sec for the original amount of the deep southern source of Antarctic bottom water is correct, a large proportion of this remains to travel east, if only 3 million m^3/sec move north as suggested in table 3.1. The eastward movement continues right around the earth, in the only zone in which continuous movement in a zonal direction is possible. The water, however, is deflected north after passing through Drake Passage, south of South America, because the South Sandwich Island Arc in this area effectively prevents deep water from moving east. The total amount of this water circulating around the southern continent varies as part of it diverges into the other oceans, or as it is replenished by water flowing into it on the eastern side of the oceans. The major loss from

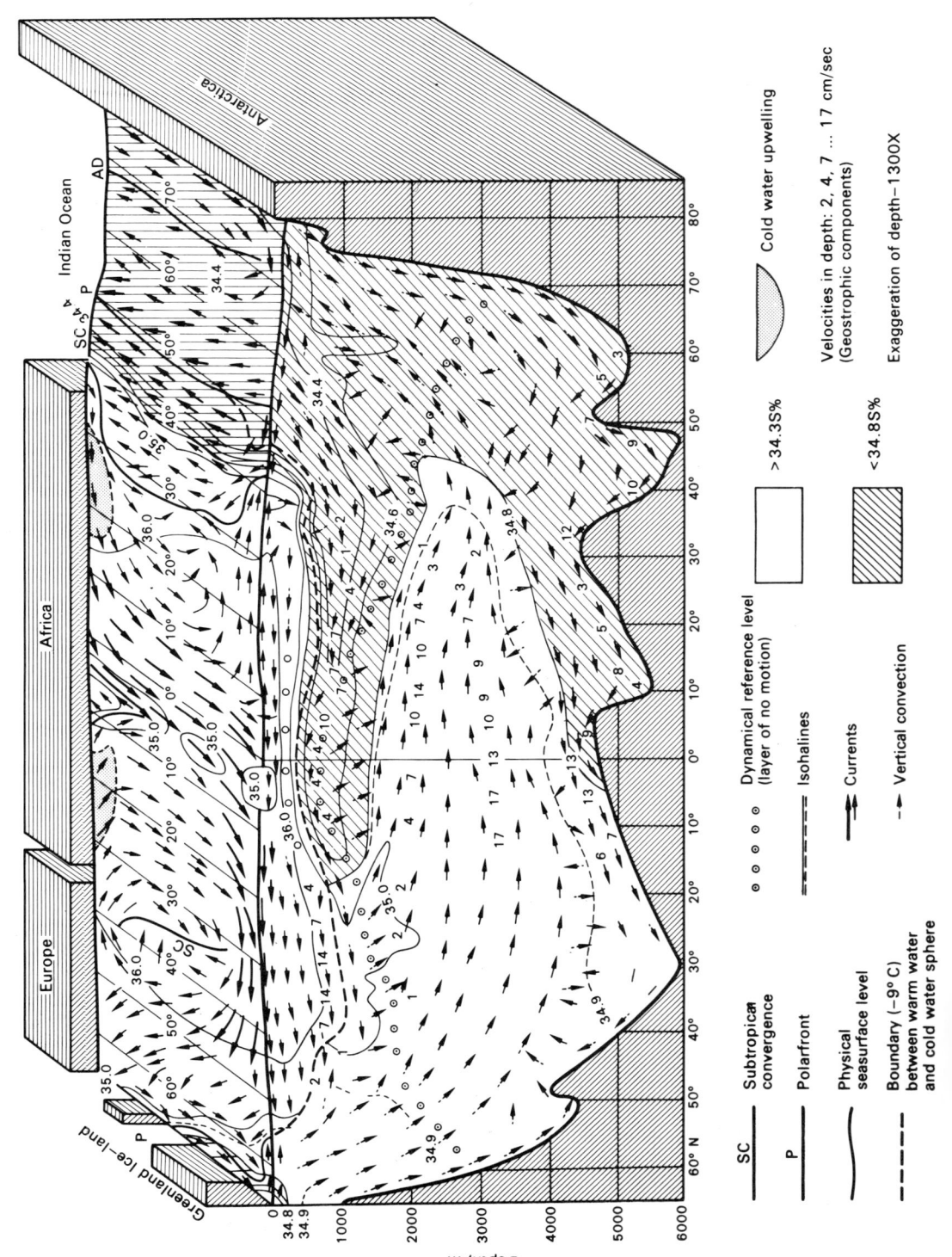

Antarctica

Indian Ocean

AD

SC ← P

34.4

60°

70°

50°

40°

35.0

36.0

30°

20°

10°

Africa

0°

35.0

35.0

35.0

10°

36.0

36.0

20°

Europe

SC

36.0

30°

40°

50°

60°

35.0

Greenland Ice-land

P

60° N

50°

40°

30°

20°

10°

0°

10°

20°

30°

40°

50°

60°

70°

80°

34.4

34.6

34.8

Depth, m

0

1000

2000

3000

4000

5000

6000

34.8
34.9

34.9

34.9

SC — Subtropical convergence

P — Polarfront

— Physical seasurface level

— Boundary (−9° C) between warm water and cold water sphere

∘ ∘ ∘ ∘ ∘ Dynamical reference level (layer of no motion)

------ Isohalines

⇒ Currents

--→ Vertical convection

◗ Cold water upwelling

Velocities in depth: 2, 4, 7 ... 17 cm/sec (Geostrophic components)

Exaggeration of depth−1300X

>34.3S‰

<34.8S‰

this circulation is into the western Pacific, where Stommel and Arons show a concentrated flow along the west side, off New Zealand.

The Antarctic circumpolar current is deep, extending to 3000 to 5000 m at the ocean bottom. For this reason it is affected by the relief. Its surface velocity is 15–20 cm/sec. Volumes of transport suggested by Kort are 190 million m³/sec between Antarctica and Africa, 180 million m³/sec between Antarctica and Tasmania, and 150 million m³/sec between South America and Antarctica.

4.3 The Pacific Ocean

The deep circulation of the Pacific is dominated by the deep western boundary current, which theory suggest flows north as far as the southern part of Japan. The much lower oxygen content of the Pacific water indicates that there is no source of deep and bottom water in this ocean. The surface character and distribution of land and sea prevent suitable conditions for water to become dense enough to sink to the deeper layers of the ocean. This applies particularly to the northern part of the ocean, which must be fed indirectly via the southern hemisphere. There is, as a result, an even lower oxygen content in the northern than the southern Pacific. Values in excess of 3·0 ml/l are found throughout the southern Pacific, but only in the lowest layers is this value exceeded in the north Pacific.

The bulk of the deep Pacific water is much more homogeneous than that in the Atlantic, and there is no evidence of the layered structure so typical of the Atlantic. There is in general a decrease of temperature with depth, the value being less than 2°C below 2000 m. There is a general decrease of salinity with depth to a level of 1000 m, the water of the upper layers being less saline than 34·6‰, while below 2000 m the salinity is only slightly above 34·6‰, except in the south where it rises to 34·7‰ and slightly above. This homogeneity suggests a very slow circulation, which is confirmed by the low oxygen values.

The water that fills the deeper part of the Pacific Ocean must have been originally at the surface in either the north Atlantic or the Weddell Sea region. The latter water has followed the shorter path to the south Pacific, where it is recognizable by its low temperature, and where the oxygen content is greater. The rest of the Pacific water probably left the surface in the north Atlantic or the Mediterranean, although in its journey to the Pacific it has been very thoroughly diluted, partly by mixing with the Antarctic intermediate water, and partly by the Antarctic bottom water. The layer of salinity minimum at the upper level reveals the influence of the Antarctic intermediate water. The exchange between the two hemispheres in the Pacific probably takes place in a northerly direction in the west in a concentrated boundary current, and to the south on the east, where the flow is sluggish.

Gordon and Gerard (1970) have used data concerning the potential temperature of deep Pacific water to establish the current pattern in the northern part of the ocean. The bottom potential temperature north of 11°S has a range of only 1·2°C, but nevertheless a pattern is discernible. The coldest water at less than 0·8°C occurs in the equatorial region from 160 to 180°W. Cold water moves north on either side of Hawaii. The water north of Hawaii is fairly warm, relative to that south of the islands, and the bottom water over and east of the crest of the mid-ocean ridge at 120°W has a large positive anomaly, suggesting geothermal heating. The basic pattern of the bottom circulation which lies below about 2500 m can be established. The sole influx of water is the Antarctic bottom water, flowing north along 175°w from 20°s to 16°n. This water is a continuation of the deep western boundary current that flows along the Tonga Trench in the south Pacific. The flow then divides into two zonal branches, the eastern one passing between Johnston Island and Christmas Island Ridge, and the western branch passing between the Marshal Islands and the Marcus–Necker Ridge. Both flows then turn north and converge in the north Pacific basin between Hawaii and the Aleutian Islands. In this zone deep upward transport is expected. They estimate that the time necessary for the water to flow from 175°w, 10°s to the centre of the convergence zone is 750 years.

4.4   The Indian Ocean

There is no source of deep water in the Indian Ocean, as in the Pacific, partly on account of the low latitude of the northern boundary. Temperature and salinity observations do, however, suggest a similar exchange in the southern Indian Ocean to that already described in the south Atlantic. There is a northward extension of bottom and intermediate water, with a return flow at deep levels. The former water-masses are the Antarctic bottom water and the Antarctic intermediate water respectively. Stommel and Arons suggest that this circulation is based on a northward-flowing western boundary current, with more diffuse southerly flow on the east. The extremely saline Red Sea water helps to give a salinity maximum to the southward-moving deep water, which has a salinity maximum at a depth of 3000 m. The Red Sea water probably penetrates eventually as far as the Antarctic. There is thus a slight transfer north across the equator in the Indian Ocean, estimated at 2 million m^3/sec by Stommel, in the deeper layers.

The deep water of the Indian and Pacific Oceans must originate in the Atlantic. Reid and Lynn (1971) have studied the part played by the Norwegian–Greenland and Weddell Seas water in providing the bottom waters of these two oceans. They base their study on the analysis of water movement relative to densities. The temperature and salinity characteristics of the bottom water of the Indian and Pacific Oceans suggest that these waters are derived from extreme sources of the high latitude Atlantic water. The density stratum that is analysed lies within the layer of high salinity that extends continuously from the north Atlantic to the abyssal Indian and Pacific Oceans. The stratum is at the sea surface in the Norwegian–Greenland and Barents Seas and in the Weddell Sea. In the north Atlantic there is a layer of water called the lower north Atlantic deep water, which is a cold less saline layer, above the bottom layer. The lower north Atlantic deep water has a higher potential density than the water above and below it, which are respectively warmer and more saline and colder and less saline. The lower north Atlantic deep water could not have formed by mixing of the waters above and below it and must have a distinct origin, probably as overflow from the Norwegian–Greenland Sea into the western north Atlantic between Greenland and Iceland. The density surface rises from 3600 m near 20°s to 3000 m near 45°s in all three oceans, extending as warm saline water from the southwest Atlantic to the Indian and Pacific Oceans. In the north Pacific it is confined to the central area and it is at a depth greater than 4000 m. The stratum rises still further as it is traced into high southern latitudes, being at 1000 m near 60°s mostly. It meets the Antarctic circumpolar water in high southern latitudes. Water of the same stratum is also formed in the Weddell Sea, where the stratum is near the surface. The effect extends eastwards into the Indian Ocean, while the salinity decreases in this direction. The saline warm water of the north Atlantic dominates the circumpolar current, rather than that of Antarctic origin. The stratum ends at abyssal depths in the north Pacific and Indian Oceans. There must be some upward flux to balance the sinking to form the deep water. Because of the large areal extent of the salinity maximum, the upward flux through the stratum is probably very small except in the north Indian and Pacific Oceans, where stability is very low at the depth of the stratum.

5   General conclusions

Although there is still considerable uncertainty about the exact nature of deep water circulation in the oceans, it is known that the deep circulation of the Atlantic is more vigorous than that of the other oceans, because only there is deep water formed to any important extent. The formation of deep water depends partly on the nature of surface currents. The Gulf Stream and north Atlantic current help to determine the occurrence of warm saline water in the north Atlantic. The cooling of this water plays an important part in the formation of north Atlantic deep water. External influences of heating and cooling and dilution or concentration of salt are also important, for example in the formation of the Mediterranean water. Excessive evaporation plays an important part in the formation of Mediterranean water. Salt concentration by ice formation increases the density of the cold Antarctic bottom water, allowing it to form bottom water by sinking to the greatest depths. Only in the Atlantic is there a marked flow across the equator at the surface, while at depth the southward flow across the equator is also more vigorous. The difference between the Atlantic and Pacific in this respect can be explained by the important source of deep water in the north Atlantic and its absence in the Pacific.

Further reading

ARONS, A. B. and STOMMEL, H. 1967: On the abyssal circulation of the World Ocean III. *Deep Sea Res.* **14,** 441–57.

BANG, N. D. 1971: The southern Benguela Current region in February, 1966: Part II. Bathythermography and air–sea interactions. *Deep Sea Res.* **18,** (2) 209–24.

BRYAN, K. F. and COX, D. 1967: A numerical investigation of the oceanic general circulation. *Tellus* **19,** 54–80.

GILL, A. E. 1971: The equatorial current in a homogeneous ocean. *Deep Sea Res.* 18 (4), 421–31.

HOLLAND, W. R. 1967: On the wind driven circulation in an ocean with bottom topography. *Tellus* **19,** 582–600.

REID, J. L. and LYNN, R. J. 1971: On the influence of the Norwegian–Greenland and Weddell Seas upon the bottom waters of the Indian and Pacific Oceans. *Deep Sea Res.* **18** (11), 1063–88.

SMITH, R. L., MOOERS, C. N. K., and ENFIELD, D. B. 1971: Mesoscale studies of the physical oceanography in two coastal upwelling regions: Oregon and Peru. In J. D. Costlow (editor), *Fertility of the sea*. Vol. II. New York: Gordon and Breach Sci. Pub., 513–35.

STEFANSSON, U., ATKINSON, L. P., and BUMPUS, D. F. 1971: Hydrographic properties and circulation of the North Carolina Shelf and slope waters. *Deep Sea Res.* **18** (4), 383–420.

STOMMEL, H. 1965: *The Gulf Stream*. 2nd edition. Univ. of California Press.

STOMMEL, H. and ROOTH, C. 1968: On the interaction of gravitational and dynamic forcing in simple circulation models. *Deep Sea Res.* **15,** 165–70.

SWALLOW, J. C. and BRUCE, J. G. 1966: Current measurements off the Somali coast during the southwest monsoon of 1964. *Deep Sea Res.* **13,** 861–88.

4   The tide

The tidal phenomena of the oceans and seas are fundamentally influenced by the moon. The gravitational attraction of the earth, moon, and sun for each other is the basic cause of the ocean tides. The movements of these bodies are complex, but regular and well known, thus enabling the tides to be predicted with accuracy. The response of the oceans to tide-producing forces is less well understood and a number of different theories to account for observed tidal phenomena have been advanced. Meteorological conditions also sometimes cause considerable differences between the predicted and actual tides.

The range and character of the tide varies greatly from place to place and in order to understand these variations, it is necessary to consider the response of the waters to tide-producing forces. In some areas special conditions give rise to abnormal tidal features, such as the double high tides of Southampton Water on the south coast of England, and the solar tides found on some Pacific Islands. Other features of interest are the tidal bores characteristic of some tidal rivers. These fairly isolated features show that special conditions must obtain if they are to develop. Tidal streams are important in influencing the movement of material and the morphology of the sea floor; they also affect waves and coastal shipping.

1   Tide-producing forces

The work of Sir Isaac Newton on gravity made possible the analysis of tide-producing forces in terms of the gravitational attraction of the sun and moon on the earth. He

E

developed the theory of the equilibrium tide, based on an ideal water-covered earth, and ignoring the effects of inertia. The equilibrium tide gives a picture of tide-producing forces. The response of the oceans to these forces is, however, an entirely different problem.

The attractive force between two bodies is proportional to the product of their masses divided by the square of their distance apart, mm'/r^2. Force is exerted in a straight line between them. Not all points on the earth are at an equal distance from the moon. The nearer ones will be attracted to a greater degree than the further. The difference in the attractive force at different points on the earth relative to the centre is the differential tidal attractive force. It is only the differential force that has any tide-raising capacity.

Figure 4.1 shows four points on the earth, two in line with the moon's direction at opposite ends of a diameter, U and U', and two at right angles to the first, V and V'. The attractive force at U, nearest the moon, will be greatest, while that at U' on the far

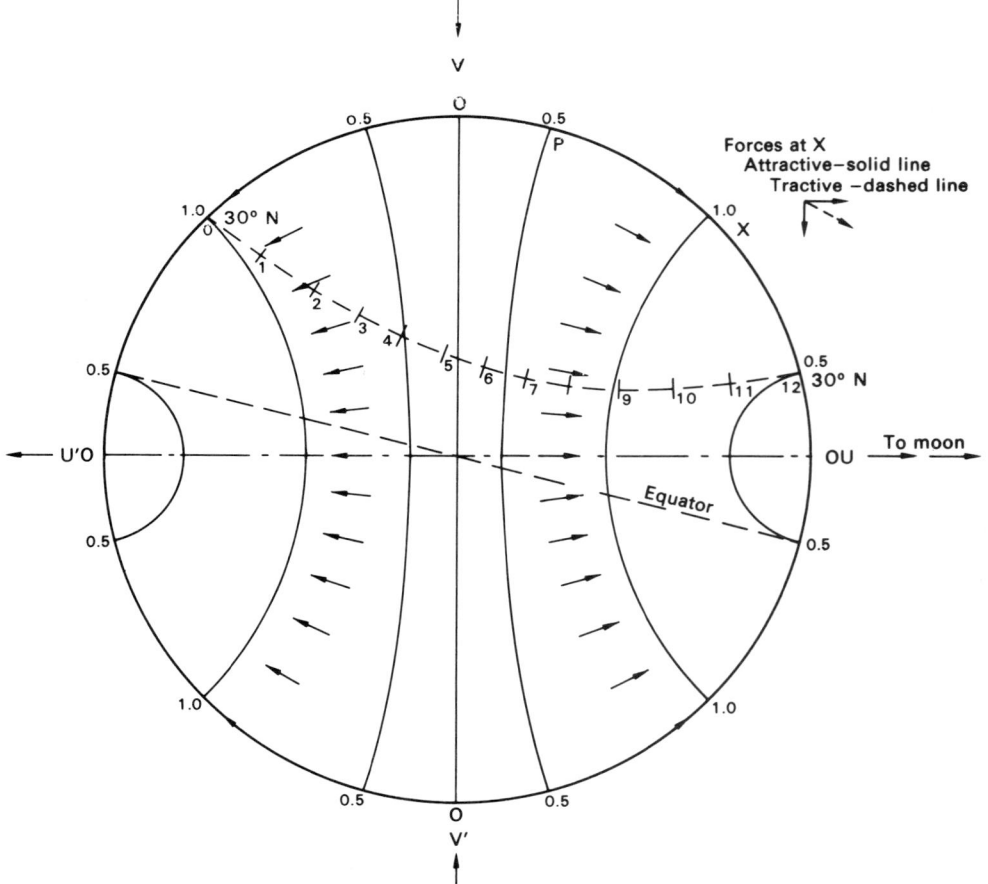

Figure 4.1 Diagram to illustrate the lunar tidal forces for latitude 30°N, with lunar declination of 15°N. The arrows at VV′ UU′ and X show the attractive forces, while the figures on the circumference of the circle indicate the relative strength of the tractive forces, whose direction is indicated by the arrows within the circle. The direction and strength at each hour on the 30°N latitude can be obtained. (*After Doodson and Warburg, 1941.*)

side will be least. The respective values are 0·000,003,493 g and 0·000,003,268 g. These forces are directed towards the moon. If, however, the differential force in relation to the earth is considered, the far side is pulled least far towards the moon, and the near side furthest. This creates a bulge under the moon and also one on the opposite side of the earth, thus accounting for the semi-diurnal character of the tidal forces. The differential forces are, therefore, directed away from centre of the earth at U and U'.

The tide-generating force at any point is the resultant of the gravitational force and the centrifugal force. The gravitational force varies both with distance from the moon and also in direction. It is the variation in direction, which reaches a maximum at the poles, when the moon is overhead at the equator, that accounts for the resultant force directed towards the centre of the earth at the poles. Figure 4.2A shows the resultant of the two forces at different points on the earth's surface. The centrifugal force is everywhere the same, and directed opposite to that of the mean gravitational force. The resultants of the two forces can be established by the parallelogram of forces. At points under the moon and opposite to it the force is perpendicularly upwards from the centre of the earth, while it is perpendicularly downwards towards the centre along a great circle normal to these points. At all other points there is an element directed along the surface of the earth. This force, acting parallel to the surface, is called the tractive force. It generates the tides. The force has the power to move particles over the surface, towards or away from the line of centres of the two bodies. The tractive force is directed everywhere from VV' towards UU', the force being zero at U and U', and on the great circle through V and V' at right angles to U and U'. It can be expressed in terms of the mass of the earth and moon and the radius of the earth, e, and their distance apart, r, and the latitude C, by the equation $3/2 \, g \, M/Ee^3/r^3 \sin 2C$. The maximum value of the force is 0·000,000,084 g at a point half-way between U and V. The cube of the distance between the two bodies comes into the equation. This means that, although the mass of the sun is much greater than that of the moon, the fact that the moon is nearer than the sun more than compensates for the moon's smaller mass. Thus the tractive force of the sun is only 0·46 times that of the moon. The tractive force reaches 10·6 m/hour after one hour and a velocity maximum of about 21 m/hour.

The strength of the tractive forces can be plotted throughout a tidal cycle on a stereographic projection, on which angles can be shown correctly. The forces are symmetrical when the moon is overhead at the equator. There are two maximums and two minimums each lunar day. The moon is not, however, always overhead at the equator. Its declination can reach 28°35', as the angle of the moon's orbit is set at an angle of 5°8' to that of the earth round the sun, which itself varies through 23°27'. When the moon's declination is not zero, which means that it is overhead north or south of the equator, the curve showing the variation of the tractive forces throughout the lunar day becomes asymmetrical and varies with latitude. The method of analysis can be demonstrated graphically by an example (illustrated in figure 4.2B) for latitude 30°N, with a lunar declination of 15°N. The varying direction of the tractive force can be determined from the stereographic projection and a diagram showing the changes in direction and strength of the tractive force hour by hour can be constructed. The curve shows that the force at hour 0 is directed southwards at its maximum value. It then swings towards southwest and between hours 5 and 6 becomes zero, followed by a slight northeasterly component, finishing up at hour 12 due

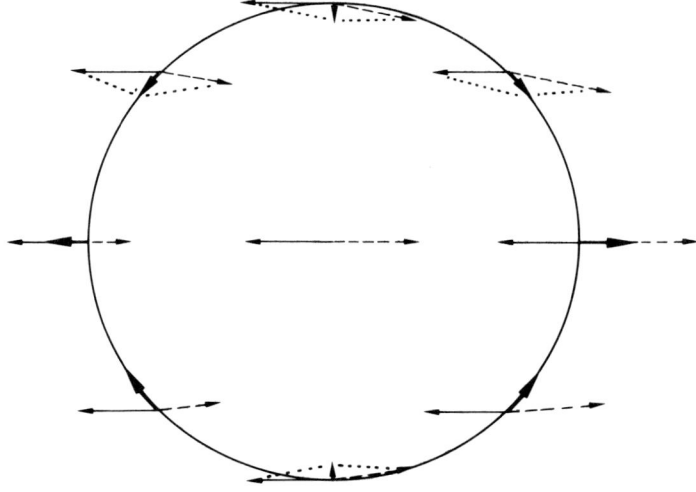

Figure 4.2A Diagram to illustrate the tidal tractive force. The thin full line is the centrifugal force. The dashed line is the gravitational force. The thick full line is the resultant tidal attractive force.

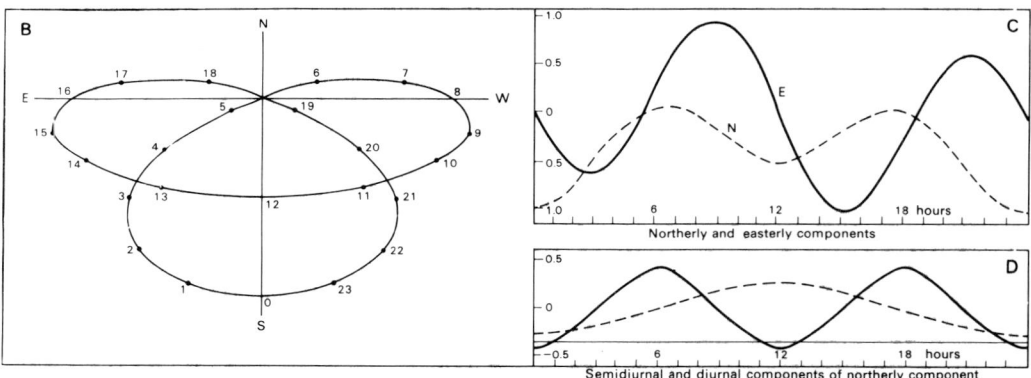

Figure 4.2B, C & D Diagrams to illustrate the lunar tidal forces for latitude 30°N, with lunar declination of 15°N. The values on the circumference indicate the strength of the tractive force, the direction of which is shown by the arrows within the circle. **B:** The direction and strength of the tractive force is shown hour by hour. **C:** The tractive force is shown split into its component easterly and northerly parts while **D** shows the northerly component split into its diurnal and semidiurnal parts.

south again, but half the value it had at hour o. The second half of the day is a mirror image of the first part.

The complex curve can be simplified by splitting it into its northerly and easterly components. These curves are still asymmetrical, but they can be divided into diurnal and semi-diurnal parts, each of which forms a smooth sine curve. The complex curve is shown to be made up of components of these periods. Thus the north component of the curve can be split into one diurnal part, which has a maximum at hour 12 and a minimum at hour o, and the semi-diurnal curve, with two maximums at hours 6 and 18.

The influence of the declination of the moon on the tractive force and its variation with latitude can be determined by the type of analysis described. The tides are symmetrical when the declination is zero, and the diurnal part of the curve is missing. The diurnal part of the lunar tide-producing force is almost proportional to the declination of the moon. The semi-diurnal part, on the other hand, diminishes in range at times of high declination. The solar tide-producing forces are similar, but of smaller magnitude and different period, as the lunar day is 50 minutes longer than the solar one.

The tractive tide-producing forces set tidal streams in motion, and these can be analysed in a similar way to tractive forces. This method of reducing a complex curve to its simple harmonic components, in the form of sine curves, is a useful technique in tidal analysis, and is called harmonic or spectral analysis. It is the basis of tidal prediction, because the complex movements of the heavenly bodies, on which the tractive forces depend, can be simplified in this way. The equilibrium tide is based on the tide-producing forces, it being assumed that there are no land-masses and inertia is ignored. These assumptions are unrealistic, but the result does illustrate the pattern produced by the tide-producing tractive forces. The response of the oceans to the tractive forces is related to the types of wave motion associated with the tidal forces and associated streams.

2 Types of wave motion in tidal theory

2.1 Progressive wave

A wave propagated in a channel of infinite length is called a 'progressive wave'. Where the wave is long compared with the water depth, the streams will extend uniformly from top to bottom. This applies to waves of tidal length, where only two crests occur around the circumference of the earth. The length of the progressive wave is defined as the distance between two adjacent crests. Its period is the time it takes to move one wave length.

Table 4.1 Wave velocity relative to depth for long waves

Depth		Wave velocity	
fathoms	m	knots	km/hour
50	91·5	58·17	107
100	183	82·27	152
200	366	117·34	217
500	915	183·95	340
1000	1830	260·15	481
2000	3660	367·90	679

The velocity of travel of the wave form is dependent only on the depth of the water in waves of tidal length, which are very long compared with the depth of the ocean. Thus the wave velocity, c, equals the square root of the depth, d, times the force of gravity, $c = \sqrt{gd}$. The variation in the height of the wave is assumed to be small compared with the depth of the water, which is true over the ocean, but in very shallow coastal water this no longer holds.

The rates of travel of the wave form can be given in terms of the water depth as Table 4.1. The actual movement of the water is, however, very much slower than that of the wave form. For example in a depth of 100 fathoms (183 m), the wave form moves at 82 knots (152 km/hour), while the water is only moving at 1·4 knots (72 cm/sec), assuming an amplitude of the wave of 10 feet (3 m). The amplitude is defined as the elevation of the wave above the mean level. This is derived from the relationship u/c = y/d, where u is the rate of the stream, c is the wave velocity, y is the elevation of the wave crest above the mean level, and d is the mean depth. The speed of 72 cm/sec is only achieved at the crest and trough of the wave, when the stream is at its maximum.

In a progressive wave the current flows in the direction of wave propagation at the crest of the wave and in the opposite direction at the trough. It becomes zero and changes sign at the mean level. The essential characteristics of a progressive wave are the dependence of the velocity on the depth for waves of tidal length, and the fact that the current flows in the direction of wave propagation and is at its maximum at the crest. At the mean water level there is no current, and at the trough the speed is at its maximum, but flows in the reverse direction.

2.2   Stationary wave or standing oscillation

The stationary wave is very important in tidal analysis. It can form in a basin of finite dimensions if the basin is tilted and the water in it allowed to rock to and fro with an oscillatory motion. The water will not leave the basin, hence the term standing oscillation. The movement of the water in the basin can be divided into four stages. First, the surface of the rectangular basin has its maximum slope, with high water at one end and low at the other. In the second stage the surface becomes flat as the water flows from the high to the low end, and the streams are at their maximum. In the third stage the streams cease as the elevations reach their maximums at the opposite ends of the basin. Finally the surface becomes flat again, and the streams reach their maximum velocity in the opposite direction. This type of wave motion differs from the first in that the streams do not coincide with the maximum elevations.

The period of the standing oscillation can be related to the dimensions of the basin. This fact is important in considering the response of the oceans to tide-producing forces. The length of the basin is L, the breadth is b, the mean depth is d, and the period of oscillation (that is the time between two successive high waters at one end of the basin) is T. The breadth of the basin does not affect the period of oscillation nor does the amplitude of the wave. The period is given by $T = 2L/\sqrt{gd}$, showing that the period depends on the length and depth of the basin only. Each basin, therefore, has its own natural period of oscillation depending on these dimensions. A state of resonance will be set up if the natural period of oscillation is the same as that of the period of the tide-producing forces. The basin will then respond to the appropriate tidal period.

The equation giving the period of oscillation also shows a connection between the standing oscillation and the progressive wave. The period of the standing oscillation is such that it equals that of a progressive wave with a length equal to twice the length of the basin, as $\sqrt{gd}$ comes into both equations. The connection between the two types of waves becomes apparent when the reflection of a progressive wave is considered.

Figure 4.3 shows what happens if a vertical barrier is placed in the path of a progressive

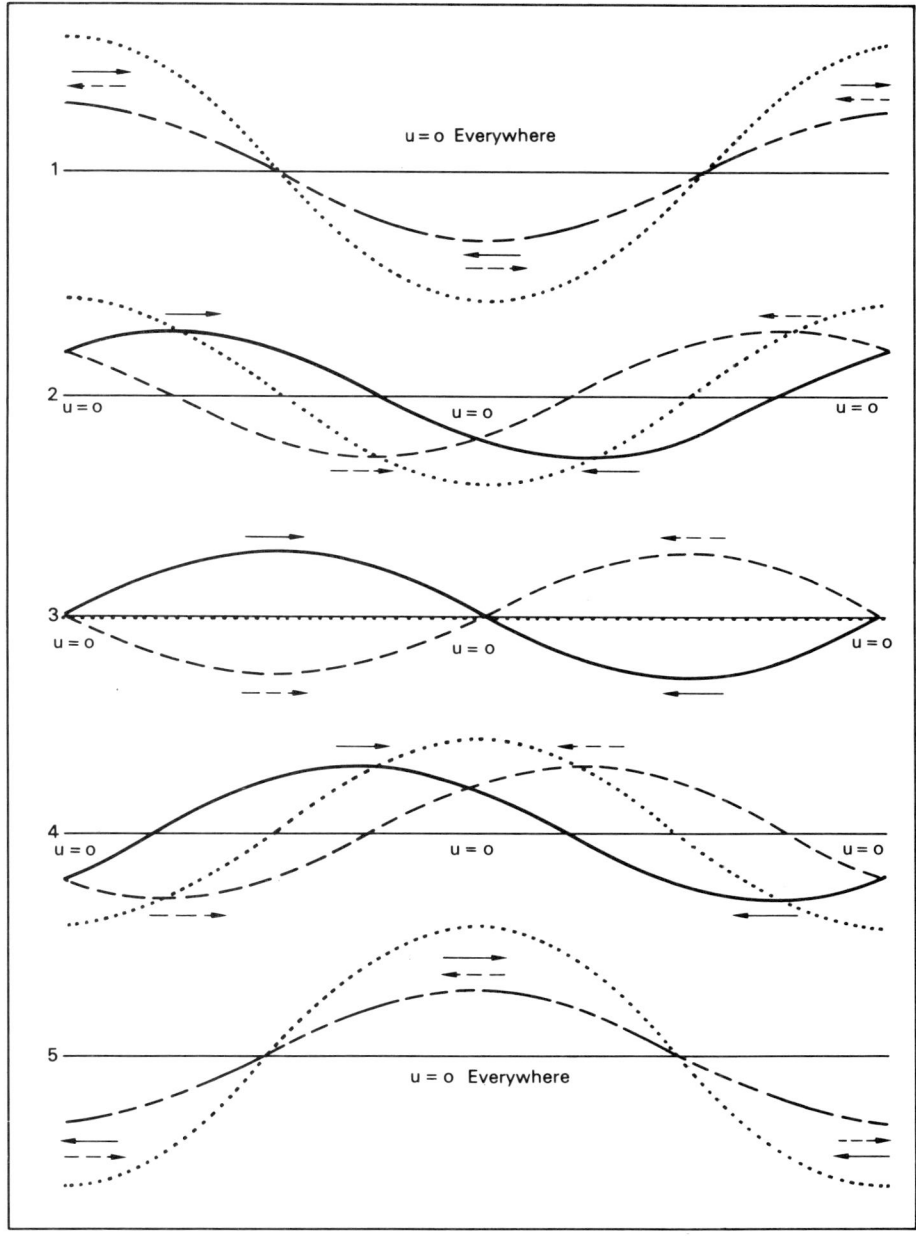

Figure 4.3 Diagram to show the formation of a standing oscillation by the reflection of a progressive wave. The dotted curve is the sum of the full curve and the dashed curve, the former being the progressive wave moving to the right, and the latter that moving to the left. (*After Doodson and Warburg, 1941.*)

wave in such a way that all its energy is reflected to form a second wave of equal size, travelling in the opposite direction to the first wave. In the first diagram the two crests coincide, which added together give the maximum elevations. The streams are flowing in the opposite directions and cancel out. In the third diagram the stage is reached when the advancing crest coincides with the retreating trough. The elevations cancel out to give a flat surface, but the streams are now flowing in the same direction and reinforce each other to give maximum velocities. A reflected progressive wave, therefore, has all the characteristics of the standing oscillation. The streams are always zero at intervals of half a wave length, while the elevation are always zero at the intermediate points. Barriers could be placed every half wave length without disturbing the motion, giving the simple oscillating system with one nodal line, as already described. The oscillating system can also be shown to operate with more than one nodal line.

3 The effect of the rotation of the earth: amphidromic systems

The importance of the rotation of the earth in the oceanic current systems has already been discussed, and it also plays a very important part in tidal phenomena. Water moving on the earth is subject to the deflecting Coriolis effect. When a particle of water is moving the Coriolis effect can either produce an acceleration to the right or create a slope up to the right in the northern hemisphere or a combination of the two effects.

The modification of a standing oscillation, by the gyroscopic effects of the earth's rotation, gives rise to the amphidromic system. When the streams are flowing at maximum velocity a gradient transverse to them is set up, causing subsidiary elevations. These will set up currents which carry the water across the basin to build up the necessary elevation on the opposite side, as the main current reverses. Figure 4.4 shows how these subsidiary elevations and streams transform the simple swing of the standing oscillation in the North Sea into a movement round three points. These points are called amphidromic points, because the tidal elevations and streams appear to move round them and there is no change of elevation at these points. The movement is in an anti-clockwise direction in the northern hemisphere and the reverse in the southern. When the elevations and streams are combined the maximum currents now occur at high water and they flow in the direction of progress of the wave round the basin at high water. These are characteristics of a progressive wave. The result of adding the effect of the rotation of the earth to a simple standing oscillation is to produce an amphidromic system, which has the characteristics of a progressive wave moving anti-clockwise round a point in the northern hemisphere.

The co-tidal lines, which join points where high water occurs simultaneously, radiate from the amphidromic points and the tidal range increases away from the points. The tidal streams associated with the system are rotatory in character, moving in an anti-clockwise sense in the northern hemisphere. Towards the edges of the basin, however, the pattern is modified and the currents are more nearly rectilinear, flowing in opposite direction at the flood and ebb of the tide, with little change of direction until they die away and then reverse in direction.

The rotation of the earth also affects a progressive wave. If this is moving along a narrow channel, the earth's rotation will lead to the development of a surface sloping up to the

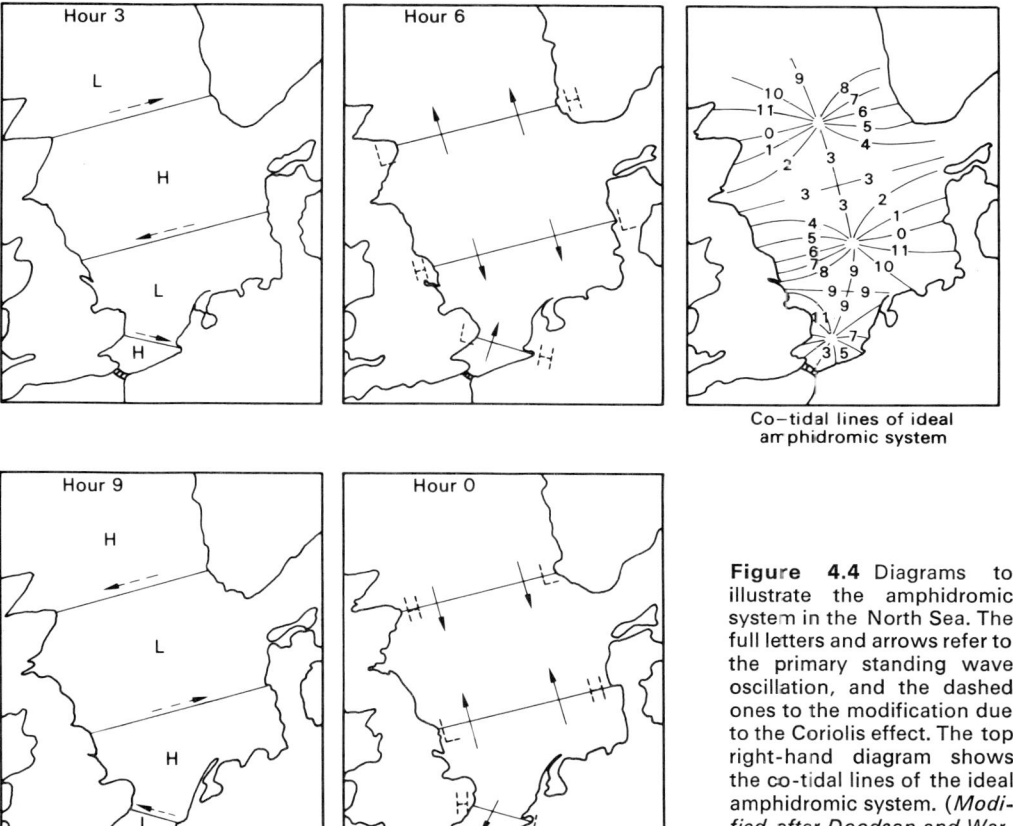

Co-tidal lines of ideal
amphidromic system

Figure 4.4 Diagrams to illustrate the amphidromic system in the North Sea. The full letters and arrows refer to the primary standing wave oscillation, and the dashed ones to the modification due to the Coriolis effect. The top right-hand diagram shows the co-tidal lines of the ideal amphidromic system. (*Modified after Doodson and Warburg, 1941.*)

right in the northern hemisphere. In an east–west channel such as the English Channel, the effect increases the range on the south coast of the channel. The tidal current flows east at high water, causing an elevation on the south side and a depression on the north side. At low water the stream flows west. The level will then be higher on the north and lower on the south. The combination accounts for the low range on the north coast and the high one on the south. This type of tidal modification is known as a Kelvin wave, and it is found in a modified form in the English Channel, where the tidal range on the coast of southern England is much smaller than that on the north French coast. The Kelvin wave is found in situations in which much of the energy of the progressive wave is lost, preventing the formation of a true amphidromic system by reflection of the wave.

Amphidromic systems are fundamental in modern analysis of tidal phenomena in the oceans, seas, and gulfs. Earlier ideas concerning oceanic tides show how the development of the idea of the amphidromic system, as described, follows the development of theories of tidal motion in the oceans.

4   The response of the oceans to the tide-producing forces

4.1   Theories of oceanic tides

The earliest ideas concerning the tide in the oceans, such as the equilibrium theory, were not very realistic, because they did not take into account the distribution of land and sea. The work of Laplace, done as long ago as 1776, took into account the rotation of the earth and inertia. He worked out a solution for a globe completely covered by water of a constant depth. It was nearly 100 years before Lord Kelvin justified the mathematics of Laplace in 1875. The data were worked out for four different depths of 2200 m, 4450 m, 8850 m, and 15,000 m. The results shows that for depths greater than 8850 m, the tide progressively approached the equilibrium value. For a value a little less than 8850 m, the tide became infinitely large. At still smaller depths it was inverted in relation to the equilibrium tide. This affects the movement of a progressive wave of tidal length around the earth in different latitudes. At the equator the circumference is such that the tidal wave would have to travel at 1400 km/hour to get round the earth in the time available. A free wave can only travel at this rate if the water depth were 20,400 m deep. This is much greater than the depth of the oceans. Tides near the equator would be inverted, but would become direct as the circumference becomes smaller in higher latitudes.

The earliest theory of the tide in the actual ocean is the progressive wave theory which developed from the idea given above. Tide-producing forces, according to the theory, generated forced tidal waves, which travelled round the southern ocean at a speed determined by tide-producing forces and the latitude. The forced wave travelled as a free progressive wave up the other oceans, according to their depth. There were some observations in support of the theory, such as the progressively later time of high water northwards along the coast of South America in the Atlantic, but there is more serious evidence against the theory. Most important is the fact that such free waves would be reflected from the edges of the oceanic gulfs, to form standing oscillations. Another point is that even the Atlantic Ocean is quite large enough to generate its own tidal oscillations. If the sea were 4800 m deep, it would only need to be 1440 km long to support an oscillation of 12 hours period. A depth of 4800 m is not uncommon.

The next important theory put forward towards the end of the nineteenth century was Harris's standing oscillation theory, based on a system of stationary waves. He stressed the importance of resonance in his explanation of the tides, suggesting that where the natural period of oscillation of a water body coincided with one of the periods of the tide-producing forces, the ocean would respond to that period. Harris produced a chart on which he divided the ocean into a series of rectangular areas of suitable dimensions to support oscillations of a particular period. The analysis explained a number of tidal anomalies by reference to the position of a place relative to the nodal line of an oscillating system. Thus, if the tide were small, the place should be near the nodal line, but if it were large it should be far from it. One of the weaknesses of Harris's theory was his assumption that two neighbouring oscillation areas did not react with one another, and his omission of the effect of the rotation of the earth. His emphasis on the importance of resonance, however, was an important step forward.

Most modern co-tidal charts are based on a series of amphidromic points. An early one

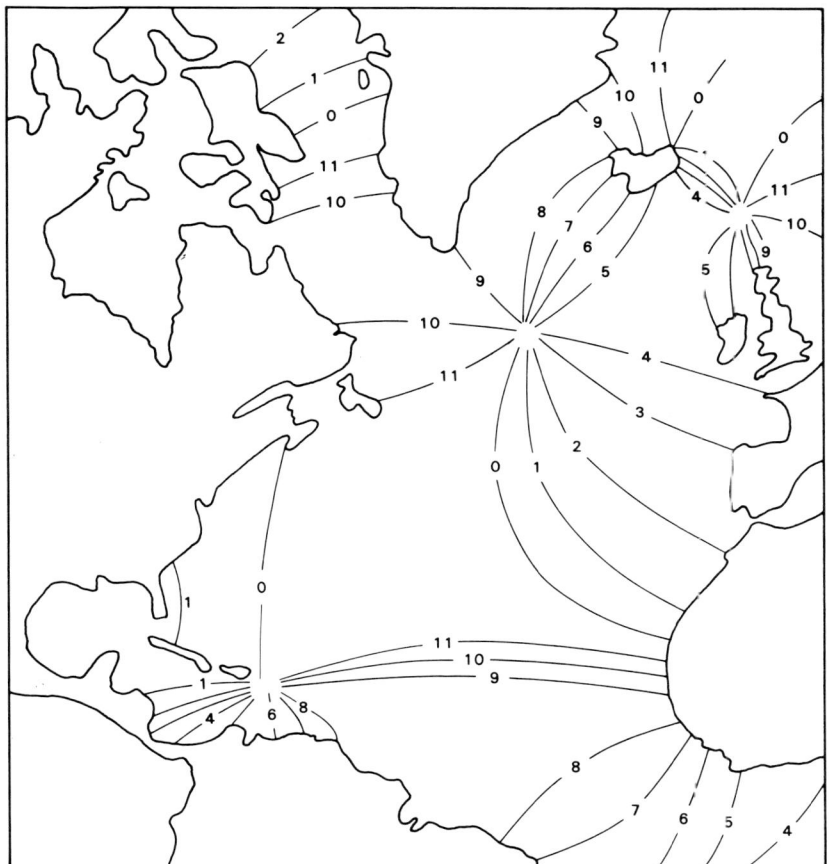

Figure 4.5 Co-tidal lines in the Atlantic Ocean after Sterneck (1920). (*After Doodson and Warburg 1941.*)

of this type was prepared for the north Atlantic by Sterneck (1920–21), and is very similar to that of Dietrich (1963). Sterneck's chart (reproduced in figure 4.5) has three amphidromic points—one between Scotland and Iceland, one in the centre of the north Atlantic about half-way between Ireland and Newfoundland, while the third is situated near the West Indies. The chart shows the semi-diurnal co-tidal lines; others are needed for the diurnal component, which because of its longer period does not necessarily react in the same way as the semi-diurnal one. The tidal pattern in oceans bounded by different meridians and of varying depths has been worked out. The results are mostly based on some pattern of amphidromic systems. Although the models differ from the actual oceans, they do suggest that these are of suitable dimensions to support amphidromic systems of varying periods according to their dimensions.

Pekeris and Arcad (1969) have computed the semi-diurnal lunar, M_2, tidal component

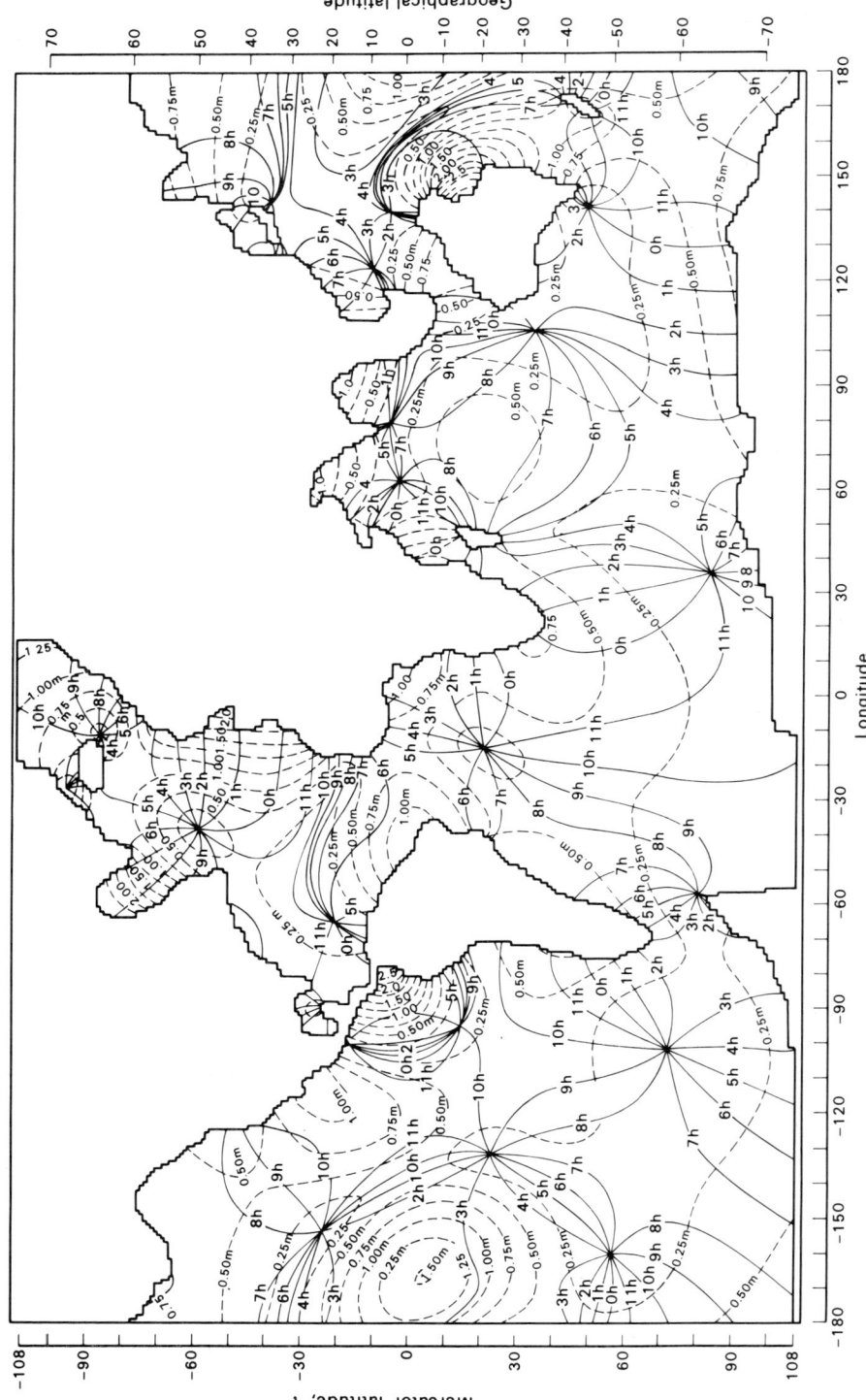

Figure 4.6 Co-tidal lines are shown in full lines and co-range lines are dashed for the M₂ tide for a computation grid of 1° latitude and longitude. *(After Pekeris and Accad, 1969.)*

in a model of the world ocean. The coastline is approximated by steps of two sizes. The finer grid is of 1° steps and the coarser is of 2° steps. A smooth bottom relief was used. Tests were run both including and excluding a frictional parameter. With the frictionless ocean model the tide was sensitive to the ocean configuration, suggesting a near resonance with the lunar semi-diurnal tide. The positions of the amphidromic points were, however, stable in all the models, although their amplitudes varied considerably. Without friction the values were too large, and numerical convergence was slow. It was concluded that tidal friction was a necessary element in the model to give reliable results.

The friction was simplified by using a linear rather than a second power relationship between velocity and frictional forces. The square law is more accurate and will be included in subsequent models. The frictional solution gave stable solutions even with the coarse coastline model. The results were not sensitive to changes in the coastline. The tidal amplitudes were reasonable, but even with friction, tidal phases were somewhat advanced compared with reality. The time difference was not, however, as great as 3 hours obtained in the frictionless model. Some of the model tests suggested that an amphidromic point was situated in the south Atlantic at 19°s, 13°w, or at 21°s, 15°w. Such observational evidence as is available agrees with the latter position. The results for the 1° fine coastal outline with friction are shown in figure 4.6.

Munk (1969) has discussed the value of deep sea tide measuring devices, two of which are under current development. They would provide data that could be used for computer-based tidal prediction. They might also elucidate problems concerning the formation of long waves in the ocean, similar to those that develop in the atmosphere.

4.2   Types of oceanic tide

There is large variety in the various forms of tide curves in different areas, due to the great number of possible combinations of the several variables. Three main groups can be identified. Firstly there are the synodic tides (such as those around Britain) where diurnal inequality is small, but the fortnightly cycle of spring and neap-tides is dominant. The latter feature is due to the relative position of the sun and moon. Secondly, there is the declinational type of tide in which the diurnal element is marked and changes in value over the spring tide period (as in parts of Borneo). Thirdly, the anomalistic tide (as in the Bay of Fundy) is dependent to some extent on the varying distance of the moon from the earth.

The tides of the Atlantic are dominated by the semi-diurnal lunar pattern, with two high waters occurring each 24 hours. This suggests that the size of the Atlantic is more suited to react with the semi-diurnal tide-producing forces. In the Pacific, which is large enough to respond to the diurnal tide-producing forces, the tide is often mixed, having one high water higher than the other each lunar day.

The occurrence of diurnal tides in the Atlantic is limited to those bays where the conditions are such that resonance with tidal forces of this period can take place, as for Chaleur Bay and Northumberland Strait. The former has a marked inequality of high water and the latter of low water. Diurnal tides are also common in the Pacific, as at Victoria in British Columbia. The curve is mainly diurnal, but the period of low water is shorter than that of high water at spring tide. At neap-tide this is reversed, giving a double low water.

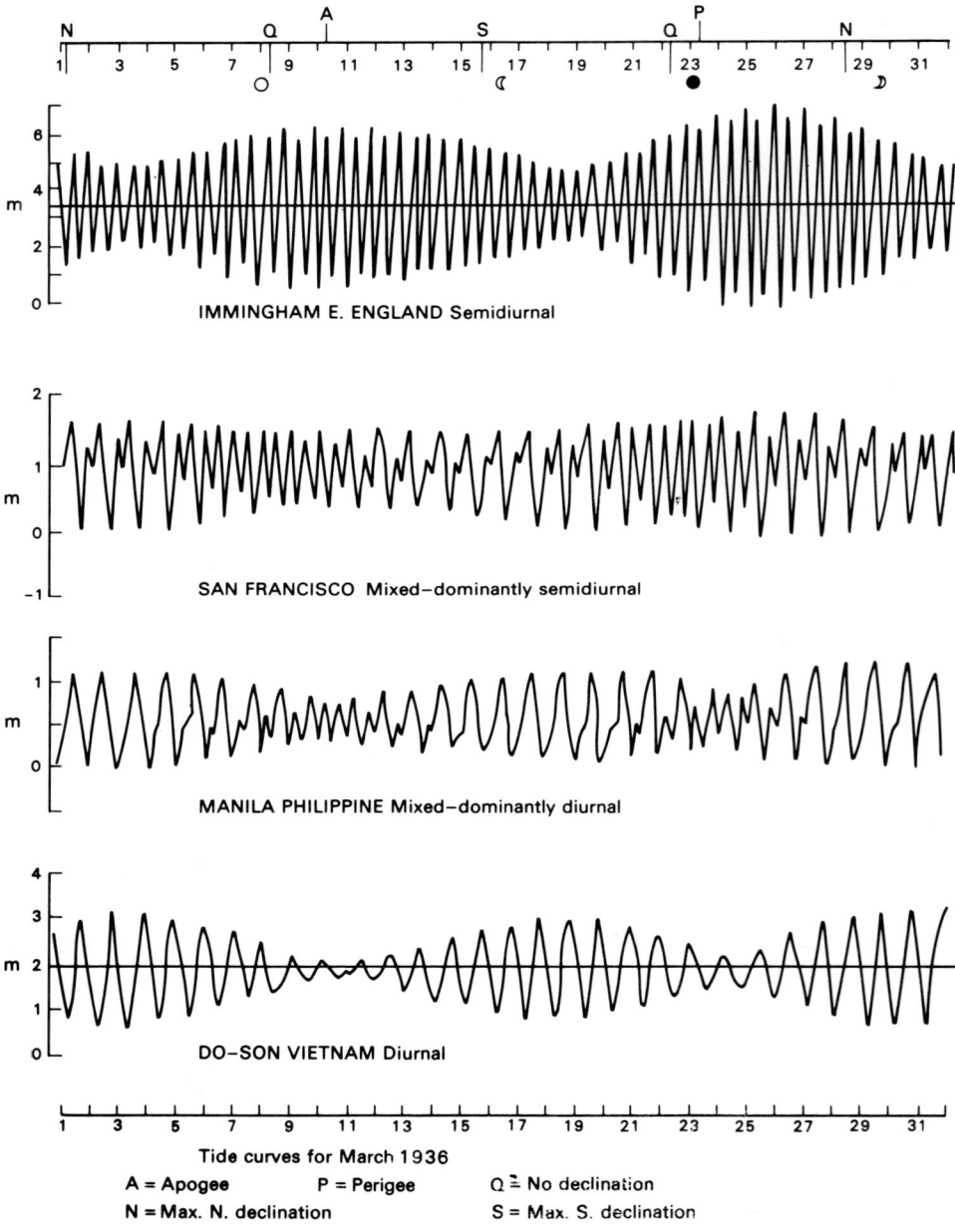

Figure 4.7 Representative types of tidal curves. *(After Dietrich, 1963.)*

The tide is also influenced by the sun and is a mixed tide. Asymmetry is also well developed on the coast of California. Some areas in the western Pacific have diurnal tides only, some a mixture of semi-diurnal and diurnal during the fortnightly cycle of neap and spring tides. Examples of some types are shown in figure 4.7. Purely diurnal tides are not common, but do occur in parts of Vietnam (see Do-Son) and the northern half of the Gulf of Mexico, where the range is small. On the north coast of Borneo, which shows the influence of the diurnal tide-producing force, the maximum range is small, being just over 1·8 m, but the tide is diurnal for a period of a week around spring tide, when its range is greatest. During neap-tide the range is small and the pattern is semi-diurnal. The range increases in this area during the solstices.

The tides of southern California are associated with regular movements of sand on the beaches during part of the summer. Grunion, *Leurethes tenuis*, take advantage of this cycle in their spawning habits. Eggs are deposited on the beach in such a position that they are covered by sand during the spring tide stage, and are uncovered again at the next neap-tide, when the eggs are just ready to hatch, after being protected in the sand during the incubation period of 14 days. There is also a longer-term seasonal change, whereby the diurnal range is considerably less at the equinox than it is at the solstices. The maximum range is 2·65 m and the mean range is 1·13 m.

The tide in some areas is more influenced by the solar forces than the lunar ones. Tahiti, for example, has high tide at the same times each day, instead of being 50 minutes later each day. This can be explained if an amphidromic point of the lunar tide-producing forces were situated near the island. The lunar tide would, therefore, be reduced in effectiveness sufficiently for the smaller solar tide to become conspicuous. The total range is not large.

Apart from the types of tide already mentioned, there are also the variations due to the distance of the moon and other factors related to the movement of the moon and sun. The fortnightly succession of spring and neap-tides occurs when the sun and moon are pulling together or against each other respectively. The various elements of which any complex tide curve is composed can be analysed by harmonic analysis. When the major constituents have been isolated various longer-term effects are left. They can be related to the movements of the sun, moon, and earth. The distance of the moon from the earth varies in a monthly cycle, being a minimum at perigee and a maximum at apogee.

The tides of the Bay of Fundy respond to the lunar distance in such a way that the spring tides at perigee are considerably greater in range than those at apogee. This type of tide is called anomalistic. At perigee the spring range is 15·4 m, while at apogee it is 12·25 m. There are also longer-term effects related to the repetition time of certain specific relationships in the orbits of the earth and moon, one of which has a period of about 18 years.

The range of the tide over large stretches of the coasts of the world is fairly small, while it is probably even smaller in the open ocean. Oceanic islands rarely have large tidal ranges. The areas where the tidal range exceeds 6 m are restricted to more enclosed seas—for example, around parts of the coast of Britain. The only section of the coast of Africa to have this range is near Zanzibar on the east coast. A few isolated areas having this range are found in the Gulf of Cambay and the Hoogly delta region of India, near Rangoon and the west side of the Malay peninsula. There are also small areas on the south coast of Borneo, the west side of the Straits of Formosa, the western part of the north coast of Australia, and the southern tip of South America. In North America ranges over 6 m occur in the inner end of the Gulf of California, parts of the Gulf of Alaska, Ungava Bay in Quebec province, and the Bay of Fundy on the Atlantic coast. Ranges over 3 m are more widespread, but most of the coasts of the world have ranges under this value. Nearly all large tidal ranges are found in gulfs and bays.

4.3 Tides in gulfs and seas

A rectangular gulf is about the nearest approach to the theoretical rectangular basin in which the character of the amphidromic system can be examined. The length and depth of the gulf determine its natural period of oscillation. If this natural period of oscillation is nearly equal to that of one of the tidal periods the gulf will respond by resonance and the tidal range will be large. There is, therefore, a critical length in relation to its depth, which will enable a gulf to support a standing oscillation.

The position of the node of the standing oscillation in relation to the mouth of the gulf will depend on its length relative to its depth. There is a critical length for each depth that will support a standing oscillation in such a way that the nodal line will run through the mouth of the gulf. There will be no change of level at the mouth and the range will increase to a maximum near the head of the gulf. If the gulf is shorter than this critical length, high water will occur throughout the gulf at the same time; but there will be a small tidal range at the mouth. If, however, the gulf is longer than the critical length high water will occur at the head of the gulf while it is low water at the mouth. If, however, the gulf is twice the critical length the nodal line will occur in the centre of the gulf and the range will be as great at the mouth as it is at the head, high water occurring at one end when it is low at the other. Under these conditions there will be no streams across the mouth of the gulf. A gulf of critical length derives its tidal energy from the open ocean and although the level does not change at the mouth, when the surface is flat at mid-tide the currents are at their maximum.

The critical length of the gulf is given by $L = \frac{1}{4}T\sqrt{gd}$. The gulf dimensions are more common than those of the enclosed sea that can operate as a self-contained system with a length twice that of the gulf. Therefore in many gulfs it is high tide throughout the gulf at the same time, and strong tidal currents may be expected across the mouth of the gulf. Table 4.2 gives the critical lengths for gulfs and enclosed seas.

So far the tide in a gulf has been considered as a simple standing oscillation, but it will be affected by the rotation of the earth, giving it the characteristics of an amphidromic sys em. Gulfs in nature also have not the symmetrical form of the ideal theoretical gulf, and other factors, such as friction, will modify the results.

The Bay of Fundy, which has one of the greatest tidal ranges anywhere, illustrates the importance of resonance. The bay averages 68·5 m depth, its critical length is 256 km, which agrees closely with its measured length of 259 km. The natural period of oscillation of the bay is about 6·29 hours, which is almost exactly the period required for resonance

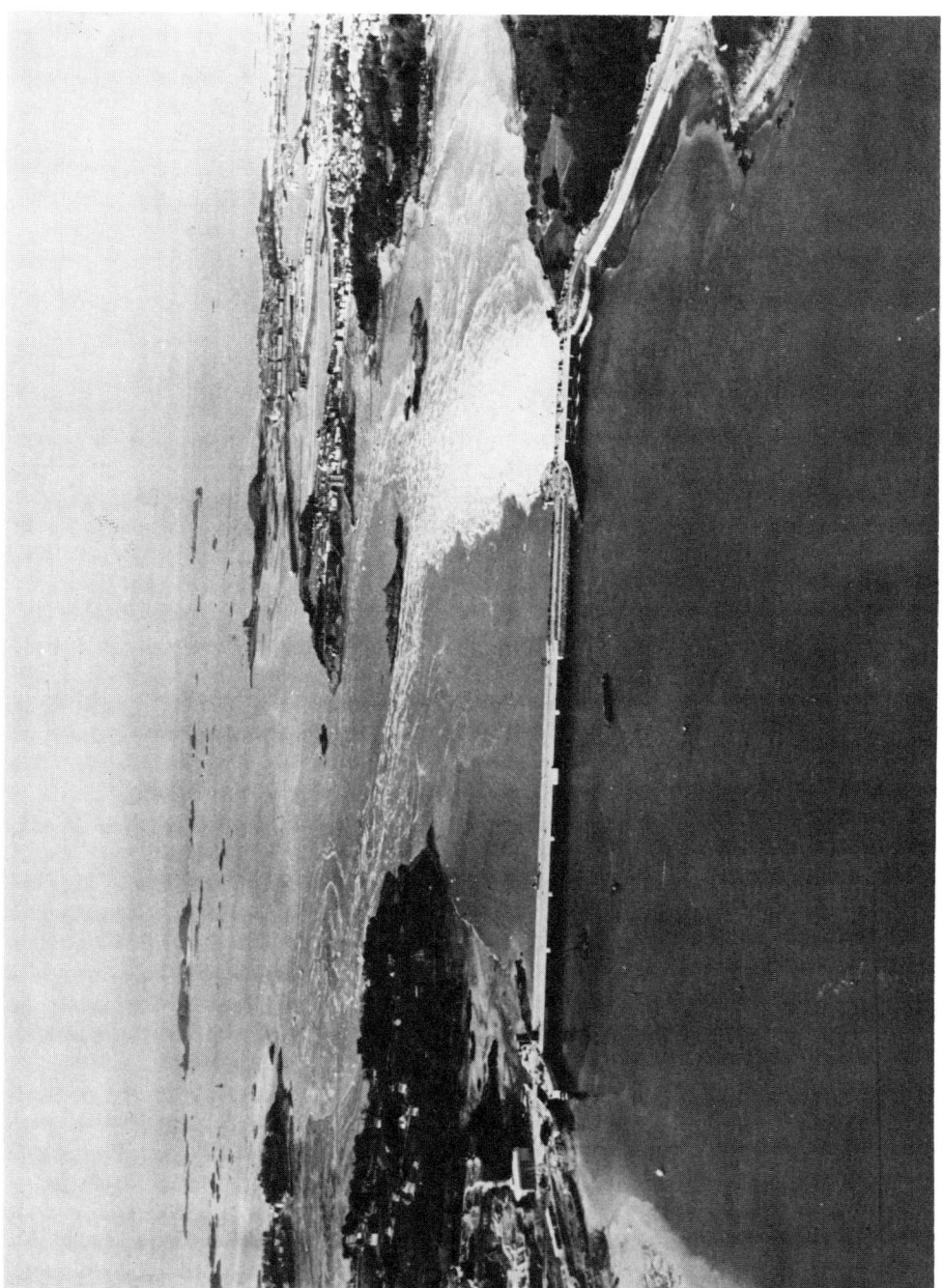

Plate 3 The tidal power station on the Rance in north France is shown, the ebbing tide flowing seaward through the generator on the right side of the barrage (*French Embassy Information Service*).

Table 4.2 Critical length of gulfs and enclosed seas

Critical length of gulfs in nautical miles and km for 12- and 24-hour periods

Depth		Period 12 hours		Period 24 hours	
fathoms	m	n miles	km	n miles	km
50	91·5	175	324	349	645
100	183	247	457	494	915
200	610	349	645	698	1290
500	915	552	1020	1104	2040
1000	1830	780	1440	1561	2890
2000	6100	1104	2040	2207	3830
Critical length of enclosed seas					
50	91·5	349	645	698	1290
100	183	494	915	987	1825
200	610	698	1290	1396	2580
500	915	1104	2040	2207	3830
1000	1830	1561	2890	3122	5780
2000	6100	2207	3830	4415	8170

to occur. It reacts, therefore, to the semi-diurnal tidal forces. The range of the tide increases towards the head of the bay as would be expected from the theory. This effect is increased by the shallowing water and narrowing and bifurcation of the bay towards its head. Spring tidal range exceeds 15 m at its head. Rotation of the earth increases the range on the southern side of the bay. The time of high water occurs almost simultaneously throughout the bay, being only 24 minutes later at the bifurcation than at the mouth of the bay.

Long Island Sound at the eastern entrance to New York Harbour also illustrates the relationship. Again the tide is high at nearly the same time throughout the basin and the range increases from 0·75 m to 2·3 m from the mouth towards the head. Over the same distance the streams decrease from 180 cm/sec to almost negligible velocities. The dimensions of the harbour are mean depth 20 m and length 128 km. The critical length for this depth is 132 km.

The English Channel exemplifies a gulf whose dimensions are more nearly those of an enclosed sea. The mean depth is about 36 fathoms (66 m) and the channel is about 480 km long, the length required for an oscillation with a nodal line across the centre. High tide occurs in the east at Dover at about the time when it is low tide off Cornwall.

4.4   The tides around the British Isles

The English Channel has already been mentioned in connection with the characteristics of a Kelvin wave, and its dimensions are such that it can react to the semi-diurnal tide-producing forces. In theory a nodal line should run near the centre close to the Isle of Wight, but friction and other variables prevent the formation of a true amphidromic system. The amphidromic point is degenerate and it lies inland, as shown by the convergence of the co-tidal lines towards the coast in this vicinity. This fact helps to account for the double tides characteristic of the area around the Isle of Wight.

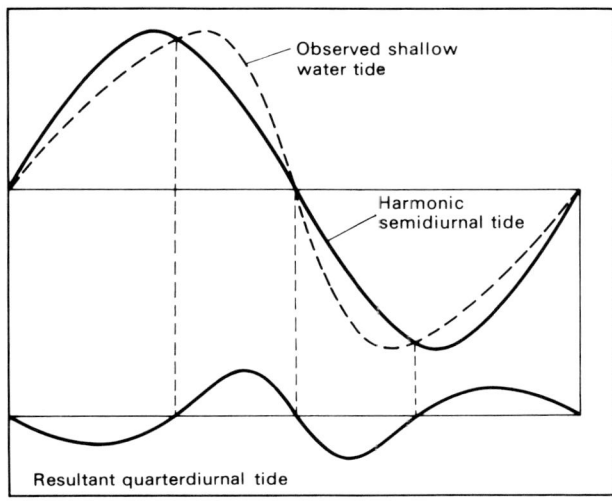

Observed shallow water tide

Harmonic semidiurnal tide

Resultant quarterdiurnal tide

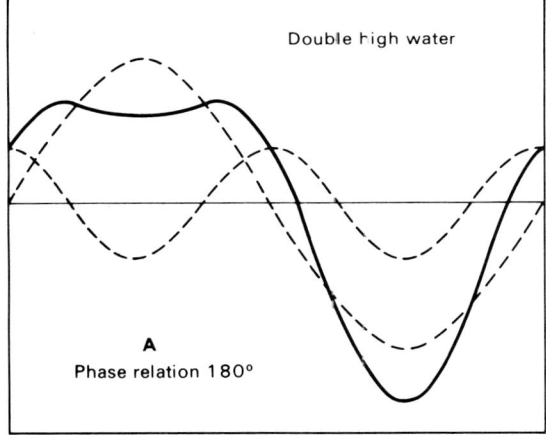

Double high water

A
Phase relation 180°

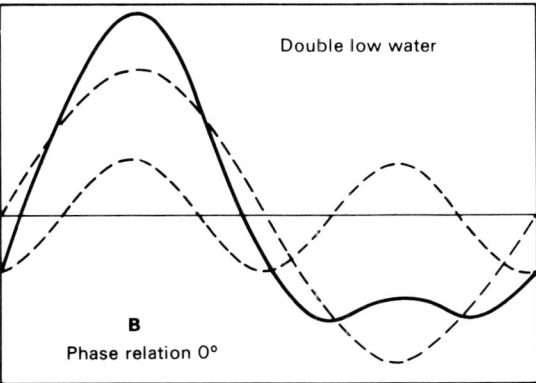

Double low water

B
Phase relation 0°

Figure 4.8 Shallow water tidal phenomena. (*After Doodson and Warburg, 1941.*)

4.4a Shallow water tides

The double tides near the Isle of Wight can be partly explained by shallow water effects. The influence of shallow water on a progressive wave is fairly simple. The speed of a progressive wave depends on the depth of water. The water is deeper under the crest, which moves faster than the trough, in which the water is shallower. The smooth sine curve of the wave form thus becomes distorted. The tide rises faster than it falls, because the crest is accelerated and the trough retarded. The asymmetrical curve can be resolved into its regular components in the same way as the complex curve of tidal tractive forces. The residual curve has two crests to one in the original curve (figure 4.8). This curve is called the quarter-diurnal tide. It is irregular and can be further split up to give a symmetrical quarter-diurnal tide and the remainder (which has three crests) is called the sixth-diurnal tide. The process can be continued. The curves when added together never give a double low or high water because their crests and troughs do not coincide. They give the distorted curve characteristic of a progressive wave in shallow water.

It is not clearly known how shallow water affects a standing oscillation, but it is reasonable to assume that the effect is somewhat similar. If the crests or troughs of the semi-diurnal and quarter-diurnal tides coincide and they have a specific range relationship, the two curves added together will give a double high or low water. When the two crests coincide the phase relationship is said to be 0°, and this will tend towards the development of a double low water. If a trough of the quarter-diurnal tide coincides with the crest of the semi-diurnal tide the phase relationship is 180°, and double high water will result if the amplitudes are suitable, as shown in figure 4.8.

The quarter-diurnal tide must be one-quarter the semi-diurnal tide to produce a double tide and the sixth-diurnal tide must be one-ninth the semi-diurnal tide. These amplitude ratios are most likely to occur where the range of the semi-diurnal tide is diminished near a degenerate amphidromic point close to the shore. These conditions are fulfilled in the area around the Isle of Wight. At Portland the phase relationship is −002°, showing that the crests of the two wave types nearly coincide. This would suggest the probability of double low water. The amplitudes are such that the semi-diurnal tide has an amplitude of 0·63 m, while the quarter-diurnal tide is 0·125 m. Four times 0·125 is not quite equal to 0·63 m, but it is near enough to produce a long stand at low water, and the higher species of tide produce the double effect. Fresh water, with a phase of relationship between the two tides of 176°, is well situated to have a double high water. The respective amplitudes of the semi-diurnal and quarter-diurnal tides are 0·615 m and 0·16 m, which are suitable to produce a double high water by the simplest means, as $4 \times 0 \cdot 16 > 0 \cdot 615$.

The case of Southampton, important from the point of view of shipping, is less simple (Macmillan, 1949). There is a very long stand at high water, which at some phases of the tidal cycle becomes a double high water. The tide here, as in other areas where tidal currents can flow in from two directions, is affected by the hydraulic gradients set up as a result of different levels at either end of the straits. At the eastern entrance to the Solent the tide has a range of 3·95 m at spring tide, while at the western end the range is only half. The analysis of the tide at Southampton shows that the phase relation of the semi-diurnal and quarter-diurnal tide, which is −279°, is not suitable for the generation of double high water. The amplitude ratio is also not suitable to produce a simple double high water. The semi-diurnal amplitude is 1·36 m and the quarter-diurnal one is 0·25 m. The sixth-diurnal tide accounts for a long stand at high water, but the eighth- and tenth-diurnal tides are needed to produce the double effect. Many complex phenomena are concerned with the double tidal effect at Southampton.

Double tidal features will occur normally a) where the phase relationship between the semi-diurnal and quarter-diurnal shallow water tides are suitable, and b) where the amplitude relationship of these tides is favourable.

Double tides also occur on the other side of the English Channel at Le Havre and Honfleur. They are, however, caused by a different process because the range of the demi-diurnal tide is considerable in this area. The reflection of the tide from the Straits of Dover appears to cause the double tide in this area. The tidal current first flows east to produce a preliminary high water, while at times the subsequent westerly stream produces a second high water. The westerly-flowing stream, affected by the rotation of the earth, has an up-gradient on the English side of the channel and this may help to produce the long stand of high water at Southampton, while shallow water effects produce the double tide.

4.4b   The North Sea

The development of a multi-nodal amphidromic system is well illustrated in the North Sea. The dimensions of the sea are such that, in the absence of the rotation of the earth, there would be three nodal lines (shown in figure 4.4). The most northerly would run from southwest Norway to Rattray Head in Scotland. The middle one would run from central Denmark to the east coast of England near Flamborough Head, and the southern one would lie between south Holland and East Anglia. The pattern of tidal streams, when the surface was flat, and the position of high tide and low tide at the different hours are indicated in figure 4.4. High and low water would alternate as shown. The energy to keep the tidal motion going is derived from the Atlantic Ocean tide to the north. As a first approximation it may be assumed that there is a barrier across the Straits of Dover.

The rotation of the earth converts the standing oscillations into three amphidromic systems. Subsidiary slopes are set up across the nodal lines when the streams reach their maximums. They are reversed six hours later. The co-tidal lines show that the three systems interconnect in such a way that the tide moves anti-clockwise continuously south down the coast of Scotland and England and northwards along the continental coast.

If the barrier across the Straits of Dover is removed and the effect of friction is considered, the ideal tidal pattern can be related to the actual tidal chart (figure 4.9). The energy to maintain this tidal wave moves anti-clockwise around the North Sea. It is derived from the Atlantic Ocean and is gradually absorbed as it travels round the shallow sea. The force of friction thus reduces the tidal energy, resulting in a diminishing range of tide towards the continental coast; the amphidromic points move eastwards towards the area of reduced range.

The northern amphidromic point lies somewhere near the coast of southern Norway, and may in fact be degenerate, as the range on the south coast of Norway is very small. The central point also moves eastwards. It lies a short distance off the coast of central Denmark, where the tidal range is less than 1·2 m. The southern amphidromic point is not moved east appreciably, partly owing to the closer proximity to the source of energy, but mainly

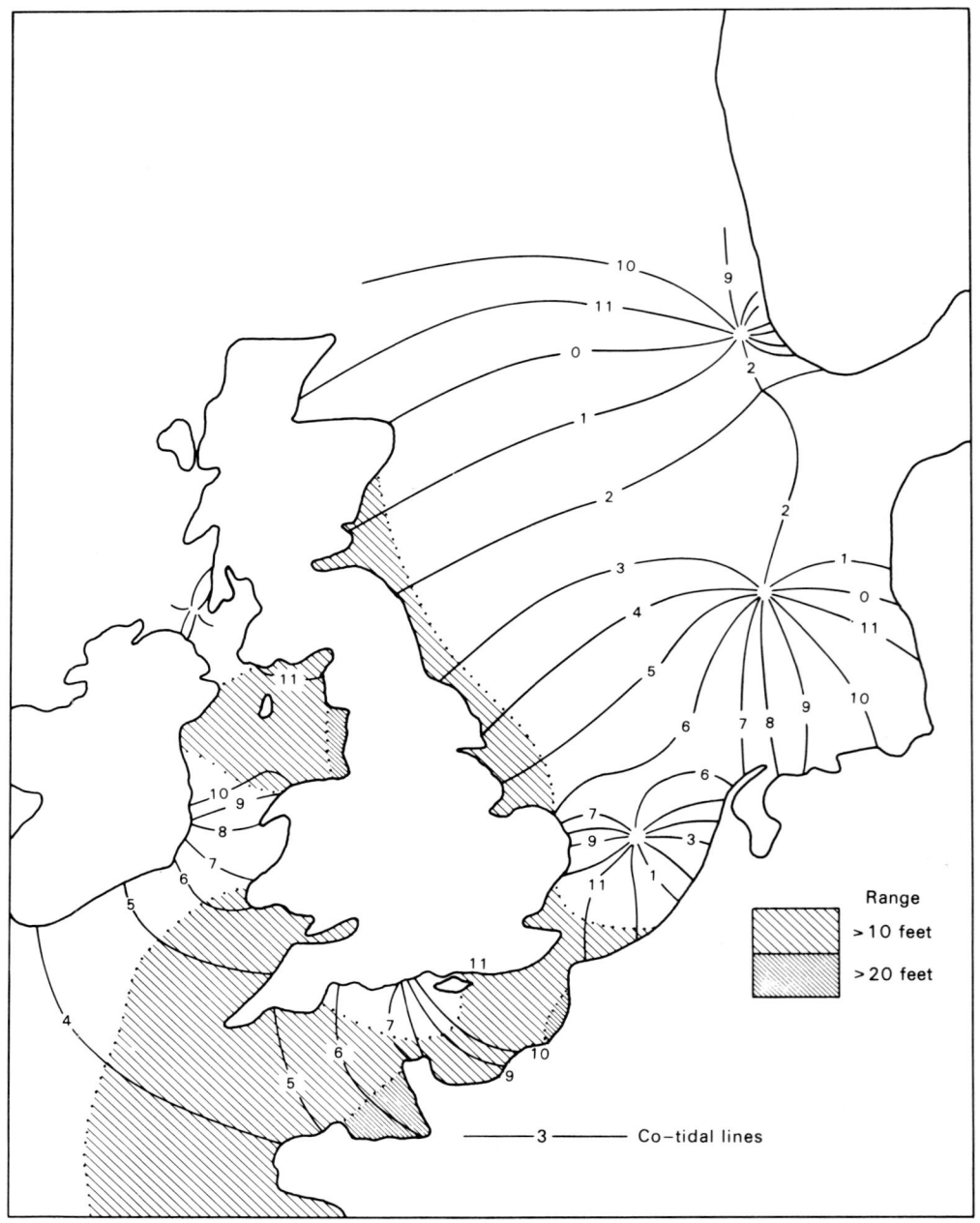

Figure 4.9 The tides around Britain. (*Based on Admiralty Chart No. 5058.*)

as a result of the influx of energy and water through the Straits of Dover. The amount of water coming through the straits is only felt in the Flemish Bight, where it swings round the co-tidal lines to a certain extent. The tidal range along the east coast of England increases away from the amphidromic points towards the heads of gulfs. It is, therefore, highest in the Wash, where the spring range exceeds 6 m.

Cartwright (1969) has examined the tides at St Kilda, where they show a strong diurnal inequality, and where the diurnal tidal streams are stronger than normal. The two phenomena may be related. The evidence suggests a non-divergent tidally-induced continental shelf wave. The diurnal tide is normally only one-tenth the semi-diurnal in British waters, but in this area it is much larger. Non-divergent waves cause currents, but little surface change of elevation. These waves are associated with sharp changes of depth and are called double Kelvin waves by Longuet-Higgins. The continental shelf waves are non-divergent; they are trapped and forced to travel parallel to the shelf edge with shallow water on the right in the northern hemisphere. Their frequency must be less than that of the local inertial frequency. They are most likely to be generated near the peak frequency of their dispersion curves, where the group velocity changes sign. The inertial frequency is 1·695 c/day in 57°35′N and the spectra show activity in the diurnal tidal range, which satisfied the requirements. The shelf wave is probably generated by some form of coupling with the normal divergent wave modes. The non-divergent wave is controlled by geostrophic forces.

5 Tidal currents and their observation

The tidal tractive forces produce a rotatory tidal force. The semi-diurnal tractive force rotates in a clockwise direction in the northern hemisphere, and the reverse in the southern, but the diurnal forces change sign at 45°N and s and at the equator. They also change sign as the moon's declination changes from north to south. In the Atlantic, however, where semi-diurnal tides predominate, the movement is mainly in a clockwise direction.

The effect of the earth's rotation produces a rotatory current. In the northern hemisphere the deflection is to the right. A current flowing east at high water is deflected to the south and will be flowing in this direction when the easterly flow ceases at the turn of the tide. At low water the current will be flowing west and will induce a northerly flow three hours later. This results in a clockwise rotation. The subsidiary currents will probably not be so strong as the main current. The line joining an hourly plot of the flow vectors would be nearly elliptical in form, the major axis being east–west.

Rotatory tidal streams will also be set up where the tidal wave is moving up a channel with gently shelving sides. The wave moves up the channel at a speed appropriate to the mean depth. Variations of level across the channel are prevented by the generation of transverse streams, which give rise to rotatory currents. On a shelving coast if the high water current is flowing east it will be at its maximum at high water if the wave is a progressive one. Three hours later the main current will be zero, but the water level on the coast will be falling rapidly, and a current will be running offshore, to the south on the northern shore of an east–west channel, and to the north on the southern shore. At hour 6 it will be low water and the level will be flat, so no transverse currents will be running, but the main stream will be flowing west at its maximum velocity. At hour 9 the tide will be rising again at the maximum rate; this will also be the time when the main current is reversing and is, therefore, zero. The subsidiary transverse current will be flowing towards the shore to build up the level. This will be to the north on the north shore and to the south on the south shore. The combination of these movements gives a clockwise rotation on the northern shore of the channel and an anti-clockwise one on the southern shore.

Several of these influences will probably be operating simultaneously, causing complex tidal streams. Hydraulic tidal streams operate in straits linking areas in which the tidal levels are different. These streams can reach considerable velocities at times. For example, at the northern end of the Seymour Narrows between Vancouver Island and the mainland, currents approaching 5·15 m/sec (10 knots) occur at some stages of the tide, when there is a 4 m difference in level at either end of the strait. This is because low water at one end coincides nearly with high water at the other end.

In general waves of tidal length are very long compared with the water depth, and tidal streams do not diminish greatly in velocity with depth. Such observations that have been made, mostly in fairly shallow water, support this contention. Streams are only appreciable in shallow water. In a depth of 50 fathoms (91·5 m), where the surface speed was 51·5 cm/sec (1 knot) the maximum velocity of about 61·7 cm/sec (1·2 knots) was observed 5·5 m (3 fathoms) below the surface. There was a very slow decrease towards the bottom below this depth. Only about 11 m (6 fathoms) above the bottom the speed was about 36 cm/sec (0·7 knots).

Measurement of tidal streams can be based on measurements over time at one point, or water paths can be traced through space. Current meters and jelly bottles are designed to measure the current at one point. The former can operate over a considerable period of time, while the latter gives a value at one position and time for each observation. Observations can be repeated hourly to give values for a complete tidal cycle. The other method that has been used extensively over the last decade is the use of the Woodhead sea bed drifter, which is described in detail by Phillips (1970). The method gives much longer-term information of water movement, as the floats move with the residual of all the currents. This technique integrates all the currents. The tidal streams are often the most important in the areas where the method has been used most extensively.

The jelly bottle (Carruthers, 1963) can be used in all depths up to 110 m, and can be adapted for greater depths. It can record currents from 13 cm/sec to 310 cm/sec (0·25–6 knots). The method was used in conjunction with the recording of dispersal of radioactive tracer on the Caister shoal near Great Yarmouth. The tracer, placed in a depth of 7 m, moved south at about 31 to 82·5 cm/sec under the influence of the flood-tide, and was not affected by wave currents. Current meter readings over a complete tidal cycle gave 108 cm/sec for the southward flood-stream, and 41 cm/sec for the northward ebb, at 1 m above the bottom, with a tidal range of 2·44 m at Lowestoft. Jelly bottle results on a slightly lower tide of 2·2 m gave a maximum stream of 141·5 cm/sec and a northward stream on the ebb of 121 cm/sec. The jelly bottle results show stronger streams than the current meter, but both show a southerly residual.

Deeper observations have been made on the La Chapelle Bank 160 km from Ushant, in the Bay of Biscay. The shallowest depth over the bank, which lies near the edge of the continental shelf with rapidly deepening water to the southwest, is 151 m. The morphology suggests strong currents. Tidal range in the area is 3 m, and direct-current measurements 12·2 m below the surface indicated a velocity of 92·7 cm/sec in a direction 17° east of north, and 108 cm/sec in a direction 17° west of south, towards the deep ocean. Currents close to the bottom were measured at 47°40′N, 7°14′ W at 155 m, 25 cm above the bottom by jelly bottles at hourly intervals for 17 hours. The tide was intermediate between mean tide and neap tide. The maximum stream was 28·3 cm/sec at 190° true, with a reciprocal direction 6 hours later of 21·6 cm/sec. At no time was the current less than 13 cm/sec. The streams were capable of moving sand, as indicated by the bed forms, which consisted of sand waves elongated normal to the maximum streams.

Further measurements have been made in the same area by Cartwright and Woods (1963) in 165 m of water above the edge of the continental shelf. The semi-diurnal tidal currents were recorded reliably. They formed current ellipses with the major axis about 30° from north, normal to the line of greatest slope of the shelf edge and to the alignment of large sand waves. The bottom currents were 69 per cent of the surface currents in magnitude, with a residual drift to the west of about 10 cm/sec in 47°40′N, 7°14′w. The observations were made by current meter. The maximum value of just over 50 cm/sec occurred at spring tide, with 41 cm/sec at mean tide and 26 cm/sec at neap tide. The currents rotate clockwise. At mean tide the following observations were obtained:

	Maximum cm/sec	Minimum cm/sec	Direction east of north in degrees
Surface	47·4	30·4	28
Surface	37·0	29·3	24
Surface	42·2	29·8	27
Bottom	25·7	20·1	37

Tidal observations in deep water are very few, but one set of data are available. A deep water oceanographic observatory has been installed at a depth of 3903 m, 160 km off the coast of California, from which tidal heights and currents can be measured (Nowroozi et al., 1968). A spectral analysis of the data was used to estimate the tidal heights and currents for the different tidal periods. The tide heights and currents correlated well and various tidal components were all represented. Thus the measured ocean currents in this area are mainly caused by the tides. The results of the observations and analysis are summarized in table 4.3. The predominant spectral component is M_2, the semi-diurnal

Table 4.3 Deep water tidal data off California

Component	Period hours	Ocean tide cm	Obs. ocean current cm/sec	Phase difference degrees	Calculated current, cm/sec
O_1	25·819	24·276	0·454	16·3	1·21
K_1	23·362	34·362	0·765	−162·6	1·72
N_2	12·658	14·087	0·721	72·1	0·71
M_2	12·421	45·430	2·406	45·4	2·27
S_2	12·000	9·651	0·613	−247·8	0·49

lunar one, with an amplitude of 2·406 cm/sec. The diurnal component is also significant. The phase difference between the tides and currents can be explained if it is assumed that two waves are travelling in opposite directions.

Drifter studies differ from those already mentioned in that the trajectories of the water particles are followed, rather than measuring the flow past a fixed point. The drifters are released at one point and they move with the water at a predetermined depth. Each drift has a serially numbered returnable card attached. The major disadvantage is that only the starting and end point are known and the speed calculated must be a minimum. Some drifters are recovered from fishing boats, particularly in the North Sea. Drifters may be used at any depth, but surface ones are liable to be influenced by the wind. Near-bottom movement of drifters is not necessarily representative of the true residual flow, as the drifters are affected by currents of many different types.

The dispersion of sea bed drifters was observed by Harvey (1967) from the Menai Straits in the Irish Sea. The drifters were dropped at hourly intervals in 12 equal batches to ascertain the movement at different states of the tidal cycle. The recoveries showed a southwesterly movement followed by a northwesterly one along the west coast of Anglesey. Eventually recoveries extended further, reaching south Wales and Blackpool after 255 and 202 days respectively.

Another experiment was made by Robinson (1968) in the North Sea. He released 240 drifters 4·8 km east of Spurn Point at the seaward end of the main Humber ebb-channel at low water. During the first 3 months 33 per cent of the returns were from Spurn spit, and 7 per cent from the south bank of the Humber between Donna Nook and Cleethorpes. The remainder came from the Lincolnshire coast south of Donna Nook. The wind direction during the experiment was mainly southwesterly, suggesting that waves were not responsible for the movement of the drifters. The tidal stream, therefore, seems to be the main agent of sediment transport in the offshore zone in this area. The drifters recovered from the Lincolnshire coast probably moved with the main tidal streams and the flood residual caused their southerly transport. The flood-stream flows south along this coast in connection with the North Sea amphidromic tidal system. The tidal streams in the North Sea and the Irish Sea, as well as in many other nearshore areas, create a distinctive tidal offshore morphology which was discussed in volume 1.

6   Tides in rivers: tidal bores

Although many rivers are tidal in their lower reaches, few develop well-marked bores. Even rivers which have bores can lose them, if the bed of the river is disturbed by dredging or other means. There must, therefore, be some critical factor on which the generation of a bore depends. It has already been shown how a progressive wave moving into shallowing water develops a steeper front slope, which leads to a more rapid tidal rise than fall. The same process is carried further when a tidal bore is generated. The rise of water is in the form of a nearly vertical wall of water, advancing up the river.

The best-developed bore in the world is probably that of the Chang Tank Kiang in China. At Haining the depth of the river before the arrival of the bore is only 1·53 m, but the bore itself is nearly 3·35 m high at spring tide, and it moves up the river as a wall of water at more than 8·25 m/sec. It carries 1·75 million tons of water each minute. The evidence suggests that at one time this bore did not exist. On the other hand, the Seine in France had a very well-developed bore, called the Mascaret, which used to move up the river as far as Rouen. Since dredging and other channel improvements were started in 1780 the bore has nearly disappeared and now only affects the lowest part of the river at the highest spring tides. Other rivers that have bores include the Hoogly, the Petitcodiac in Canada, and the Severn and Trent in England. The bore on the latter has been reduced by river improvements.

A delicate balance between certain forces is necessary to initiate and maintain a bore, and they only affect one tributary of a river system in many cases. The Trent for example, develops a bore, locally called the Eagre, while there is no similar feature in some of the other rivers flowing into the Humber although the Ouse has one. A bore cannot be accounted for solely by the development of a more and more steeply sloping front to the flood profile of the waves; otherwise bores would not be the local phenomena that they are. Champion and Corkan (1936), in their study of the Trent bore, showed that the slope of the river bed is an important factor in the generation of a tidal bore.

In order to understand the generation of a bore it is necessary to consider the movement of water in a sloping channel. If the velocity of the current, moving up the river, is less than the rate of movement of the wave up the channel, the depth of water will increase down the channel. If the velocity of the current is greater than the critical value, which is the velocity of the wave, $\sqrt{gd}$, then the depth will increase upstream, causing an upslope of the water surface. The mechanism of the bore can be considered by examining the effect of the bed slope in relation to the velocity of the current.

Where the velocity is very low the surface will be flat and the bed slope will not influence it. If the current is not zero it can be shown that the gradient of the surface relative to the channel slope is given by $\dfrac{i}{1 - u^2/gd}$, where i is the gradient of the channel, u is the velocity of the stream, and d is the depth. Again the relationship between the velocity of the stream and that of the rate of propagation of the free wave is critical. When the velocity of the stream approaches that of the wave form, the surface slope may become very great. As this value depends on the depth, it is local in its effect. For example, in the Trent the tide rises up a steep slope, the depth is reduced rapidly and the point where $u^2 = gd$ may be reached. Over a very short distance there is a large change of surface elevation. This would tend to produce a steep front, which might appear as a wall of water. Where u^2 was less than gd the surface slope would be downwards upstream. This state would occur in the deeper water, as it does in normal tidal motion as the tide rises.

A steep slope in the river bed is important to the generation of a bore in two ways. Firstly, it accentuates the gradients of the water surface, and secondly, it accelerates the attainment of the point where $u^2 = gd$, where the surface slope must change sign. A bore is also more likely to occur at high spring tide as both the depth and currents are greater at this state of the tide. A narrowing of the channel will tend to produce similar results. Where the critical velocity is reached other factors come in, and the relationships do not hold accurately. Instability is set up and this results in the development of waves, which so often follow the bore as it moves upstream. They are called whelps on the Trent.

Once formed, the bore will move upstream travelling as a free wave. The rate at which the bore is propagated upstream depends on its height and the mean depth. This may be modified by the speed of the current flowing downstream in the river, above the bore. The bore travels faster than the progressive wave, at a speed given by

$$c = \left(1 + \frac{1}{2}\frac{B}{d}\right)\sqrt{gd} - U,$$

where c is the speed of the bore, B is its height, d is the mean depth, and U is the river current. The Trent bore moves upstream at about 24 km/hour or 670 cm/sec. Substituting appropriate values into the equation gives

$$c = 1 + \frac{1}{2}\left(\frac{1\cdot83}{2\cdot74}\right)\sqrt{288} = 7\cdot025 \text{ m/sec.}$$

B is 6 feet (1·83 m), as the depths before and after the passage of the bore are 6 and 12 feet (1·83 and 3·66 m) respectively, d is therefore 9 feet (2·74 m). The calculated value agrees well with the observed value. The value of U (speed of the current) is 1·83 m/sec.

The river improvement works, which normally aim to increase the depth of the river, therefore, are detrimental to the formation of bores, because they prevent the critical velocity being reached. Bores are usually associated with spring tides, because the water is especially shallow at low water, when the flood-tide is advancing up the river, and deep as it passes, giving a higher value of B, the height of the bore.

The Trent bore illustrates the close relationship between the gradient of the river bed and the generation of the bore. The lower part of the Trent, when the bore was studied by Champion and Corkan, could be divided into three reaches (figure 4.10). The lowest part between the Trent Outfall into the Humber and Burton Stather had a very steep slope. In the second stretch, between Burton Stather and Walkerith, the river sloped less steeply, and the gradient decreased still further in the third stretch upstream of Walkerith. The critical relationship is set up in the second section, where the bore is well marked. Its rate of movement and size increase to its maximum at Walkerith. Upstream from here it travels for some considerable distance, sometimes as much as 32 km.

The bore is, however, very variable and more recent observations, made at various times up to 1952 (Barnes, 1952), show that it has changed its character since the river works were undertaken. It appears to have slowed down as it has diminished in height. Its average upstream speed was found to be about 16 km/hour, although at its maximum it reached nearly 19·2 km/hour, or more on some occasions. The conclusion may be reached that bores are very individual features, liable to variation in speed and character, sometimes even disappearing, with only minor changes in the nature and gradient of the river bed up which they move.

Further reading

BELDERSON, R. H. and KENYON, N. H. 1969: Direct illustration of one-way transport by tidal currents. *J. Sedim. Petrol.* **39,** 1249–50.

CARTWRIGHT, D. E. 1969: Extraordinary tidal currents near St Kilda. *Nature* **223,** 928–32.

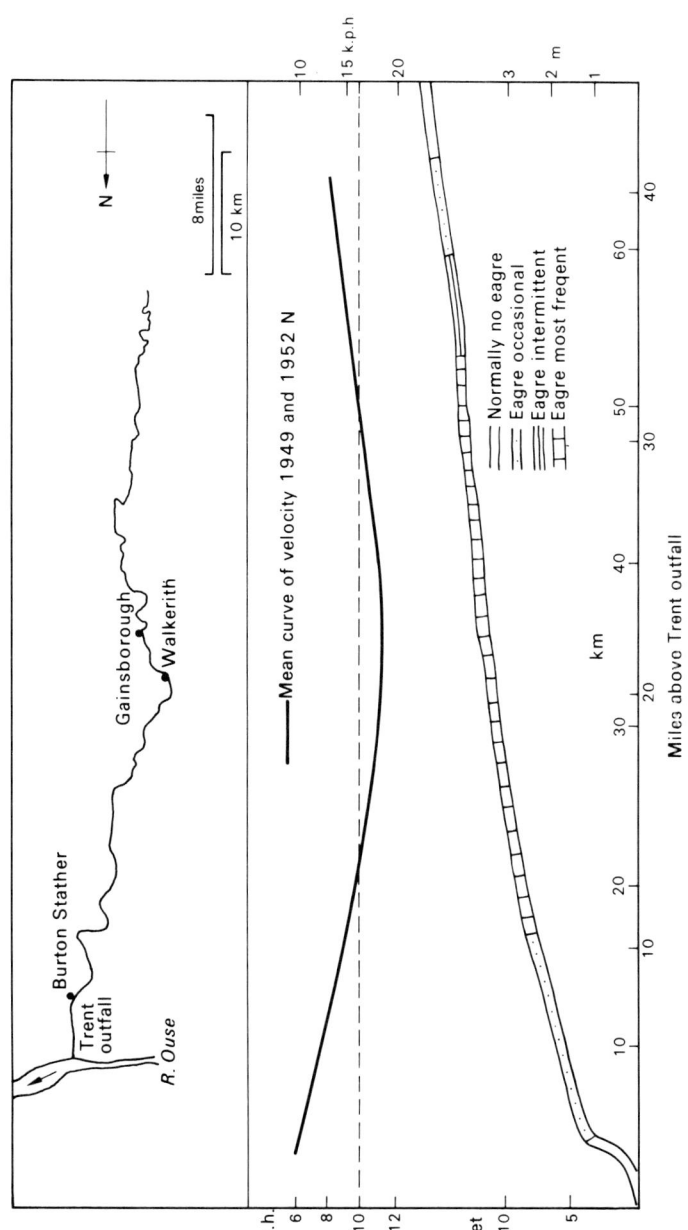

Figure 4.10 The Eagre on the Trent, showing the formation of the Eagre in relation to the slope of the river bed and the mean speed of the progress up the river. (*Modified after Barnes, 1952.*)

DALTON, F. K. 1951: Fundy's prodigious tides and the Petitcodiac's tidal bore. *J. Roy. Astron. Soc. Canada* **45,** 225–31.

DOODSON, A. T. and WARBURG, H. D. 1941: *Admiralty manual of tides*. London: HMSO.

HARVEY, J. G. 1967: Drifter studies in the Irish Sea. *Liverpool Essays in Geography*. London: Longmans, 137–56.

MacMILLAN, D. H. 1966: *Tides* London: C. R. Books.

MUNK, W. H. 1969: Deep sea tides. In M. Sears (editor), *Progress in oceanography* **V**. Oxford: Pergamon, 67–70.

MUNK, W. H. and CARTWRIGHT, D. E. 1966: Tidal spectroscopy and prediction. *Phil. Trans. Roy. Soc.* **A 259,** 533–81.

PEKERIS, C. L. and ACCAD, M. 1969: Solution of Laplace's equations for the M_2 tide in the world oceans. *Phil. Trans. Roy. Soc.* **A 265,** 413–36.

ROBINSON, A. H. W. 1968: The use of sea bed drifters in coastal studies with particular reference to the Humber. *Zeit. für Geomorph.* **NF 7,** 1–23.

SNODGRASS, F. E. 1964: Precision digital tide gauge. *Science* **146,** 198–208.

5 Waves

Perhaps the most obvious movement of the water in the oceans is the disturbance of the surface caused by the generation of wind waves. Waves formed by several other processes are also generated in the ocean. These include the rare, but often destructive, seismic waves or tsunami, which though inconspicuous in the open ocean, may become very large near the coast. Other types of waves include surf beat, surges, microseisms associated with wind waves and internal waves, which can be generated in several ways. The wind, however, is responsible for the formation of the oscillatory waves, which move over the surface of the sea everywhere and all the time, and which play such an important part in the modification of the coastline. The surface of the sea in a storm appears to be in a state of chaotic confusion. It is, however, possible (to a certain extent at least) to sort out the different wave trains present.

Ideal oscillatory waves in the ocean change as the waves move from deep water into shallow water. Deep water is defined as greater than half the wave length. In deep water a wave moves independently of the bottom, but in shallow water its form and movement are affected by the bottom.

1   Ideal waves in deep water

Waves in deep water can be defined by their length and height. Particularly important from the point of view of their effect on the beach at least, is the relationship between these two dimensions, the ratio of height to length, giving the steepness of the wave. The wave length is defined as the distance between two successive crests, measured perpendicular to the wave crest. The period and velocity of the wave form are closely associated with the length. The period is the time it takes the wave form to move through one wave length, and the velocity is the speed of advance of the wave form. The three variables are related by $L = CT$, where L is the length, C the velocity and T the period. The wave velocity depends only on the wave length, the two variables increasing together. The relationship can be expressed by $C = \sqrt{(gL/2\pi)}$, or in the form $L\ (m) = 1{\cdot}56T^2$ (secs). This is useful because it is easier to measure the period in deep water than the length. The relationship assumes a low amplitude, but can be used also for low waves of finite height without much error. The equation given applies to deep water, but in shallow water the depth effect must be included. The modified equation shows the significant difference between the movement, on the one hand, of wind waves, and on the other, of the much longer waves, including tide waves and seismic waves. This equation is

$$C = \sqrt{\frac{gL}{2\pi} \tanh \frac{2\pi d}{L}},$$

where g is the force of gravity and d is the water depth. In deep water d/L is large and tanh $(2\pi d/L)$ is almost unity, giving the deep water equation. If d/L is 0·5, that is the depth is half the wave length, than tanh $(2\pi d/L)$ is 0·9963, so that for most purposes a wave may be considered to be in deep water when the depth exceeds half the wave length. When, however, the ratio d/L is small, less than 0·05, tanh $(2\pi d/L)$ approaches the value of $(2\pi d/L)$. The equation then simplifies to give $C = \sqrt{gd}$. This is the equation already frequently used in the discussion of tidal phenomena, in which the velocity of the wave depends only on the depth, because the length is great compared with the depth.

The wave height is not fixed in relation to the length, but depends for its value on the wind generating it. Theoretically a wave cannot have a height-to-length ratio greater than one-seventh, or it becomes unstable and breaks. This value is very rare in the ocean, and most natural waves are much less steep, the ratio H/L often being below 0·02.

Various conceptions of the ideal surface form of waves have been suggested. The equations already given were developed by Airy in 1842 and Stokes in 1847, the latter extending the theory to waves of finite height. They suggested that the waves were sinusoidal in form, except when they are of finite height, when their form is nearly trochoidal. The nature of the motion is irrotational. Stokes' theory of wave character shows that the water particles do not come back to exactly the same place as each wave passes, but they progress slightly in the direction of wave propagation. This implies that there is a slow transference of water in the direction of wave movement. This is called mass transport. From this point of view their theory differs from that of Gerstner and Rankine, who suggest that the wave form is exactly trochoidal and the orbits exactly circular. As mass transport has been shown to exist, the theory of Stokes (which has been proved convergent by Levi-Civita in 1925 and applied to water of finite depth by Struik in 1926) is the more realistic.

The mass transport of water by wind waves is closely related to the nature of the actual movement of water within the wave form. Clearly the water particles cannot move with the velocity of the wave form, which for a 10-second wave would be 15·6 m/sec. They do, however, describe almost circular orbits, each orbit progressing slightly in an open circle. The particles move up as the wave crest approaches and forward on the crest, then down as the trough approaches and back under the trough, thus completing the orbit. The size of the orbit depends on the height of the wave, while the speed of movement depends also on the wave period, as the orbit must be completed each period. Beneath the water surface there is a rapid reduction in the size of the orbit. It is in geometrical progression as the depth increases in arithmetical progression. Thus for every one-ninth of the wave length downwards the orbit is halved, with the result that at a depth equal to the wave length the orbit is only 1/535 of its surface size. Wind waves, therefore, only affect a very shallow surface layer, compared to the whole depth of the ocean.

The volume of mass transport also depends on the height and length of the waves according to the equation $H^2\sqrt{g\pi/32L}$, as the square of the wave height is involved, but only the square-root of the wave length, the amount of mass transport falls off rapidly with decrease of wave height, being nil for waves of no finite height. The surface velocity of mass transport is proportional to the square of the steepness of the wave and is given by $(\pi H^2/L^2)C$. This property of waves is of considerable importance to an understanding of their effect on bottom material in shallow water.

The height and length of the wave also determines its energy, which is equally divided in two forms. The kinetic energy results from the movement of the water particles and the potential energy is due to the elevation of the wave crest above the still water level. Again the amount of energy depends on the square of the wave height, as given by $E = wLH^2/8$ where w is the weight of 1 ft³ of sea water; this can be expressed also in the form $E = 0.64wH^2T^2$ or $E = 41H^2T^2$. E is given in ft/lb/foot of wave crest/wave length. Thus the energy also depends mainly on the wave height.

The relations given above apply strictly only to waves of low amplitude, and must be modified by the addition of another term for waves of finite height, although this has little effect except under extreme conditions. The full equation is

$$E = \frac{wLH^2}{8}\left(1 - 4.93\frac{H^2}{L^2}\right).$$

In an isolated train of waves, moving through calm water, the kinetic energy remains stationary, while the potential energy moves at the wave velocity. The total energy of the wave train, therefore, moves at half the speed of the wave form. Waves appear to die out in front of the train and form in the rear, each individual wave seeming to travel through the wave train.

The ideal waves have long continuous crests, running perpendicular to their direction of movement. This is not often the case in reality, however, as there is rarely only one wave train present. The interference of wave trains moving in different directions, and of different lengths, result in the formation of short-crested waves. Even waves moving in

F

the same direction may not have the same length. This gives rise to the usual variability of wave height in the open sea, and of breakers on the coast.

2   Wave observations

2.1   Deep water

Until recently the methods used to forecast waves depended on visual observations, but methods have been developed which enable waves to be instrumentally measured both in deep water and in shallow water. A satisfactory ship-borne wave recorder has been developed at the National Institute of Oceanography (Tucker, 1956). This apparatus, which can be fitted into the hull of a ship, is designed to measure the height of the water surface above the point in the hull where it is fitted, and to relate this height to a reference level below the ship. The addition of the two values gives the wave height. The first amount can be found by the change in pressure due to the increase of water depth, but the second is not so easily measured. It is derived from a measure of the acceleration of the vertical component of wave velocity by double integration. The result shows the fluctuations of the surface, which can be analysed to give the wave spectrum. The instrument is fitted at a level 3 m below the water line, which cuts out waves of periods shorter than 4 seconds. The records have an accuracy of ± 10 per cent.

A number of different types of wave recorders have been developed. Some record waves as variations in the water level at a fixed point with respect to time. Instruments in this class include pressure-sensing devices fixed below the surface, wave poles, wave wires, ship-borne wave recorders and pulsed sound beams pointing upwards from the bottom (inverted echo-sounder). The ship-borne recorder is particularly useful, and it has the added advantage that no fixed installation is required, as is the case for wave wires.

The second type of wave record covers an area of the sea surface in three dimensions at one instant in time. Stereophotogrammetric techniques produce a contour map of the sea surface at a point in time. Low-flying aircraft can provide maps covering areas of 815 m by 540 m, as shown in the work of Cote *et al.* (1960) in connection with the project SWOP.

Other devices, such as the splashnik, which records wave acceleration, have been developed. Wave buoys developed by the National Institute of Oceanography record wave heave, surface slope, and slope differences. The curvature of the wave form can be established in this way. The NIO wave buoy is described by Longuet-Higgins, Cartwright, and Smith (1963). It floats freely in the open ocean. The vertical acceleration combined with the two angles of pitching and rolling provide the first five Fourier coefficients of the angular distribution of energy in each wave frequency band. These values can then be used to calculate the weighted average of the directional spectrum with respect to the horizontal azimuth. Other useful data provided include the total spectral energy, the mean direction of energy, the angular spread of the energy and an indication of the shape of the energy distribution. Measurements of this type can be used to test the validity of different theories of wave generation. The instrument provides a two-dimensional wave spectrum rather than the earlier type of observations that normally yielded a one-dimensional one.

Time series analysis is used to obtain data concerning the significant heights and periods of the waves in conjunction with other statistical methods to establish the wave spectra. The one-dimensional spectrum can be analysed by Fourier analysis to give the frequencies in the spectrum and their relative amounts of energy. For some statistical purposes, however, it is more convenient to obtain only a height and period from the record.

The mean wave height and period may be obtained with a suitable attachment to an electrical wave meter. The full wave record is needed for spectral analysis and the photo-electric analogue analyser is a satisfactory means of analysis. Newer digital methods are, however, rather more accurate. Digital computer methods are used to analyse the cross-spectral records (Tucker, 1963).

Bonnefille, Cormault, and Vlembois (1967) describe an ultra-sonic inverse wave recorder, which consists of a piezo-electric transducer. The instrument is set vertically and is con-nected by cable to the land, where the recorder is located. The results consist of punched paper tape in a state suitable for computer analysis. Each record consists of 20 minutes of observations and can be programmed to give the statistical distribution of crests and the period of the swell, the spectral energy, the significant and maximum values of both wave height and period, and a graphical output showing the percentage energy in each height and period band.

Despite the elaborate equipment now available there are still problems in measuring waves accurately at sea and in providing a consistent statistical description of them. The difficulties of wave measurement in the ocean stem from both sampling problems and the problem of calibration (Kinsman, 1965). However, much useful information has been obtained; some examples are given below.

Draper and Fricker (1965) recorded waves in 60 m of water on the Sevenstones light-vessel 32 km southwest of Land's End. Twelve-minute records were taken every three hours with a ship-borne wave recorder. The sum of the distances of the highest crest and lowest trough from the mean water level, H_1, the mean zero-crossing period T_z, the mean crest period T_c, the significant height H_s, the most probable height of the highest wave in the period H_{max} are given. Waves were much higher in winter and the period was lower in summer. These results as well as the wave steepness values are shown in figure 5.1.

Further wave records have been published from data obtained by the weather ship *India* stationed at 59°N 19°W in the Atlantic Ocean (Draper and Squire, 1967). These results are presented in the same form as those already given for Land's End. Two thousand records have been analysed spread over a period of 13 years. Together they may be considered to represent an average year. The wave heights are presented in terms of percentage exceedance of stated values. The Atlantic in winter has a high percentage of large waves. Heights exceeded 4·6 m for 41 per cent of the time in winter. The most com-mon wave conditions developed when a significant height of about 2·1 m was com-bined with a zero-crossing period of about 9 seconds. The higher waves were almost always associated with the longer periods, but some long periods had small heights, when the waves came from a distant storm. Wave steepness values rarely were lower than 0·056; the steepest wave recorded was 0·115, compared with the theoretical maximum of 0·14. This wave had a height of 10·65 m and a period of 7·65 seconds. The highest wave recorded during the period was 20·4 m, but the likely maximum was 24·4 m.

Darbyshire (1959b) has analysed the relative frequency of wave height and period in

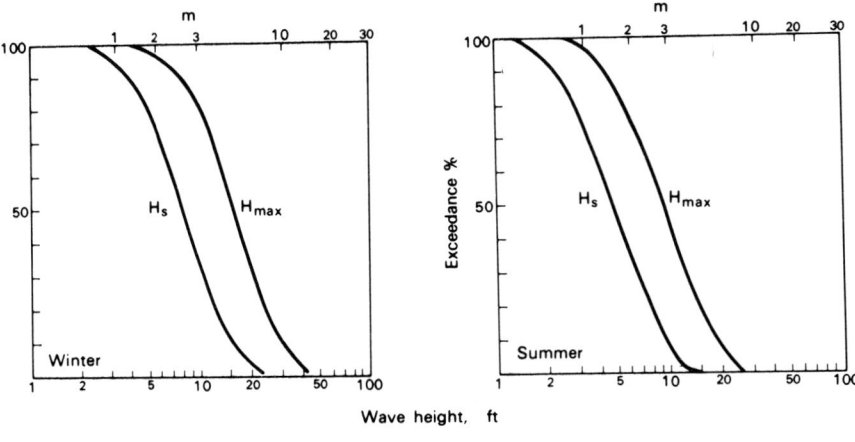

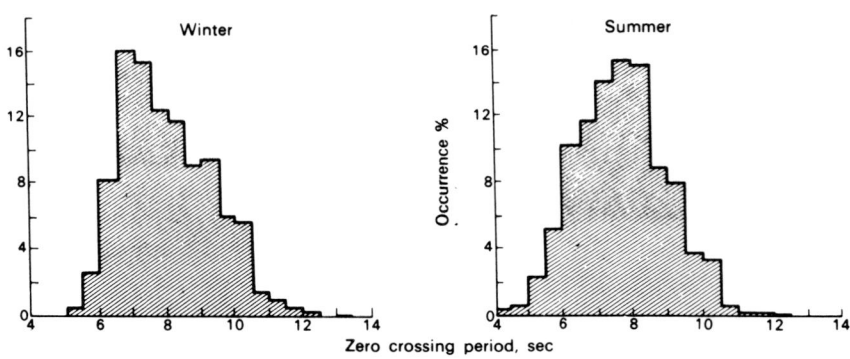

Figure 5.1 Waves off Land's End at Sevenstones. The significant and maximum wave heights are shown for winter and summer. (*After Draper and Fricker, 1965.*)

the Atlantic and in the Irish Sea, as well as at various coastal stations. This provides useful information concerning the dimensions of waves likely to be experienced in different situations. The waves which were used for the analysis in the Atlantic were in the generating area. The height of the crest above the mean water level was found for each period band of 1 second and related to the force of the wind. For force 4 winds in the Atlantic, the most common period was 6 seconds, and the height above the mean water level averaged about 30 cm. For force 6 the values were 7 seconds and 76 cm; for force 8 they were 11 seconds and nearly 3 m; and for force 9 they were only a little greater.

The largest wave for which there is reasonably reliable evidence of a visual type is that observed by USS *Ramapo* in the north Pacific in 1933. This wave was 34·2 m high. Stationary random process statistics suggest that a wave twice the mean height occurs once in 23 waves, one three times the mean height occurs once in 1175 waves, and one four times occurs only once in over 300,000 waves. A freak high wave results from the addition of a large number of different period waves in the spectrum that happen to coincide at their crests; such waves only last for a minute or less as the different period waves separate again.

2.2 Wave measurement in shallow water and wave spectra

Waves in shallow water can be measured by the ship-borne wave recorder, installed in light-ships that are continuously on station, thus providing a long-term record. For wave observations close to the shore some type of pressure wave gauge, connected to the shore by cable, is probably most useful. A simple mechanism is the flexible rubber bag held below the surface with the pressure recording device on shore and connected to the instrument by air line. The pressure changes are recorded by pen on a revolving drum. The maximum distance that the instrument can be from the shore recorder is 100 m. Wave direction can be measured by an array of three or more wave recorders, which must be fixed about a wave length apart. Interpretation is complicated by the refraction that waves undergo in shallow water. Four instruments exist that can measure wave direction (Draper, 1967b).

Wave records are now available for waves in the Irish Sea (Darbyshire, 1958) and the North Sea (Darbyshire, 1960). Observations were made with wave recorders fitted in light-ships. Records show that waves in shallow enclosed seas are generally much shorter in period and lower in height than those generated in the open ocean. Wave records in Morecambe Bay at a position 53°55'N, 3°29'W indicate that the maximum waves recorded in this part of the Irish Sea were 8·55 m high with a period of 7 seconds on 1 November 1957, but waves of this height are very rare. For observations covering a whole year even 3 m waves only occur about 2 per cent of the time, while waves occurring about 20 per cent of the time are only 30 cm high. The higher waves are all concentrated in the winter half of the year (Darbyshire, 1958). By far the most common significant period was about 5 seconds, occurring about 30 per cent of the time. This period was exceeded mainly in December and January, when it reached about 7 seconds. There was a close correlation between wind direction and wave height, the higher waves being generated by winds blowing over the largest fetch. For a given wind speed the waves were shorter than in the open ocean, and they required a shorter fetch to reach their maximum size.

These results may be compared with those given for observations in the open Atlantic and on the coast of Cornwall. In February 1953 the average height in the ocean was over 7 m, while for coastal observations in the same month of 1946, the mean height was just over 3·4 m. The difference was rather greater in winter, but was noticeable throughout the year. The significant periods in the two areas also differ. The average period of the waves was much longer at the coastal station than in the open ocean, although again the observations were made in different years. Periods over 15 seconds were much more common on the coast, where they may occur for as much as 45 per cent of the time in some months, while they never exceeded 5 per cent of the time in mid-ocean even in winter.

Darbyshire suggests that the lower waves near the coast are the result of less efficient wave generation in shallow water. The longer periods recorded at the coastal station may be the result of the attenuation of some of the shorter, higher waves, generated in the open ocean before they reach the coast, where the long-period swells predominate. Wave

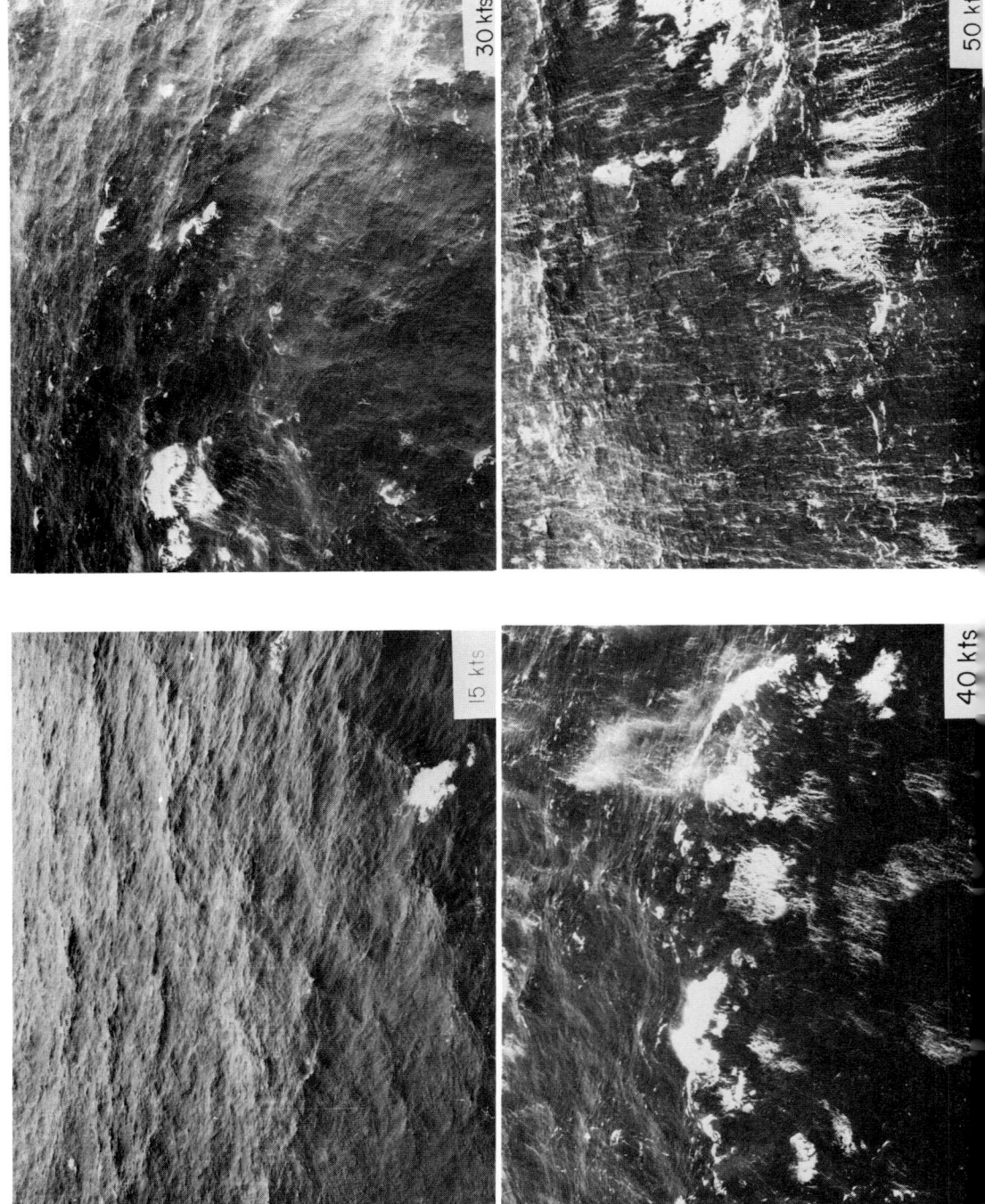

15 kts

30 kts

40 kts

50 kts

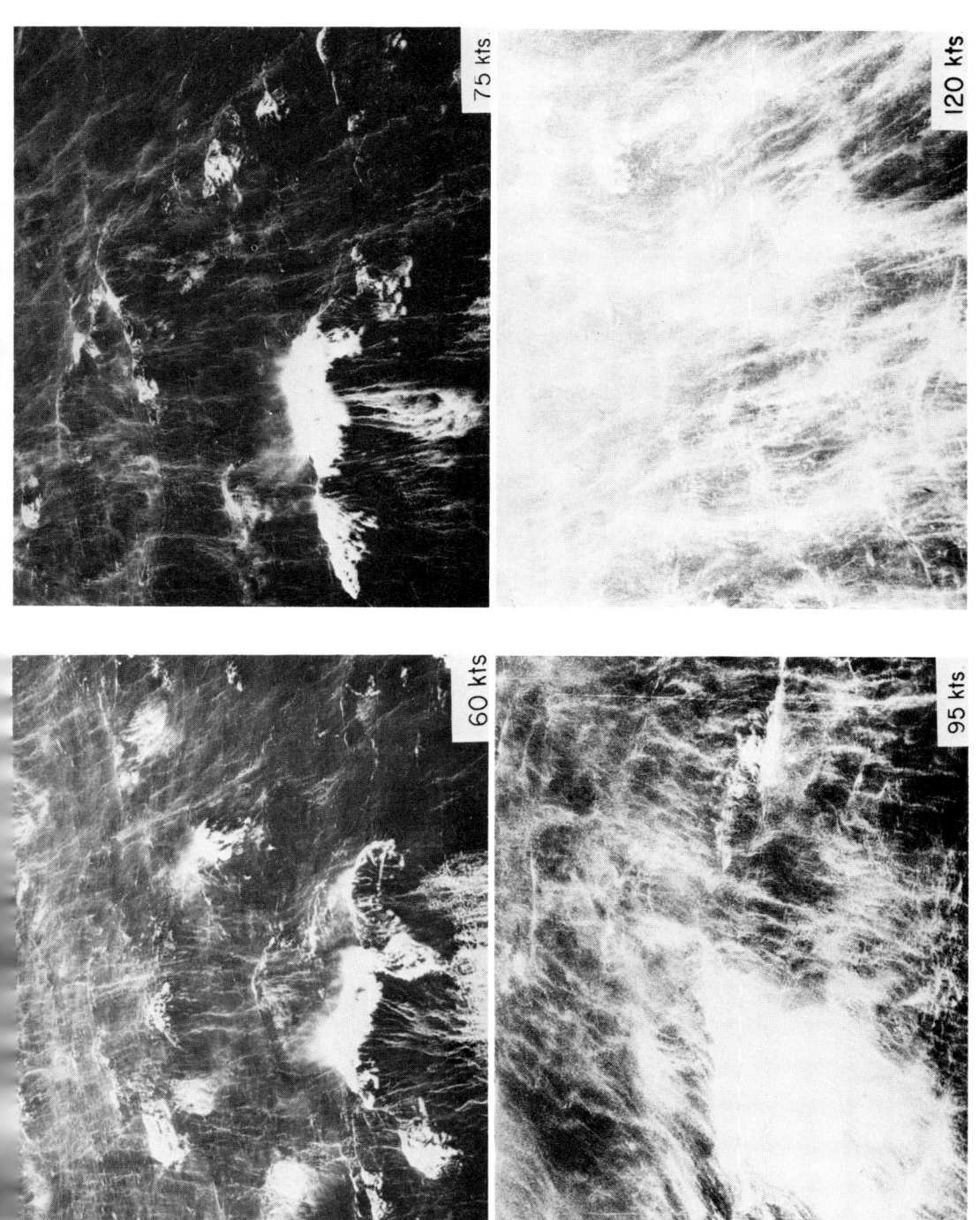

Plate 4 Waves: sea state. A set of photographs to show waves generated by winds of different velocity (*G. Neumann and W. J. Pierson*).

sizes seem to follow a fairly similar pattern in different areas, with the highest waves being extremely rare, while a particular height is dominant in each region. Thus for the Atlantic weather-ship data this height was about 3 m for all the observations, although waves of 15·3 m were sometimes recorded. At Perranporth, on the other hand, although waves over 10 m occasionally occurred, the most common height was about 1·2 m. A somewhat similar value held for observations at Casablanca in North Africa.

A useful summary of the available wave records at coastal sites in the United States has been given by Darling and Dumm (1967). Surface and submerged types of instruments were used and records were analysed to give the significant wave height and period for a 7-minute run each 4 hours. A spectrum was analysed for a 20-minute run each day. Fourteen stations were available on the Atlantic coast, four on the Gulf coast, and five on the Pacific coast. In addition there were a number of visual (27) observing stations, at coastguard stations. These provided details of surf characteristics given in the form of number of observations and percentages of heights for each breaker period band. Direction of approach, breaker heights, and periods were recorded.

The North Sea, Camp Pendleton in California, and Sekondi, Ghana, are compared below as examples of waves in different environments. The mean significant wave measured at the Smith Knolls light-ship in the North Sea in winter was about 1·2 m, and the maximum in a 3 hour period was nearly 2·5 m, while the highest recorded wave was 7·3 m with a period of 8·4 seconds. Waves of 5·5 m height had an exceedance time of 5 per cent. The most common conditions were those with a significant height of 61–90 cm, with a zero-crossing period of between 5·5 and 6 seconds. In winter the significant wave height exceeded 1·8 m for 20 per cent of the time. In summer the value was only about 1 m, while the modal period increased to 6–6·5 seconds. At Camp Pendleton in California the most common situation was a wave about 45 cm high with a period between 13 and 14 seconds. The highest wave recorded was 3·5 m high, and the calculated maximum was 4·4 m high. These waves are, therefore, very much flatter than those characteristic of the North Sea. In Ghana at Sekondi, waves of significant height exceeding 1 m occurred for 70 per cent of the time, but they only exceeded 1·2 m for 20 per cent of the time. The most common wave was just over 1 m high and the period was 10–11 seconds. These waves may be compared with those recorded in mid-ocean at the weather-ships in the Atlantic. Records at weather-ship *Juliet* at 52·5°N 20°W (Draper and Whitaker, 1965) show that the commonest wave was between 1·2 and 2·4 m high with a period between 8 and 9 seconds. In winter the significant height exceeded 4·6 m for 53 per cent of the time, at which season the modal period was 10–10·5 seconds. In summer the modal period fell to 7·5–8 seconds. The light-ship in Morecambe Bay in the Irish Sea had a maximum wave height of 8·6 m and 7·6 seconds period. The most common wave was 30–60 cm high with a period of 4–5 seconds. Winter modal period was 6–7 seconds, and the summer value was 4–5 seconds. At the Varne light-ship in the Straits of Dover the maximum wave recorded was 7·6 m high with a period of 6·4 seconds. The most common wave was 61–90 cm high and 4·5–5 seconds period, with a winter mode of 4·5–5·5 seconds period and a summer mode of 5–5·5 second period.

3   Wave spectra and wave generation: sea

Waves in the open ocean and on the coast are much more complex than the ideal waves discussed earlier. The great complexity of waves in the open ocean, particularly in the generating area where they are known as 'sea', has led to the development of the method of spectral analysis, by which complex wave profiles are split into their constituent parts. The method is similar to harmonic analysis used in tidal work. The relative energy in each frequency band present in the wave pattern at any time is given.

Figure 5.2 shows the wave spectrum derived by Fourier analysis from a wave record. The energy of the wave is usually concentrated around a fairly narrow frequency band. The frequency, $1/T$, is the inverse of the period. The significant wave height is defined as the mean height of the highest one-third of the waves, and the significant period is the period of these waves. The significant height is $1 \cdot 6$ times the mean wave height, while the maximum height is about $2 \cdot 4$ times the mean height.

Although much of the wave energy is concentrated in a fairly narrow frequency band, analyses show that the total wave spectrum usually covers a wide range of periods. Neumann (1953), who has used the wave spectrum technique to relate waves to the winds that form them, has shown a typical spectrum in which the frequency varies from less than $0 \cdot 06$ to over $2 \cdot 0$. The periods covered by this frequency range are from 20 seconds to 5 seconds. Within the range, however, there is a marked peak, which Neumann suggests becomes more marked as the wind force rises and the energy of the waves increases. Thus in a sea generated by a 20-knot wind he shows a low peak with a frequency of $0 \cdot 124$ or a period of $8 \cdot 1$ seconds. When the wind force is 40 knots, the maximum energy is concentrated strongly around the frequency of $0 \cdot 06$ or 16-second period. In this band, therefore, the highest waves would be expected to occur.

Darbyshire (1959b) has analysed observations made in the north Atlantic by a weathership. He uses the spectrum technique to relate the various dimensions of the waves, as actually observed and analysed, to the winds generating them. Provided the fetch is large, he found that the wave spectrum has only one form, even though the generating wind force varies. The spectrum he gives is generally of the same form as that suggested by Neumann. There is a sharp peak in which the wave energy is concentrated and the waves highest. This frequency he calls f_0 and from this value the energy falls off more rapidly to the high period or low-frequency side and less rapidly towards the lower periods.

A useful result of his analysis is the relationship which has emerged between the different variables on which the wave size depends, and the relationship between the form of the spectrum and the significant wave dimensions. The significant period is related to the period which has the most energy in the spectrum by $T_f = 1 \cdot 14 T_{1/3}$; thus the period with the highest energy is only slightly larger than the significant period. The wave height which is dependent on the wave energy in the spectrum, can be related to the square of the wind speed, assuming a large fetch.

Several theories have been put forward to account for the initial generation of waves and a variety of techniques have been developed to forecast or hindcast wave dimensions from meteorological data. Many of the theories put forward to explain the transference of energy from the air to the wave have been shown to be incorrect in some respect. Phillips'

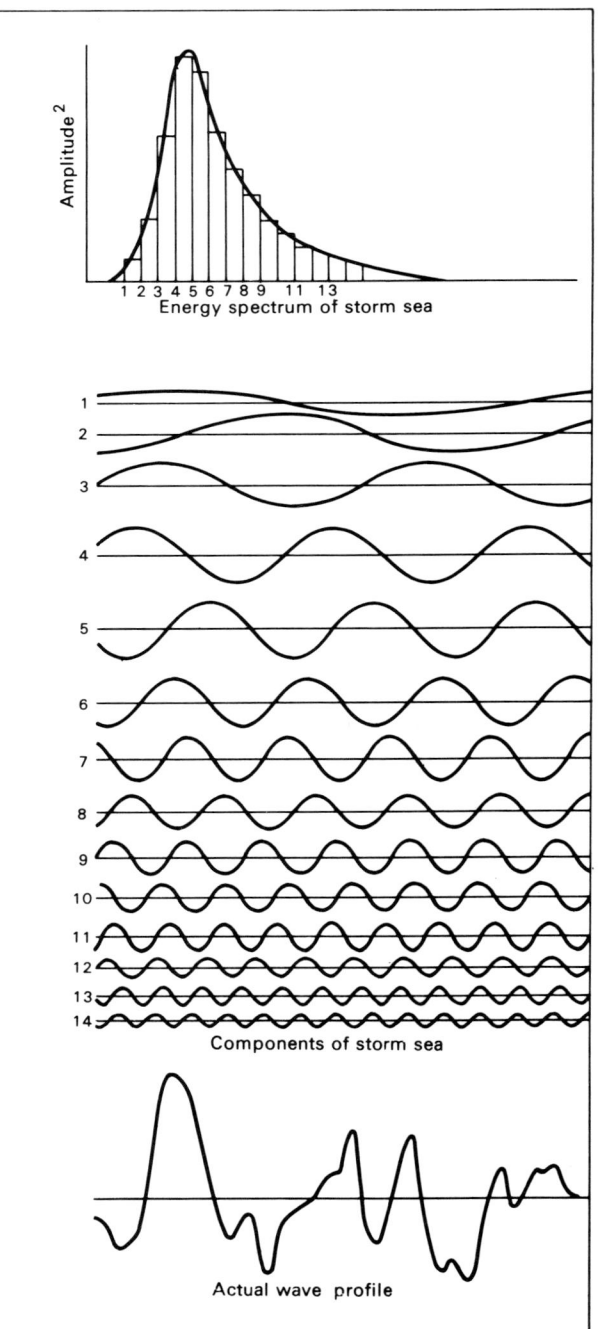

Figure 5.2 Diagram to show the addition of fourteen different components of a wave spectrum to form the actual wave profile. The proportion of energy in the different parts of the spectrum is indicated in the upper part of the diagram. (*After Deacon, 1960.*)

(1957) theory appears to be rather more successful, although there are still difficulties in the establishing a complete explanation.

One of the early theories was put forward by Jeffreys (1925), who suggested that the air flow was laminar over the windward slope of the wave and turbulent on the lee slope, which is partially sheltered by the wave crest, giving lower velocities there. An important factor in his theory was the sheltering coefficient, which is related to the proportion of the windward slope of the wave offering resistance to the wind. The value he deduced for this coefficient seems to have been too great. His theory also cannot account for the original generation of waves from a flat surface, although it does seem to apply better in the early stages of wave generation than when the waves become large.

The work of Sverdrup and Munk (1947) shows that the tangential stress of the wind on the water was an important aspect of wave generation. This varies with the square of the wind velocity, above a certain threshold. When the water particles move in the same direction as the wind, they acquire energy from the wind, but energy is lost when the particles move in the opposite direction. Because of mass transport there is a net gain of energy. The wave velocity is much greater than the water particles velocity, which would allow the wave form to move faster than the generating wind velocity. When the wave velocity exceeds the generating wind velocity, energy is transferred in two ways—the waves gain energy by wind stress, when the particles of water are moving in the same direction as the wind, but they lose it by pressure of the wave form on the wind. In the earlier stages the increase of energy builds up the height of the wave more than the length, but when the wave form is moving faster than the wind, most of the energy increases the wave length and velocity.

The more recent theories of wave generation by Eckart (1953) and Phillips (1957) consider that the flow of the air is fully turbulent and that the normal pressure, due to gusts of wind, is randomly distributed. It appears that the pattern used by Eckart was not sufficiently random, and Phillips' results appear to be more accurate. Phillips assumes that a gusty wind starts to blow on a surface previously at rest. This causes pressure fluctuations on the surface, which are both normal and tangential. Eddies are caused by the wind and these tend to develop slowly and to move at a speed related to the wind velocity. These local pressure centres are carried in the air-stream and when their rate of movement is similar to that of a developing disturbance on the water surface, a state of resonance will be set up. This initial disturbance may develop into a growing wave, because if it is moving at a speed similar to the wind pressure pocket, they move together and the wave can grow. This, however, assumes that the pressure pattern is not changing and is, therefore, an over-simplification. When the wave dimensions increase other factors come in and the waves grow in a different way. According to this analysis there does not seem to be a minimum wind velocity necessary to develop waves. Even the lowest winds which cause some disturbance on the surface may form waves, although these may not grow. The smallest wind velocity which is likely to generate resonance waves is about 23 cm/sec.

He suggests that with light winds the wave spectrum is narrower, the waves appearing more uniform than with strong winds, when a wider wave-period band is present. It has been shown by Phillips that the wave height grows, after the initial stages, at a rate that is uniform with time, at least until the wave height exceeds a certain point. His theoretical results agree fairly closely with some observations over a restricted range of wind duration.

The main problem of wave generation is related to the distribution of pressure on the moving, random water surface under the influence of a turbulent wind. The pressure fluctuations are of two kinds: one is produced by turbulent eddies in the wind, and the other is induced by air flow over the irregular surface. The total pressure pattern is the sum of these two. Turbulent pressure provides energy input over a wide spectral range, while directly induced pressure provides a selective 'feedback' which allows certain components to grow rapidly. Phillips (1966) in a more recent study again showed the importance of resonance in the relationship between the free surface waves and the turbulent pressures that cause them.

Miles (1967) dealt with an alternative process of wave generation. His theory considered the flow of air over a fluid boundary with a small sinusoidal displacement. He then considered the pressure on the water surface resulting from the perturbation of the original flow. The initial disturbance of the water surface may be due to turbulent pressure as described by Phillips at the lower frequencies, although pressure fluctuations are weaker than first estimated. In the main stage of wave growth the new idea put forward by Miles suggests that shearing flow is effective, leading to the observed reduction in angular spread of the wave spectrum. Turbulence does appear to have a profound effect on the induced pressure distribution, particularly when the wave speed is comparable with the wind speed. As waves grow in amplitude they become distorted and their crests become sharper than their troughs. If the spectrum has a sharp maximum, secondary peaks may sometimes be found at multiples of the frequency of the primary peak. There is also energy transfer among the different wave components. The process involves resonances among particular groups of wave components. The result causes a gradual spread of energy beyond the high spectral region. There is a trend towards a more uniform directional distribution of spectral density as a result of resonant interactions in gravity waves, but this effect is sometimes obscured by others.

The growth of waves under the influence of the wind cannot continue indefinitely according to Phillips (1966). The interaction among the waves is not capable of transferring energy from a given wave-number band as rapidly as it is supplied by the wind. The size of the waves is limited by the requirement of stability of the water surface. Instability results in the breaking of the wave locally; energy is lost and stability is restored. Breaking will often occur when two wave crests run together. When waves are generated by a uniform and steady wind the typical spectrum has a steep forward face, where the frequencies are low, rising to a sharp maximum, and falling at higher frequencies in the equilibrium part of the spectrum. As time passes the spectral peak occurs at lower frequencies. It seems likely that Miles' mechanism operates as a primary means of wave growth in conditions of relatively short wind duration and short fetch.

Bretschneider (1965) has summarized the state of knowledge on wave generation. The random velocity fluctuations of the air flow over the sea surface provide the mechanism for Phillips' theory of wave generation. The stresses are both normal and tangential. The unstable eddies are carried forward in the wind stream at the mean wind velocity (convection velocity), developing and interacting as they move. These pressure fluctuations are responsible for the early wave generation. Phillips does not consider the tangential stresses. As the ratio of wave velocity to wind velocity approaches unity other processes come into operation, including sheltering effects and variations of shear stresses. Phillips has shown that for high-frequency waves the energy varies as f^{-5}.

Miles' theory takes over as the waves grow. His model for wave generation is based on the instability of the interface between the air flow and the water. Whereas Phillips' theory predicts the growth of waves as proportional to time, Miles' theory shows an exponential growth rate. Phillips' model is uncoupled, in that the air flow is assumed independent of the response to it, while Miles' theory is coupled, in that the coupling can lead to instability and rapid wave growth. It seems very likely that both mechanisms occur in the ocean, with the different models being dominant at different frequencies. The advantages of both Phillips' and Miles' theories are that no unknown constants enter the equations and both yield theoretical wave spectra which can be compared with measured ones. Phillips' theory gives both the frequency and direction of travel. There is still, however, much to be learnt concerning the dissipation of energy in growing waves before accurate predictions can be made. This loss of energy is seen in the 'white horses' that accompany storms at sea where the waves are being actively generated.

The larger waves are so complex that all the theories break down before the largest waves are reached. These waves have great significance for behaviour of ships at sea and the attack of waves on coastal defences. It is, therefore, necessary to develop empirical relationships between the main factors on which wave growth and wave characteristics depend. The most important variables are wind strength, wind duration, and the fetch, which is the stretch of open water over which the generating wind is blowing. In an enclosed sea it will be limited by the nearest shore in the direction from which the wind is blowing, while in the open ocean it will depend on the meteorological situation, which determines the distance over which winds of constant direction can blow.

The effect of fetch can readily be seen where the wind is blowing offshore. Near the shore the waves are very small, but they gradually increase away from the coast as the fetch lengthens. There are in fact two distinct zones. In the one near the shore, the waves increase in size away from the coast because the fetch is increasing in this direction; but each wave maintains its size, which is limited by the fetch. In the second zone, further offshore, the fetch does not impose a limit to size of the waves beyond a certain point, which varies with the strength of the wind. The waves in this zone will all be of equal size at any one time, but will continue to grow until they reach the maximum size for that wind speed and duration, assuming that the wind continues to blow in the same direction.

Different authorities differ greatly in their interpretation of the fetch needed to develop waves of maximum size. Darbyshire, for example, considers that a fetch of about 160 km is sufficient to develop waves to their maximum size, even when the winds are strong. Neumann and Pierson (1957) consider that the fetch necessary to produce the maximum size waves depends on the strength of the wind, which itself largely determines the wave size. They suggest that much longer fetches are necessary than those given by Darbyshire (1957a). Thus a 63 km/hour wind would require 740 km fetch to produce the maximum waves possible, while a 93 km/hour wind needs 2620 km and 69 hours wind duration. The latter situation is likely to be very rare, so that it is difficult to test the upper part of the curve. The data of Bretschneider also indicate that waves require a very long fetch to reach their maximum size.

The depth of water over which the wind is blowing also exerts some influence on the wave growth, and Darbyshire has found that different equations apply to shallow and deep water data. Further complications are introduced by the instability of the air immediately in contact with the sea, or the air–sea temperature difference. A decrease of $11\,^{\circ}$c of the air temperature relative to the sea temperature caused the wave height to double for the same wind speed. It has been stated that a 25 per cent increase of wave height results from a $5\cdot5\,^{\circ}$c temperature difference between air and water. Other observations in the Irish Sea have suggested that the effect of instability is marked only when the waves are not fully developed (Darbyshire, 1958).

All these variables make it very difficult to assess the waves generated by any given storm, particularly as there were probably some waves present before the disturbance developed. The most favourable regions for the generation of really large waves will be those parts of the ocean where the winds blow strongly and constantly in direction over wide stretches of open sea. These conditions are most nearly fulfilled in the westerly wind belt of the southern ocean. Winds in the stormy latitudes of the north Atlantic are much more variable in position and direction. It is the very long waves, with their great energy, that can be traced furthest from their source across the ocean.

4　Wave attenuation: swell

When a wave moves out of the generating area into calmer waters it is modified to become 'swell'. Waves of high energy can travel for very long distances before losing their identity. As soon as swell travels out into calmer water, the shorter waves tend to die out, because they lose their small amount of energy fairly quickly. The swell thus becomes more uniform and longer crested. A spectrum of swell waves is usually deficient in the short wave lengths, although local winds may superimpose a series of small waves on the swell. Many observations of swell on the coasts of the continents in the southern hemisphere give a mean length of the swell generated in the southern ocean of 300 m, with a period of 14 seconds.

Some authorities suggest that the swell lengthens as it moves away from the generating area, but most workers agree that there is no marked change in the wave length of the individual wave trains as they move out from the generating area. The waves move as individual trains, each travelling with a velocity appropriate to the wave length. They therefore spread out, with the longest waves outstripping the shorter. As the waves move out they become longer crested and more uniform and sinusoidal in form, with smooth rounded crests. Their height is reduced as they move away from the generating area, a process accelerated by encounter with a head wind. Reduction in wave height is due to three possible causes: 1) dispersion, 2) loss of energy due to air resistance, and turbulence, and 3) divergence.

Darbyshire (1957b) has studied the attenuation of waves in the north Atlantic over distances of 640–2560 km from the storm area, from records of waves taken instrumentally on the weather-ship *Weather Explorer*. The analysis of the records showed that the reduction of wave height was independent of wave period and could be expressed in the form $H_T/H^0_T = (300/R)^{1/2}$, where H_T is a function of the wave height at the recording point, It is the square-root of the sum of the squares of the peaks for each one-second interval on the wave spectrum; H^0_T is in the same form, but refers to the wave height at the edge of the generating area. This was arrived at by the use of the equation for wave generation and may, therefore, not be quite accurate. R is the distance from the recording point to the centre of the storm. Despite the uncertainty of the wave height on the edge of the generating area, the results are very consistent. They seemed to be independent of the speed of the following wind in the decay area. The equation shows that wave height is reduced fairly rapidly with distance from the storm, being halved in about 1600 km, but decreasing much more slowly thereafter (figure 5.3).

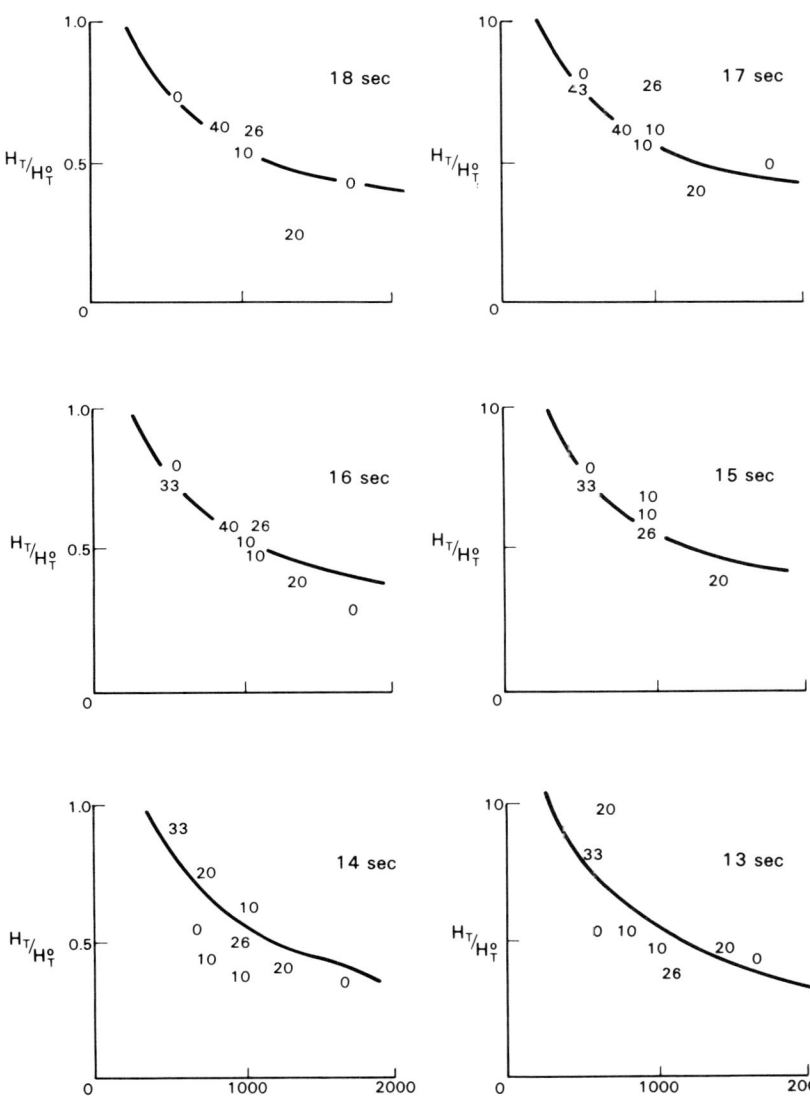

Figure 5.3 Reduction of wave height in relation to distance from a storm centre and strength of the following winds. (*After Derbyshire, 1957.*)

Distance from storm nautical miles N

Causes for this reduction were analysed. The first, dispersal, resulted from the spreading out of the wave trains as each moved out from the generating area with its own appropriate velocity. The longest waves get furthest ahead, thus spreading out the zone over which the energy is distributed and leading to a reduction in wave height. Within any one period band, such as was used for this analysis, the effect would not be marked and could not account for the observed wave height reduction. It is probably this factor, however, that accounts for the apparent increase in wave length, which some authorities associate with swell waves. The second factor is the loss of energy, but too little is known as yet concerning the method of transfer of energy from water to air, and vice versa, for a reliable estimate of this effect to be made.

The third effect can, however, account in considerable measure for the observed result. Divergence is the spreading out of a wave front laterally as it moves out from the storm area. This must result in a loss of energy per unit width of crest and, therefore, of wave height. Darbyshire considers that this factor is important in causing the reduction of wave height in the decay area. He has used the relationship successfully in comparing predicted and observed swell heights. In fact the attenuation of waves seems to be better understood than their growth, as forecasts of swell are usually more reliable than those of seas in the generating area.

One of the best analyses of swell waves used a record of swell waves measured by a wave gauge at Pendeen in Cornwall (Barber and Ursell, 1948). This record was analysed to give the wave spectrum, from which the progress of swell of a particular frequency band could be traced on successive records. Their analysis was based on the assumption that wave trains travel across the ocean at speeds equal to half the wave velocity. By projecting the waves backwards in time and distance from their point of arrival, the position and time from which they set out could be found. Where these propagation lines meet is the most likely centre of the storm. Its distance away can then be read off from the scale, and the time determined. Examples of wave spectra and propagation diagrams are shown in figure 5.4. Meteorological charts can then be studied and the storm, from which the swell travelled, can be identified from these data.

The swell in the example analysed first reached the coast of Cornwall at 1900 hours on 30 June 1945, when it had a period of 18 seconds (figure 5.5). Later observations showed a steady decrease in period as the shorter waves arrived, but the height increased. The origin of these swell waves was traced back to an intense tropical hurricane, which reached its maximum intensity on 26–27 June, while situated off the coast of North America between Cape Hatteras and Nantucket, Mass. This was about 4320–4800 km away from Cornwall. Other swells recorded on this coast have been traced back to storms in the southern hemisphere near the Cape of Good Hope. Fluctuations in the value of the wave period have been explained by interference of tidal currents, as their periodicity agreed with that of the reversals in the tidal streams.

Wave records in California have shown that swell waves can travel even longer distances than those given without losing their identity. Wiegel and Kimberley (1950) have concluded that swell reaching California during the summer months must have originated in the south Pacific Ocean between 40° and 65°s and 120° to 160°w. This is about 11,200 km from California. The swell, generated in the winter in the stormy latitudes of the southern ocean, arrived with periods between 12 and 18 seconds, sometimes reaching

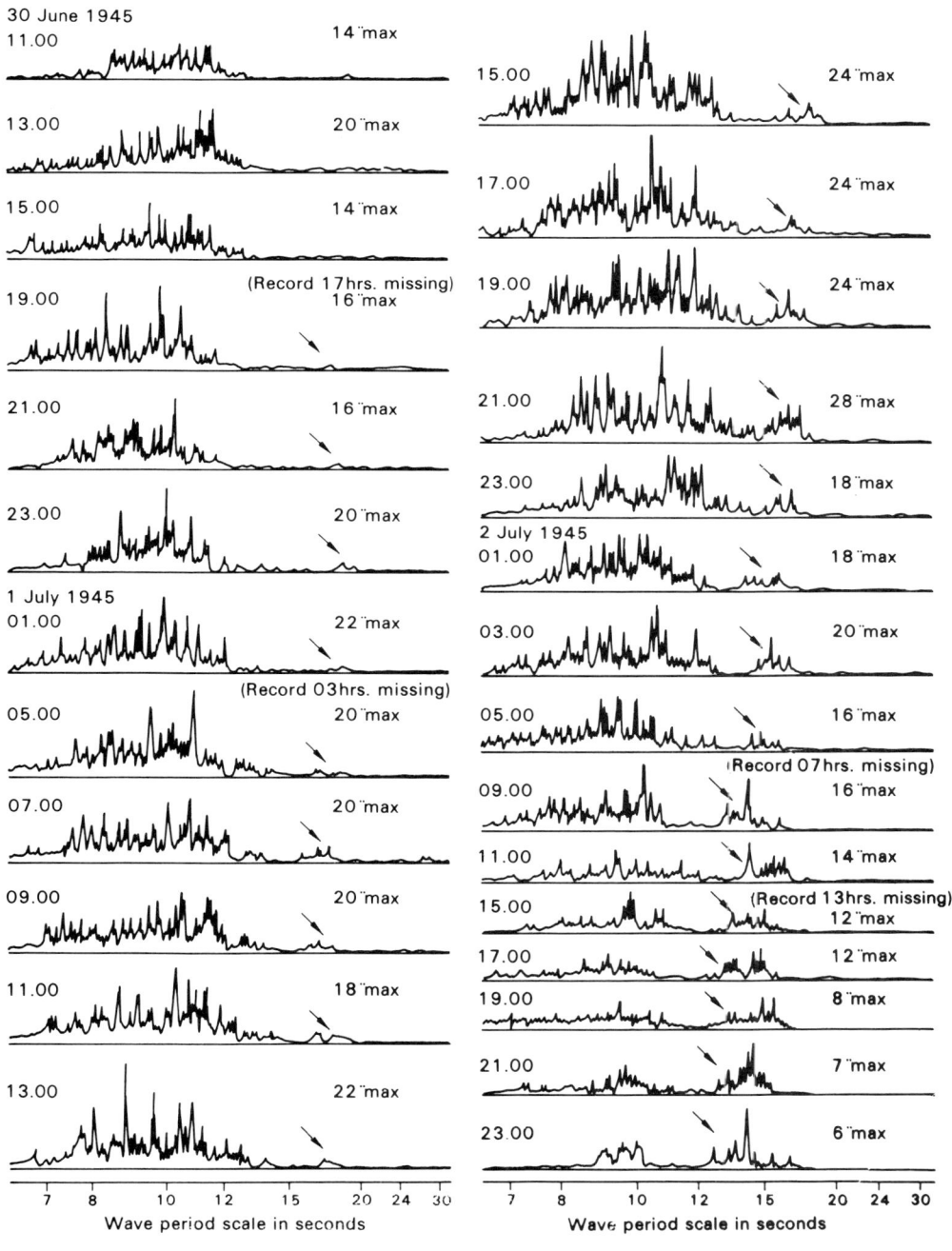

Figure 5.4 Wave spectra at Pendeen, Cornwall, 30 June to 2 July 1954. (*After Barber and Ursell, 1948.*)

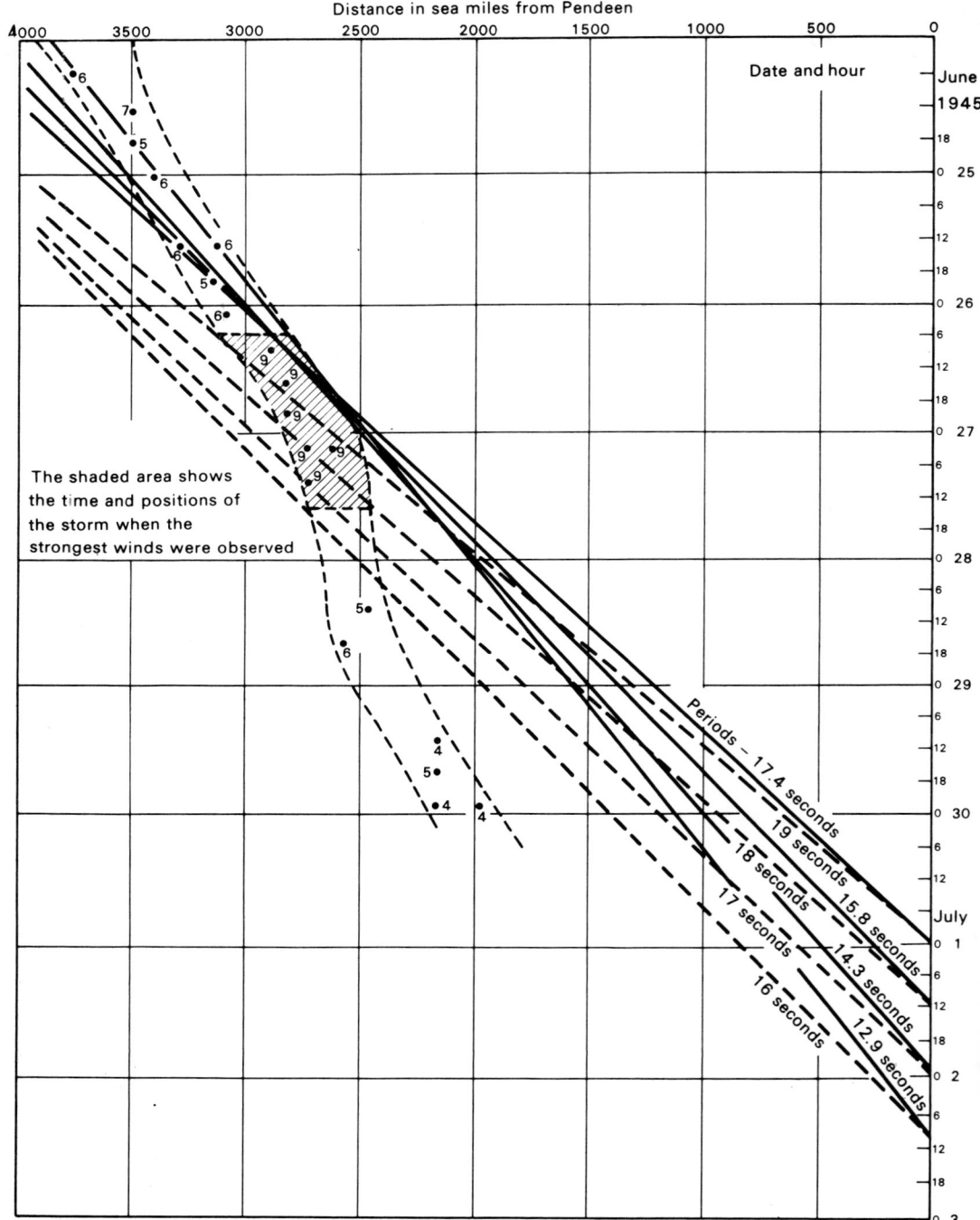

Figure 5.5 Propagation diagram for the waves reaching Pendeen between 24 June and 2 July 1954. (*After Barber and Ursell, 1948.*)

values as high as 22 seconds. Its usual height was about 0·6 and 1·8 m, but at times it reached a maximum of 3–3·7 m, suggesting that the original waves must have been very large indeed. They could only have been generated in the region of strong and persistent westerly winds and unlimited fetch.

Swell waves must always be taken into account in analysing waves arriving at any point, and their source of origin may be many thousands of kilometres away across the ocean. This fact in large measure differentiates the wave records of areas in enclosed seas from those to which the long swells have access on the shores of the open ocean. It is also the long swell waves that are most modified by refraction as they enter shallow water because of their great length. They feel the bottom for this reason long before the shorter waves generated locally by less strong winds.

Since the valuable work by Barber and Ursell, more records of the distant travel of swell waves have been collected. Analysis is now mainly by digital computer rather than by analogue machines. Cartwright (1967) shows that water waves are dispersive, their speed of travel depending on their period; a 24-second wave propagates at 19 m/sec or 14·5° latitude/day. Munk *et al.* (1963) studied the propagation of swell across the Pacific Ocean towards the coast of California. The swell was recorded by Vibroton, which supplies data suitable for digital computer analysis by means of a vibrating wire in a vacuum capsule. Three Vibrotons were placed in a triangle in 100 m of water so that the direction of wave approach could also be determined. The spectra are illustrated in the form shown in figure 5.6, which shows equal energy density contours plotted against time for different wave frequencies. The figure shows the gradual increase of the frequency of waves with time for any one set, the direction of which is indicated by the straight lines. Each main set of waves is represented by a ridge on the diagram. The time and origin of the linear events can be given accurately in terms of bearing and angular distance. Swells from the antipodes can reach the west coast of the United States through a window south of New Zealand, the travel distance being 165–175° from their source between Kerguelen and Madagascar. The higher wave frequencies or shorter periods above 50 c/ks are heavily attenuated. In the generating area in deep water there is much energy in the 100 c/ks (10-second period) band. The loss of energy in this band could be due to opposing winds in the trade-wind belt, causing breaking of smaller waves on the swell and a loss of energy. Scattering of waves due to non-linear interaction within the spectrum itself could also cause attenuation. Munk set up six wave recording stations on a great circle between New Zealand and Alaska. The records were obtained by FLIP (Floating Instrument Platform) in the gap between Honolulu and Yakutat. Spectral plots were obtained for twelve major storms. There was no evidence of loss of energy in the trade-wind belt; indeed at times the energy increased. The attenuation of waves with frequencies above 50 c/ks in the first 20° from the generating area was much greater than that over the whole of the rest of the distance, probably as a result of scattering.

5 Wave forecasting

Accurate wave forecasting requires detailed knowledge both of the theory of wave generation and a detailed picture of the weather situation over the ocean. Owing to the lack of

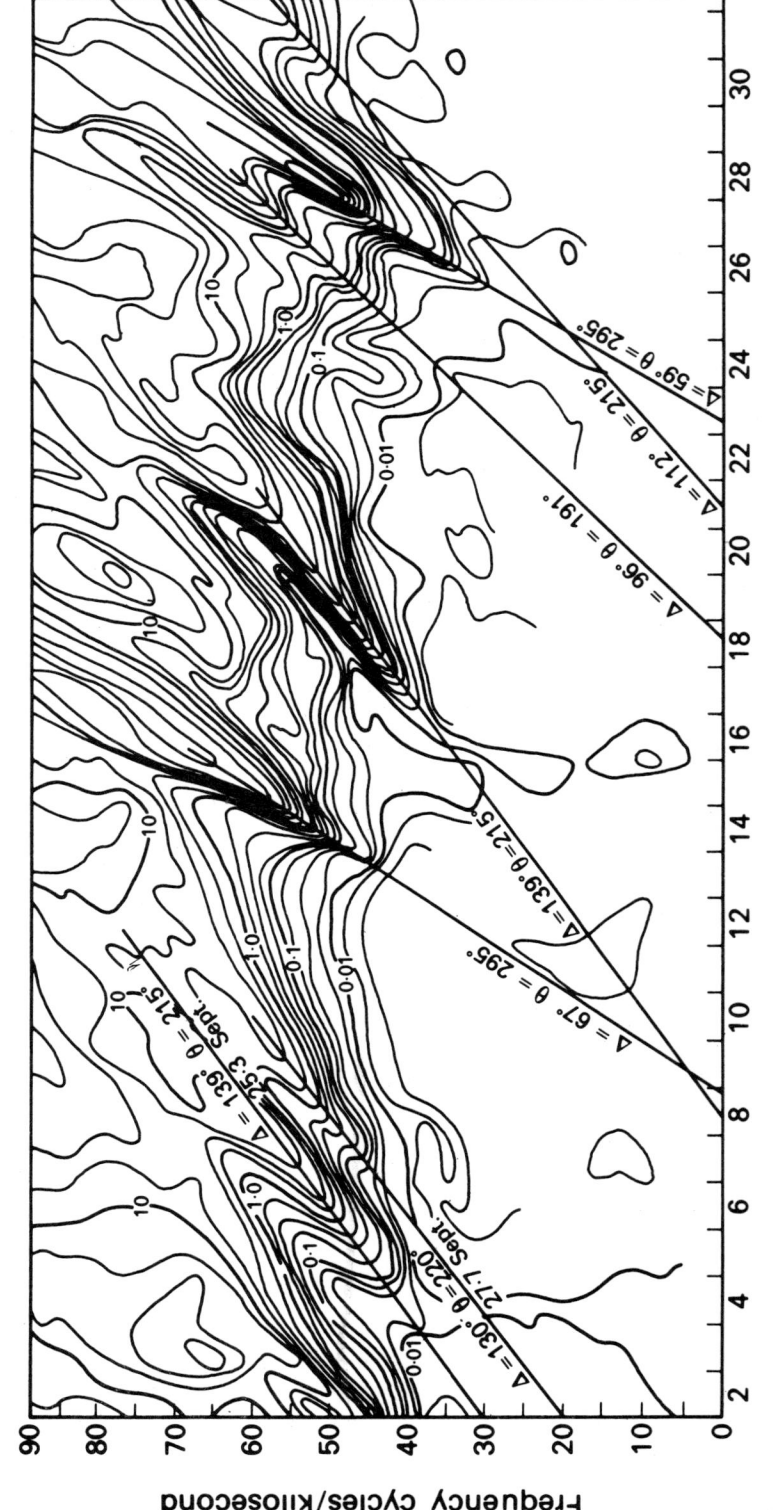

Figure 5.6 Contours of spectral energy density plotted against time for waves at San Clemente Island during October 1959. (*After Cartwright, from Munk et al., 1963.*)

such information, the methods that have been developed, therefore, are mainly empirical in character. They seek to relate observed wave dimensions to the winds generating them. The results are often expressed in the form of curves, from which the wave dimensions can be read off from observed data of wind strength, duration, and fetch. An example is shown in figure 5.7. Some of these curves are based on visual wave observations and cannot hope to be more accurate than the information on which they are based. Others, for example those of Darbyshire, are based on instrumentally observed wave data. Forecasting procedures have been devised by Sverdrup and Munk, and modified by Bretschneider, called SMB for short, secondly by Pierson, Neumann, and James (PNJ), and thirdly by Darbyshire (D). The results obtained by the different methods vary significantly, as shown by the following examples.

With a fetch of 93 km and a wind speed of 55 km/hour, the significant height according to SMB would be 3·5 m; PNJ give 2·0 m and D 2·5 m, for a 74 km/hour wind the figures are respectively 6·0 m, 7·0 m, and 4·3 m. The SMB method does not make use of the spectrum technique, which is used by the other two methods. The figures given above are for a very short fetch. When the fetch is longer the results are rather different, because Darbyshire considers that wave height is independent of fetch when this exceeds about 160 km, while the other methods allow the waves to continue to grow as the fetch increases. For higher values of the fetch, therefore, the wave heights given by Darbyshire are considerably lower than those given by SMB, and the PNJ values are very high indeed.

Darbyshire has given data which suggest that the PNJ equation for wave height results at times in a 100 per cent over-estimate, when compared with observed significant wave heights, while his own values are much nearer the observed ones. The fetch in this particular instance was 1600 km. With a fetch of only about 240 km the PNJ results seemed too small. In many examples the results given by the SMB method are even greater than the PNJ ones. One difficulty in comparing the different curves is that Darbyshire uses the gradient wind, in the earlier equations at least, while the other methods use the surface wind, measured at an elevation of 10 m.

Charnock (1958) has suggested that it would be better to use the friction velocity, but this is not so readily obtained. Because of the difficulty of application of the gradient wind, Darbyshire (1959b) used the wind as measured at the weather-ship from which the wave observations were made. On the assumption that the relationship between the gradient wind and the surface wind is 3/2, his results are very similar to the earlier ones, giving respectively $H = 0·0038U^2$ and $H = 0·0036U^2$ for the earlier and later data, where H is the square-root of the sum of the squares of all the peaks in the spectrum in feet, and U is the gradient wind in knots. The newer relationship between the wind at the surface and the significant wave height is $H_{1/3} = 0·0133W^2$. According to his analysis, therefore, the wave height varies as the square of the wind speed. The wave period containing the peak of the wave spectrum energy, T_f, was related to the square-root of the surface wind speed by $T_f = 1·94W^{1/2}$ this value held until the wind force was very strong, when the result became more accurate if the equation was given as $T_f = 1·94W^{1/2} + 2·5 + 10^{-7}W^4$. These equations assume that the waves are fully developed and that no limits are imposed by fetch or wind duration.

The SMB technique provides curves which enable a variety of different situations to be analysed. Not only are deep water conditions allowed for, but different conditions that

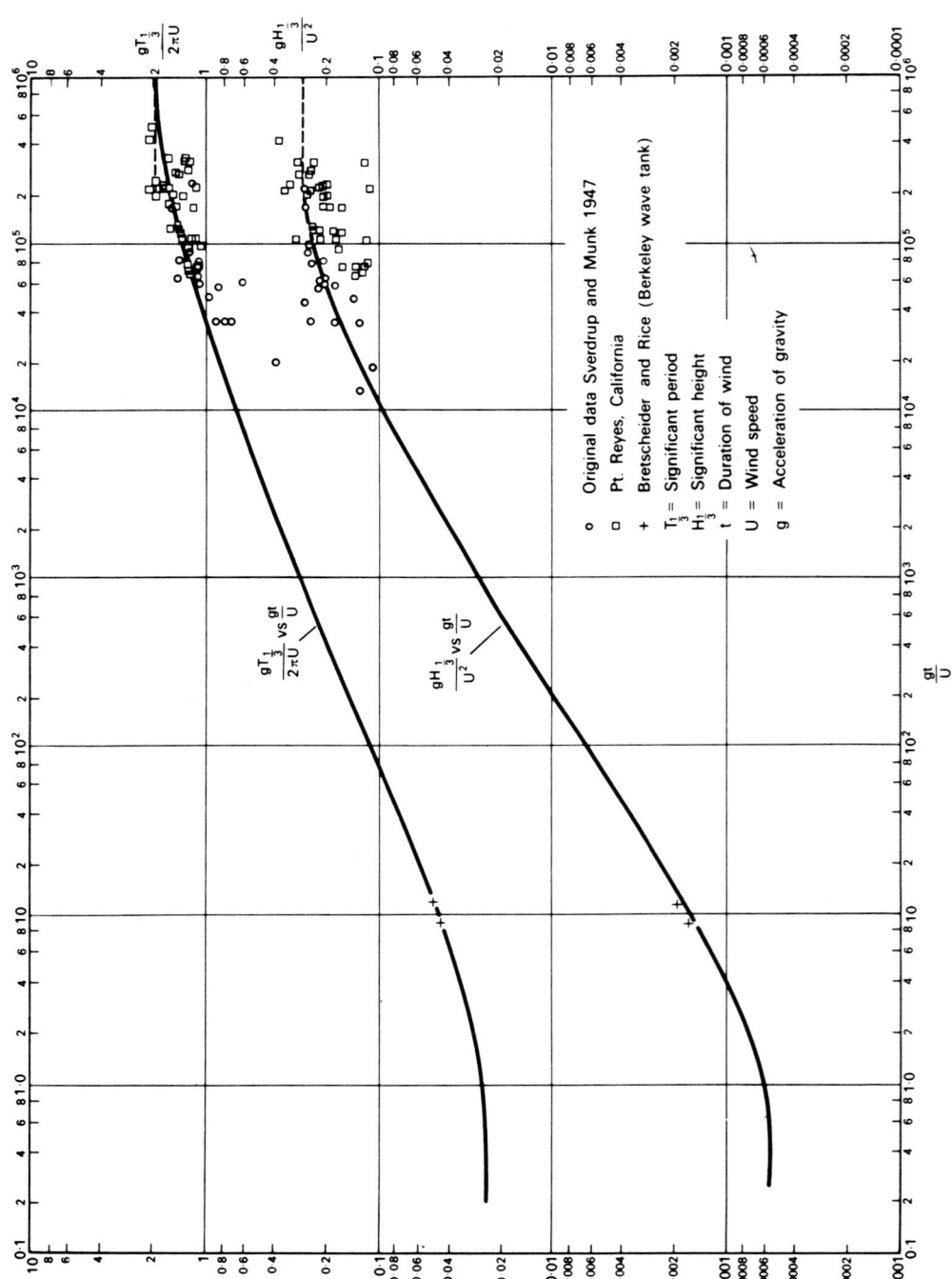

A: Duration graph.

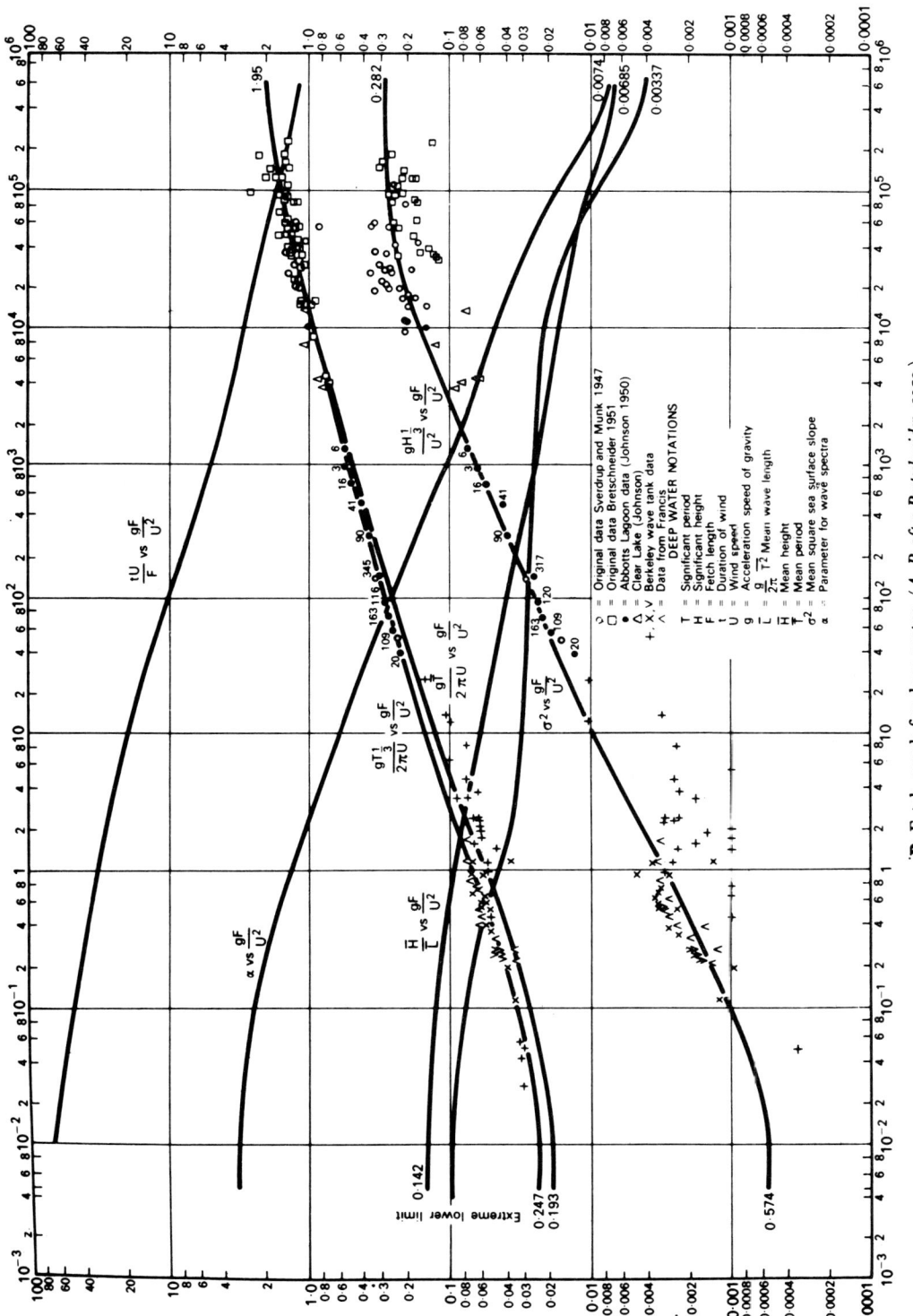

B: Fetch graph for deep water. (*A, B after Bretschneider, 1952.*)

Figure 5.7 Deep water wave forecasting curves, given as a function of wind speed, fetch length, and wind duration. (*After Bretschneider, CERC, 1959.*)

apply to wave generation in shallow water are also taken into account. One graph, from which deep water wave characteristics of significant height and period can be obtained, is plotted with the wind speed on one axis and the fetch length on the other (figure 5.7). The minimum duration necessary to obtain full size waves is also given. The curves for wave height show a steady increase as the fetch increases. For example, taking a 55 km/ hour wind in a fetch of 160 km, a wave just over 3 m high would be expected. When the fetch increases to 1600 km the wave height should be over 6 m, and would require a duration of 55 hours, instead of only 9 hours in the smaller fetch.

Curves are also given which allow the decay of waves to be estimated. Again there is a difference between the methods, in that the significant periods are held to increase with decay distance in the SMB graphs. Complications arise owing to the fact that the storm generating the waves is itself moving, and this movement becomes very important in the instance of a hurricane, which is treated in the revision of the SMB technique by Bret-schneider (1959).

The PNJ method (1955), by providing a series of additional curves, allows for the angular spreading of the waves as they move away from the generating area. Allowance is also made for the fetch over which the storm waves are blowing, for a situation in which the storm is moving at about the same speed as the waves generated by it, and finally, for the result of the rapid cessation of the storm winds. Other factors that must be considered are air stability, given by the air–sea temperature differences, the effects of shallow water, tidal currents, and wave refraction, all of which affect the wave dimensions.

Accurate prediction is not yet possible from theory of the wave spectrum. It is also difficult to measure the wave spectrum accurately at sea, especially in storm conditions when waves are being actively generated. Wave forecasting must be done in three stages. Firstly, the wave spectrum for the particular wind velocity, duration, and fetch must be predicted. Secondly, the effect of wave attenuation must be taken into account as the sea changes into swell. Thirdly, shallow water effects must be considered as these change the waves before they arrive at the coast.

Moskowitz (1964) has attempted to test the various forecasting methods. He selected sets of comparable wave spectra in terms of wind speed for winds of 37, 46, 55, 65, and 74 km/hour, and tested them statistically to ascertain whether they were likely to have come from the same population. The results showed that some variability in the waves and weather conditions had not been removed, and that spectra chosen for the same wind, but without allowing for other criteria, were nowhere near coming from the same population. This is one of the difficulties in testing the various empirical wave forecasting methods. In Moskowitz mean spectra, the area under the curve is proportional to the square of the wind speed, and so the wave height is proportional to the square of the wind speed. Darbyshire's predicted heights fit the measured waves better after allowing for a recali-bration of the wave recording device on which the results were based. Darbyshire's results then agree reasonably with those of the PNJ and SMB methods. Bretschneider (1965) has shown that the generation of waves in shallow water is less efficient than in deep water because of dissipation of wave energy by bottom friction. This factor must be combined with the deep water wave forecasting curves to provide an estimate of waves generated in shallow water. The best agreement between wave data and the numerical calculations was found when the friction factor was taken to be $f = 0.01$. Limitations of fetch and duration can be allowed for. Bretschneider's curves provide good agreement between wave spectra for Atlantic City, New Jersey and the shallow water hindcast wave spectrum, which contains much less energy than the deep water hindcast spectrum.

Wave refraction complicates wave forecasting in shallow water. Mehauté and Koh (1966) consider that the most convenient method is based on the principle of conservation of energy flux between wave orthogonals. This provides a good approximation for very gentle slopes. Partial reflection occurs where there are abrupt changes in depth, which complicate the analysis. Bottom friction is not negligible. There may be energy transference across the orthogonals where the relief is complex owing to variations of wave height along the wave crests. The relationship for angle approach at the break point and the deep water steepness at the break point approximate to $\alpha_b/\alpha_0 = 0.25 + 5.5 H_0/L_0$ for $0 < \alpha_0 < 50°$ for the linear theory and a horizontal bottom. The effect of slope is seen in the modified relations $\alpha_b/\alpha_0 = 0.23 + 5.2 H_0/L_0$ for $20° < \alpha_0 < 40°$ on a $1/10$ slope, and $\alpha_b/\alpha_0 = 0.06 + 7.5 H_0/L_0$ for a $1/20$ slope. Non-linear effects are more important for larger initial wave steepness values. The maximum angle of the breaking wave occurs when the deep water wave angle is $45°$, whatever the deep water steepness. This angle will also produce the maximum longshore current.

6 Waves in shallow water

6.1 Modification of ideal waves in shallow water

As a wave begins to feel the bottom nearly all its characteristics are modified. The most consistent dimension, which does not alter as the water becomes shallower, is the wave period. The wave velocity and length decrease as the depth decreases according to the equation given at the beginning of the chapter. These relationships were used during the 1939–45 war to measure the gradient of enemy-held beaches from air photographs. The length of the waves in deep water could be measured and the wave period obtained. Then by measuring the decreasing wave length as the waves approached the shore, the depth could be determined (Williams, 1947). The reduction of wave length accounts for wave refraction, by which the wave crests turn to lie more nearly parallel to the bottom contours.

Wave height does not vary so systematically as length. The wave height increases slightly as the wave first enters water shallower than half the wave length, until d/L is about 0.06. At this point the value of H_b/H_0 is about 0.9. Thereafter there is a rapid increase in wave height until the wave breaks. The increase is larger for initially flat waves. A very flat wave may double its deep water height before it breaks, while a steep wave may increase relatively little. Owing to the decrease in wave length and increase in height in very shallow water, the wave steepness will increase very rapidly near the break-point. This increase, however, is rarely sufficient to cause the wave to reach the limit of stability, when $H/L = 1/7$.

As the steepness increases the wave form changes, the crest becomes sharply defined and narrower, while the trough becomes wide and flat. This change is particularly noticeable in a long, low swell, which may be inconspicuous in deeper water, but is rejuvenated as it enters shallow water and frequently is the most conspicuous wave train to break on the

Plate 5 Waves in shallow water: The wave pattern approaching an embayed coast shows refraction of a long swell. Concentration on headlands and dissipation in the bay of the wave energy can be seen. The increasing sharpness of the wave crests as they get into shallower water is shown, together with their reduction in length. Short-period locally generated waves are superimposed on the long swell (*CERC*).

beach. The orbits of the water particles also change. Instead of following open circular paths, they now follow open ellipses, the long axis being parallel to the bed. The particle velocity is also no longer uniform. There is an acceleration under the short, sharp crest, while the seaward flow under the trough takes place more slowly but for a longer time. This asymmetry becomes more marked near the break-point. On the bottom the ellipse becomes flattened into a to and fro movement, with a short, rapid acceleration under the crest and a slower seaward flow under the trough.

6.2 Mass transport in shallow water

The pattern of mass transport also changes as the waves enter shallow water. This can be defined as the net movement of water particles which results from the difference between the landward and seaward components of movement. It can be well demonstrated in a narrow model wave tank, by following the distortion of a thin thread of dye, as shown in

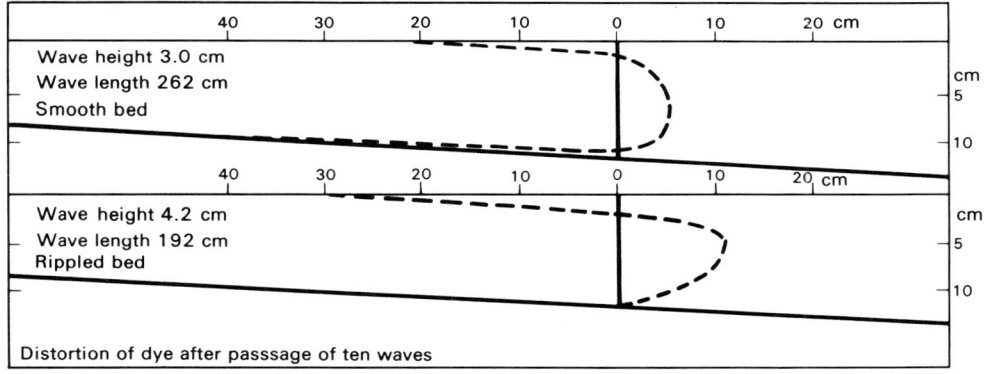

Figure 5.8 Water movement in a model wave tank to show the forward thrust on the bottom and surface with seaward drift in the centre.

figure 5.8. There is usually a landward movement at the bottom and on the surface, with a net seaward movement taking place in the centre. If the bottom is rippled, the turbulence breaks up the thread of dye and the bottom forward thrust is not apparent. This movement has been studied experimentally by Russell and Osorio (1958), who have confirmed the theoretical work of Longuet-Higgins (1953), on the mass transport of water in waves in shallow water. The forward bottom movement took place on a horizontal bed and on a gently sloping one. The velocity of the forward thrust on the bottom increased as the water became shallower for any one wave length.

Russell has suggested that there is a variation in the pattern of the mass transport in waves of different steepness, which would help to explain the very important differences in the action of steep and flat waves respectively on a mobile beach. In deep water there is nothing to hinder the continuous landward progress of mass transport, but near the shore the forward movement of water, as the depth decreases, increasingly piles up water on the shore and would raise the water level unless counteracted by a seaward return drift. Hence the seaward flow in the central part of the depth profile is accounted for.

In a flat wave in shallow water, forward movement of mass transport takes place uniformly throughout the water depth. The return drift is superimposed on this and is largest at the surface, steadily decreasing to zero at the bottom. The net result of superimposing these two movements gives a slow seaward movement throughout the bulk of the water, but a narrow layer near the bottom has a strong forward thrust towards the land. Nearly all the sediment under flat waves moves near the bottom, and so is influenced by the landward movement, creating a constructive effect. Russell suggests that under steep waves the landward component of mass transport is greatest at the surface and falls to zero at the bottom, while the return drift is at a maximum in a seaward direction at the surface and also falls to zero at the bottom. When these two curves are combined the resulting net movement is landwards in the upper part of the water, but in the lower part it is seawards, falling to zero at the bottom. Under these conditions the water flow is likely to be turbulent, and material raised from the bottom will drift seawards. Steep waves are often actively generated by local onshore winds. These winds will exaggerate the seaward return flow near the bottom by blowing the surface water landwards, and thereby intensifying the compensating seaward drift on the bottom. This effect can be shown experimentally to take place both inside and outside the break-point. A similar effect has been observed off the Northumberland coast by means of a current meter. The surface drift was 2·24 km/day to the south by west; at a depth of 40 m, the drift was 4·5 km/day to the east by north, under the action of the wind. The destructive effects of steep waves can be greatly enhanced by the action of strong onshore winds.

From the point of view of coastal geomorphology, waves in shallow water are of fundamental importance to an understanding of the character and development of coastal landforms. Analysis of mass transport provides further explanation of the well-known destructive effect of storm waves in eroding beaches, with the exception of storm shingle ridges. The constructive effect of flat, calm-weather waves in building up the beaches is also explained more fully. Destructive waves are steep and are often accompanied by strong onshore winds, while the constructive ones are long, low swells that have been reduced in height on their long journey from their place of origin to the beach on which they finally break.

6.3 Wave refraction

Before the waves finally come to the break-point, they undergo very considerable change in their direction of movement. The fundamental cause of wave refraction has already been mentioned. It is due to the reduction of wave length and velocity with depth. The parts of the wave front that are in deeper water will advance faster than those in shallow water, with the result that the wave crest will become more and more nearly parallel with the bottom contours. The main significance of this bending of the wave crest is to be found in the redistribution of wave energy, to which it gives rise. In deep water, where the wave crests are assumed to be straight and parallel, the energy is evenly distributed along the wave front. The energy between equally spaced points is the same. If these points are continued as lines perpendicular to the wave crest towards the shore as the wave crests bend, so the spacing between the lines no longer remains constant. The lines, which are called wave rays or orthogonals, as they are at right angles to the crests, will tend to converge in some areas and diverge in others.

It can be assumed as a first approximation that the energy between the wave rays, which are equal in deep water, will remain so as they are traced into shallow water. Where the orthogonals converge there will be a concentration of wave energy, and the wave heights will be greater. Where they diverge the reverse will apply, and the waves will be lower. The concentration of wave energy on headlands and its dissipation in bays can be explained on the assumption that the bottom contours are parallel to the coastline. The wave crests are bent in such a way that the orthogonals converge on the headlands and diverge in the bays. The waves are, therefore, higher and steeper, and more destructive on the headlands than they are in the bays. This and other types of wave refraction are shown in figure 5.9.

If the coastline is straight, but the bottom contours are not, the zones of convergence and divergence of energy bear no apparent relationship to the shore. Over a submarine

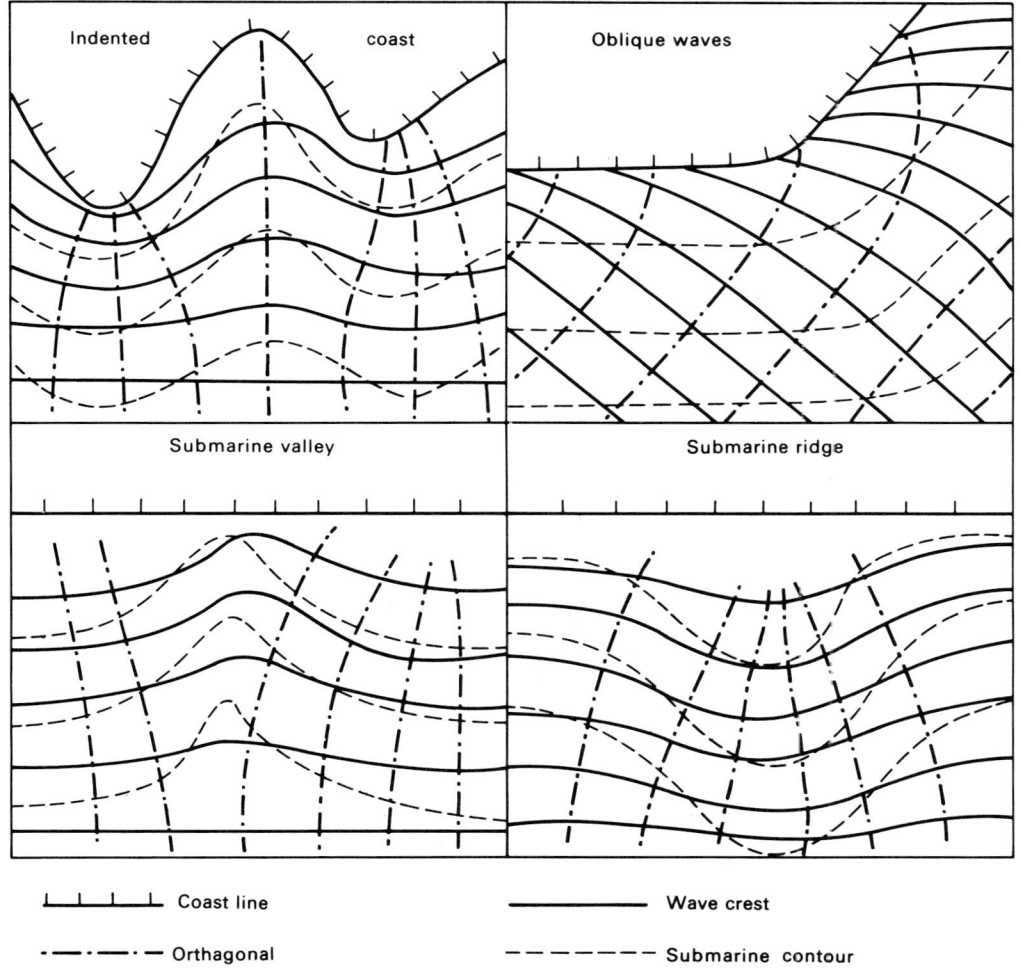

Figure 5.9 Diagrams to illustrate different wave refraction patterns.

valley at right angles to the coast there will be a zone of divergence, and abnormally low waves, while over a submarine ridge the orthogonals will tend to diverge, giving waves higher than usual. Where waves are approaching the coast obliquely in deep water, they will tend to turn to reach the coast more nearly parallel (see figure 5.9). This effect will be more marked for longer waves, which feel the bottom first. Shorter waves will not be seriously affected and may finally break on the coast at a considerable angle. Thus long waves will not be associated with longshore drift in the same way as short waves, because longshore currents depend to a considerable extent on the angle which the waves make with the shore as they break in shallow water.

Where the submarine relief is complex, wave refraction will be similarly complex, and in some situations orthogonals may cross. Such a situation may occur when waves are refracted round both sides of a shoal or island, to meet on the far side from both directions. Such patterns of refraction account for the formation of some tombolos, which tie islands to the mainland or to each other.

Wave refraction may also account for the orientation of wave-built structures. Sand berms and sand barriers (built up by long, constructive swells) are particularly affected by the refraction of the waves, because these waves are very long and, as a result, are much refracted before they reach the shore. The smooth curves of such beaches, for example those discussed by Davies (1959) in Tasmania and south Australia, fit the pattern of the refracted wave crests. The beaches respond to the refraction of the waves rather than the reverse. The submarine relief is probably the most important factor in conjunction with the direction and length of the dominant constructive waves in determining the plan of the coast under such conditions. Beaches of this type will be orientated to face the refracted wave direction, which may well be at a very considerable angle to the deep water direction, where the dominant swells are long.

Because wave refraction can have such a marked effect on the wave height, it is useful to determine the amount of refraction waves of any given period and direction of approach will undergo. Refraction diagrams can be drawn graphically if the submarine relief is known. One type of diagram shows the pattern of refraction for a considerable stretch of coast for one wave length and direction of approach in deep water. The other shows the effect of refraction at one place for waves coming from different directions and of different lengths. When the waves are forecast in shallow water it is necessary to take wave refraction into account. The pattern of wave refraction is complicated by the fact that waves usually arrive in a spectrum, each element of which is refracted differently. If two waves of slightly different frequency are travelling in the same direction in deep water they will form a long crested pattern. When they are refracted on entering shallow water, each one will respond differently. They will now travel in different directions, with the result that the pattern becomes short crested. Waves of the same length approaching from slightly different directions in deep water will tend to become longer crested as they enter shallow water, because each turns more nearly parallel to the bottom. Waves reaching the beach will normally, therefore, differ considerably from those in deep water.

Particularly complex patterns of refraction occur where long waves approach over an irregular bottom relief. Wave rays commonly cross in this situation causing a caustic, where theoretically the wave energy becomes infinite. The interpretation of the effects of wave refraction is complicated by this occurrence. One method of dealing with complex refraction phenomena is discussed by Hardy (1964). He constructed wave refraction diagrams for the Norfolk coast for swell waves coming south down the North Sea from between Scotland and Norway. He found that nearly all the orthogonals crossed in a very complex manner owing to shoals and channels offshore. He therefore calculated the distance apart of orthogonals adjacent in deep water when they arrived at the coast, and determined the incoming energy on this basis. The increments of energy at any stretch of coast were then summed to give an indication of the total incoming energy at that point. The results are necessarily tentative owing to the complexity of relief offshore and the uncertainty of the physical results of crossing orthogonals. The areas of concentration of these constructive swells did, however, agree with the areas of greatest accretion on the beach for three wave directions and periods. The conclusion drawn from studies of this type must be tentative until more accurate refraction diagrams can be constructed, which will be difficult owing to the complexity of bottom relief. The effect of crossing orthogonals also requires further study.

6.4 The reflection and diffraction of waves

If a wave approaches a vertical cliff or sea-wall, the foot of which is in deep water, it will not break and its energy will not be absorbed. As with tidal waves under somewhat similar circumstances, the wave will be reflected from the vertical barrier. Under ideal conditions this will lead to little loss of energy, and an equal and opposite wave will travel away from the face if the primary wave approaches parallel to the barrier. The reflection will set up a standing oscillation similar to that already described in tidal movement. It is sometimes called a deep water clapotis. If the incident wave arrives at an angle to the barrier, the reflected wave will lie at an equal and opposite angle to it. The two waves will interfere with each other in such a way that their crests and troughs sometimes coincide and sometimes cancel each other. The result is a diagonal pattern of high crests alternating with low troughs.

The phenomenon of diffraction accounts for the presence of waves in areas which might be expected to be sheltered from them. This is particularly important with regard to wave action in harbours. If waves are travelling towards a break-water, with their crests parallel to it, as shown in figure 5.10, those in deep water beyond the end of the break-water will continue to move forward in the same way; but as they pass the break-water they will spread out and become lower in height in its shelter. In plan they will form the arc of a circle, centred on the end of the break-water. As these circular fronts derive their energy from the wave crests advancing in deep water, they are half the normal height along a line at right angles to the break-water in the direction of wave movement. The reduction of height increases along the curved arcs as energy is transmitted along the wave crest, as shown in the figure.

In this way it is possible to explain why areas in the geometrical shadow of an obstacle, such as a break-water, are still affected by wave action. Although the theory of diffraction, as applied to light waves, is not strictly applicable to sea waves, experiments have shown that results worked out from it are not far wrong, usually giving slightly higher waves than those actually recorded. The amount of height reduction can be calculated theoretically and depends on the nature of the barrier. The wave length and depth at the end of the barrier must be known and diagrams have been prepared by CERC (1966) for

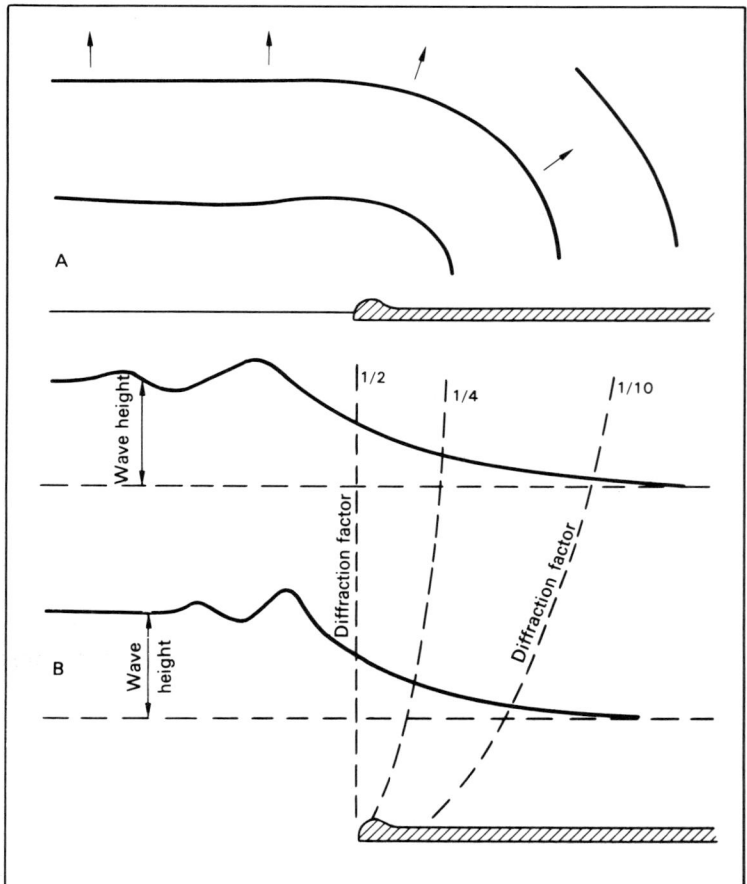

Figure 5.10 Diagram to illustrate wave diffraction. **A** is the plan view of the wave crests. **B** shows the height and diffraction factor for $\frac{1}{2}$, $\frac{1}{4}$, and $\frac{1}{10}$ of the open sea wave height. (*After Russell and Macmillan, 1952.*)

waves approaching from different angles to the barrier. Wave heights are given in terms of the ratio to the incident wave height, and curves are drawn for equal values of K′, the ratio of incident wave height to diffracted wave height. The diffraction through gaps of different widths relative to the wave length between two barriers is also calculated theoretically and illustrated graphically. The amount of diffraction increases as the angle between the wave crests and the barrier increases, resulting in rapidly decreasing wave heights for large angles.

Mobarek and Wiegel (1967) have studied a two-dimensional method of analysing wave diffraction behind break-waters, based on two-dimensional wave spectra. A laboratory study was used to test the theoretical results. The two-dimensional spectrum was treated as though the waves came from five directions at angles 30° apart. The calculations were repeated for each frequency band. The energy for each direction and frequency was found for selected sites in the lee of the break-water and the individual results summed to give the diffraction value at that site. The results agreed with the laboratory tests for the total amount of energy reaching the area in the lee of the break-water by diffraction. The peak energy was also about the same. However, it occurred at a lower frequency for the spectrum measured than the theoretical one. The evidence supported the assumption of linearity in the theory of diffraction.

6.5 Offshore and longshore wave currents

Water particles reach their maximum velocity in the turbulent zone of breaking waves, but the particle velocity may also be appreciable at a considerable depth in storm-generated waves. Hadley (1964) has analysed the velocity of wave-induced bottom currents in the Celtic Sea. The waves analysed were those generated by a force 10 gale, according to the generating curves of Darbyshire. The estimate of wave energy may be 20 per cent in error. The values in terms of V_{50}, the calculated velocity for the largest wave in a set of 50 for different depths are given in table 5.1. The relationship between V_{50} and $\sqrt{\overline{V_b}^2}$,

Table 5.1 Wave-induced bottom current velocities

Depth, m	V_{50}, c/sec	Depth, m	V_{50}, cm/sec
18·3	448	109·6	103
36·6	283	128·0	82
54·8	206	146·4	72
73·2	165	164·6	62
91·5	118	183·0	52

the mean square bottom velocity, is given by $V_{50} = 3 \cdot 2 \sqrt{\overline{V_b}^2}$. The increase for V_{100} is 7 per cent, for V_{500} is 20 per cent, and for $V_{50,000}$ is 50 per cent. These wave-induced currents are oscillatory. Storms capable of generating them are rare, only occurring for about one or two weeks in the year. The results do, however, indicate that waves cause currents throughout the continental shelf under extreme conditions.

Longshore currents can also be generated by waves. They exert a strong influence on the movement of material alongshore, which is one of the fundamental reasons for coastal erosion and accretion. One of the causes of longshore currents is wave refraction. Zones of high waves are likely to alternate with zones of low waves, particularly where the offshore relief is rather complex, the coastline indented, and the waves long. There is greater mass transport where the waves are higher and this should raise sea level slightly at these points, compared with the zones of low waves. The water surface will slope from the zones of high waves to those of low waves and longshore currents will flow in these directions.

Currents due to wave refraction have in fact been observed by Shepard and Inman (1950). Their observations were made off southern California, where the offshore relief is complicated by submarine canyons, which come right up to the beach. Currents flow from headlands to the bays, because wave concentration is greatest on the headlands. When the waves are long, and therefore much refracted, zones of divergence and convergence are well developed and currents flow alongshore from the zones of convergence, reaching 50 cm/sec at times. Currents are also induced by the landward movement of water by mass transport and by the oblique approach of waves. The wind also exerts an effect. Some of the longshore currents were closely associated with the seaward transport of water, which at times becomes concentrated to form rip-currents. Strongly developed rip-currents are supplied by longshore feeders. The conclusions reached from observation of currents off southern California indicated that the direction of the currents along the shore was dependent mainly on the wave period, height, and deep water direction of approach. The last factor had least effect when the waves were longest, as these were the most modified by refraction. At points where longshore currents meet, which may be near points of divergence of orthogonals, they often turn seawards as rip-currents.

When shorter waves approach the shore they are less refracted and the direction of approach plays an important part in determining the direction of longshore currents, which often follow the direction from which the waves come. Rip-currents then tend to be more numerous, but much smaller. When the short waves approach oblique to the coast the longshore currents will be almost continuous in one direction, with speeds often exceeding 50 cm/sec. Under calm conditions, as obtained when the observations were made by Shepard and Inman, they found that by far the greater number of observations showed the same movement on the surface and at depth.

Rip-currents are most conspicuous on shores on which the very long swells are common. They are localized and may change position with changing wave conditions. They form where the return drift is concentrated into a fast-flowing seaward current. The rip is often fed by strong feeder currents flowing alongshore. The higher the waves, the stronger will be the rip-currents, because the landward mass transport is increased. Rips may be identified on the beach because they reduce the breaker height, where they break through the surf zone. They do not, however, extend very far seaward of the breaker zone, beyond which they dissipate rapidly. They may be dangerous to bathers, who may be carried offshore in their strong flow. It is usually possible, however, to swim in again alongside the rip-current, where the breakers are higher and the mass transport on the surface is landwards. Rip-currents are variable in velocity, being faster after a series of high waves, when the water level on the shore has been raised temporarily.

Observations by McKenzie (1958) in New South Wales have shown that rip-currents on the sandy beaches are stronger, but much less numerous, during periods when the waves are exceptionally large. These large rips were fed by strong currents, up to 800 m long and 2·44 m deep. Such strong currents are responsible for cutting channels across the beach. He also confirmed that smaller waves tended to produce smaller, but more numerous, rips. Rip-currents have been measured by Draper and Dobson (1965) at Holywell in west Cornwall. Speeds of nearly 250 cm/sec were observed on the surface, where the rip broke seawards through the surf. The waves were 3 m high and the rip extended 275 m seaward.

Inman and Bagnold (1963) suggest the importance of rip-currents in their theoretical approach to longshore current analysis. They take continuity of flow into consideration, and give a relationship

$$\bar{U}_1 = q \frac{l_1}{d_b{}^2} \tan \beta \sin \alpha \cos \alpha,$$

where $\bar{U}_1$ is the mean longshore current velocity, q is the gross incoming wave discharge in unit time per unit length of wave crest entering the surf zone at angle α, l_1 is the separation between rip-currents, β is the beach slope and d_b is the depth at the break-point. The power that is needed to accelerate the surf to the velocity $\bar{U}_1$ and to overcome bed drag and internal resistance to flow comes from the longshore component of the flux of wave momentum. At a distance l_1 from a previous rip, the acceleration would cease and the longshore discharge break outwards to form a new rip-current. The longshore current must next be evaluated in terms of the wave characteristics, and the relationship

$$V = 4\sqrt{\frac{\gamma l_1}{3T}} \tan \beta \sin \alpha \cos \alpha$$

is given, where γ is H/d and T is the wave period, H is the wave height and d the water depth.

Twelve different equations have been proposed to give longshore current velocities (Galvin, 1967). These equations are in three classes. The first is based predominantly on the conservation of energy of momentum, and include those of Putnam *et al.* and Eagleson (1965). The second is based on the conservation of mass, and includes that of Inman and Bagnold just mentioned. The third is based on the empirical correlation of data and includes those of Brebner and Kamphuis (1963) and Harrison and Krumbein (1964). Galvin points out an objection to the conservation of momentum equations by showing that the alongshore component of the breaker velocity, which is assumed to drive the current, must exceed the current velocity with which it is equated in the theory. In fact the longshore current velocity must have a component in addition to the alongshore component of breaker velocity.

Harrison (1968) has suggested an empirical relationship for longshore current velocity based on 98 observations of velocity, $\bar{V}$, beach slope β, breaker height H_{bs}, wave period T_b, angle of wave approach α_b, crest length and trough depth made on an Atlantic coast beach in Virginia. There was an offshore bar parallel to the shoreline for two-thirds of the study. The analysis is based on linear multiple regression. The results differed significantly from those of Harrison *et al.* (1965), which were based on deep water data, while the more recent values are for breaker characteristics. The results show that angle of approach is the most important variable, followed by the wave period and wave height. The predictive equation is

$$\bar{V} = -0{\cdot}170455 + 0{\cdot}037376\alpha_b + 0{\cdot}031801T_b + 0{\cdot}241176H_{bs} + 0{\cdot}03923\beta.$$

In view of the unreliability of theoretical prediction equations, the empirical approach provides a useful alternative. It can, however, only apply strictly to the conditions in the field in the area investigated.

Galvin and Eagleson (1965) made model experimental studies of the characteristics of breaking waves and the longshore currents to which they give rise. The waves broke on a smooth concrete beach with a slope of 1 in 10 over a 6 m test section. The results showed that most of the fluid in the surf zone remains there and that the longshore current increases downstream from an obstacle. The energy that maintained the longshore current was less than one-tenth of the energy brought into the shoaling zone by waves. The variables measured in the wave tank included wave height, speed and form, breaker position, angle and run-up limit, change in mean water level due to the waves, and the longshore current velocity, which was measured by miniature current meters and floats. The longshore currents were measured in both vertical and longitudinal distribution, giving bottom, mid-depth, and surface values. The test area was bounded by training walls.

At the upwave end of the beach water was rapidly drawn into the longshore current and the longshore component increased with distance from the upstream training wall, while the net onshore water motion decreases with distance from it. No strong net offshore movement occurred except at times near the downwave training wall. The surface velocity showed the most consistent shoreward trend. Conclusions from this study suggest that most of the water injected into the surf zone when a wave breaks has been drawn from the surf zone. It therefore already has a longshore velocity, V_b, when it becomes part of the breaker. The velocity of the longshore current increases first at an accelerating rate downstream from an obstacle, and then at a decelerating rate, approaching a constant velocity at a large distance from the obstacle. The mean velocity of a uniform longshore current, using field and laboratory data, is given approximately by $V = g\beta T \sin 2\alpha_b$.

The study of longshore currents can be made theoretically, in the model wave tank, and in the field. All methods are helpful in establishing the relationship between the currents and the forces that create them. Much work remains to be done in this important and complex study, which is multivariate in character.

6.6 Breaking waves

Waves expend most of their energy in breaking. The point at which waves break is deter mined mainly by their height. A wave will break in water that is about four-thirds its own height at the break-point. Thus a three-metre wave at its break-point will break in four metres of water. The deep water height of low waves, however, may be almost doubled before they break.

Friction is not the primary cause of the breaking of waves, although waves may lose some energy in this way if the slope is very gentle. The change in the orbital velocity and size of the orbit explains the process of breaking. As a wave approaches shallower water, its orbit enlarges to become elliptical in form, but the time it has to complete its orbit remains the same, as the period of the wave is not affected by the shallowing water. At the same time the velocity of the wave form is decreasing, and there comes a time when the velocity of the water particles in their orbit approaches that of the wave form. When the orbital velocity exceeds the wave form velocity, the water will overtake the wave form and the wave will break. This accounts for the possibility of surf riding on a wave that is just about to break, for only on the crest of a wave in this state does the water itself move at the same speed as the wave form. Another factor that leads to the breaking of a wave is the fact that as the orbit increases in size, the amount of water in the wave form is decreased by the reduction in length, with the result that there is not enough water to complete the orbit. The front of the wave becomes hollow and the crest, being unsupported, crashes down into the trough, as the wave breaks.

Breakers can be of two types, depending on the nature of the wave and the character of the bottom. In plunging breakers the crest collapses into the trough, enclosing a pocket of air and completely destroying the wave form in the process of breaking. In spilling breakers the wave crest advances as a line of foaming water, moving forward at the speed of the wave form. On a flat beach there may be several surf waves, advancing at the correct speed for the water depth in the surf zone. This type of wave is used for surf riding, as the breaker does not lose its identity, but gradually gets lower and lower until it becomes a thin layer of swash on the beach.

Plunging breakers are best developed on a steep beach where waves were fairly flat in deep water, while spilling breakers occur on a gently sloping, usually sandy beach, where the waves in deep water were fairly steep. There is, however, a gradual transition between these two main types, which vary with the factors on which their character depends. Much of the energy of a wave is lost as it breaks, but the remaining energy moves water up the beach as swash. This returns down the beach, under the action of gravity, as the backwash.

6.7 The effect of waves on structures

Waves can at times do considerable damage to break-waters and sea-walls, as well as to natural cliffs of solid rock. If the water is sufficiently deep in front of a sea-wall or break-water, the whole of the wave energy is reflected and the only force exerted on the structure will be the hydrostatic head of the rise and fall of the water level. If the water is deep enough to prevent the wave breaking against the structure, most of the energy will be reflected. This will form a shallow water clapotis, in which it is possible to see the retreating water wave moving through the advancing one. This can often be observed when the waves are advancing against a sea wall at high tide. Again this type of reflection will not cause any damaging pressures on the sea-wall. Damage is much more likely to occur where the water is shallow enough for the wave to break against the structure. In this situation the particle velocity within the wave form is equal to that of the wave form and water moving in this way may exert considerable pressure on the sea-wall. It has been observed that the speed of the water at the crest of the wave may in fact be equal to twice the wave velocity at times.

The most intense shock pressures will, however, be set up if, when the wave breaks, it encloses a pocket of air against the structure as it does so. This air pocket becomes intensely compressed, and it throws up the surrounding water with great violence, setting up extreme shock pressures. Such pressures are difficult to measure, as they take place over a very short period of time. The pressure due to breaking waves is both short-lived and localized, as experiments have demonstrated. These high pressures were recorded for periods varying from 1/20 to 1/200 second. One wave produced a shock pressure of 12,700 lb/ft², but the total time the pressure exceeded 6000 lb/ft² was only 1/100 second. Experimental work by Bagnold has also shown that shock pressures are very short-lived. He found that even in the most uniform waves that could be generated in a model tank the shock pressures varied greatly and only occurred when a pocket of air was enclosed. They were much greater when the pocket of air was very thin. The total amount of pressure each wave produced was fairly uniform, but its distribution varied. Thus extreme pressures only lasted for a brief time, while less severe ones lasted longer. The observation that the thinnest pocket of air produces the highest pressures suggests that the air pocket acts as a cushion, separating a vertical wave front from a vertical wall against which it is brought instantaneously to rest and its momentum lost. It is such shock pressures that are most liable to do serious damage on the coast. Their rarity and brief action in time is balanced by the more frequent and longer-lived, but less intense, forces exerted in other ways.

6.8 Solitary waves

When the depth-to-length ratio is less than 0·05 the velocity of the wave depends only on the depth, and waves move independently of each other with a velocity of $C = \sqrt{gd}$.

Such waves are called solitary waves. Each one consists of a crest separated from its neighbours by flat water, in which the water particles are at rest. The particles under the crest move in the direction of wave propagation. In order to maintain a uniform water level, a slow seaward flow is superimposed on the landward movement. The movement of particles in such a wave is up and forwards as the crest approaches, and downwards as it passes, while the superimposed seaward movement converts this latter part of the orbit into a seaward flow.

Solitary waves are of interest because theoretical velocities on the bed can be calculated for different wave heights. Experimental work designed to measure these velocities has shown that the results agreed more closely with theoretical ones, using the solitary wave theory, than those worked out from the Airy–Stokes wave theory for low progressive waves. Solitary waves of this type probably exist in very shallow water in nature. They may occur where waves reform after breaking on a submarine bar.

7 Long waves

7.1 Surf beat

There are a number of waves in the ocean whose length is considerably longer than the normal, wind-generated waves, the periods of which do not usually exceed about 24 seconds. A type of long wave, sometimes called surf beat, was recorded by Munk (1949) and discussed by Tucker (1950). These waves had a period of about 2 minutes and a height of about one-eleventh of the normal waves. They sometimes give trouble in harbours and have been recorded in Tema, Ghana, and Perranporth, Cornwall. At Ghana waves of this type had periods between 1 and 10 minutes, 2–4 minutes being most common. Their height was about one-twentieth that of the normal waves, reaching a maximum of about 13 cm. Those recorded at Perranporth had heights about one-twelfth of the normal waves and their period was also 1–5 minutes.

Groups of extra high waves appear from time to time in a complex wave spectrum and these cause surf beat. Mass transport increases rapidly with wave height and near the break-point, resulting in an abnormally great volume of water moving towards the shore as the groups of larger waves break. The increase of mass transport in the breaker zone sets up a shoreward wave which is reflected from the shore; in this way the long wave or surf beat is initiated.

The surf beat is a forced wave, but it can become modified into a free wave close to the shore. Cartwright (1967) has shown that in surf beat the primary waves are destroyed on the beach, but that the free waves are released into the ocean. They can echo back and forth for days. These waves account for the energy in the spectrum below 30 c/ks. They can also cause range action in some harbours that resonate with this particular frequency.

Adams and Buchwald (1969) draw attention to spectral waves with a frequency of 9 days in summer and 5 in winter. The associated variations in level correlate highly with pressure. They are called shelf waves and are generated on a sloping shelf by the geostrophic wind. The longshore component of the geostrophic wind accounts for the shelf waves. Waves of an amplitude of several cm travel north along the east coast of Australia at 350 cm/sec, and south along the west coast at 300–600 cm/sec. They also move along the west coast of the United States at 250 cm/sec. Long waves with periods of a few minutes have also been reported on the wide shelves of the Argentine. The Argentine shelf waves are a form of seiche activity. A critical intensity can cause sea-level fluctuations of up to a metre or more in height. They are probably due to meteorological causes as they frequently occur when sharp pressure fronts come from central Patagonia (Inman, Munk, and Belay, 1962).

7.2 Microseisms

Another type of wave associated with ocean storms is worthy of brief mention. Under some conditions it is important in forecasting the approach of damaging storm waves or swell (Iyer, 1959; Darbyshire, 1960). Careful comparison of wave records and microseisms records of very small seismic disturbances show that the microseisms are related to storm waves. Both have a spectrum of waves. Microseisms frequently arrive some time before the waves reach the coast. If it were possible to relate the direction from which the microseisms travel to the direction in which the storm waves are travelling, they would provide a very valuable warning system, especially in those areas where weather observations out to sea are scanty.

The main cause of microseisms is the interference between two wave trains of equal period, but travelling in opposite directions. This situation can arise in a small depression, in which winds blow in opposite directions, one either side of the disturbance. In some instances the interference between waves of the same period could be achieved by reflection from a steep coast. When waves of equal period but different height meet head-on, Longuet-Higgins has shown that they will exert an effect on the bottom even in deep water, and in this way microseisms can be set up. They will travel much more rapidly in the earth's crust than the waves on the ocean surface. The point of origin and the direction of approach of microseisms is difficult to locate because they, like ordinary waves, suffer from refraction and also a number of different periods of seismic waves are involved.

The reason that microseisms can penetrate to the bottom, unlike normal wind waves, is that they are generated by standing waves, caused by the interference of two swell waves moving in opposite directions. Observations have established that microseisms have periods equal to half that of the generating waves. The crests and troughs change places twice for each complete cycle and this causes the centre of gravity to change slightly in phase; pressure is exerted on the bottom. The change in pressure accounts for the microseism period being half that of the sea waves from which they are derived. Fluctuations are in phase and therefore build up. A secondary type of microseism has been identified near the coast, but only contains 1/100 of the energy of the main microseisms.

Microseisms are long waves and are affected by depth and geological structures, which cause some refraction. The rays by which they travel can be constructed as refraction diagrams. It is found that the rays converge and diverge at different points, there are some barriers across which microseisms are not propagated. These barriers are the result of geological shadow zones or refraction. As a result of these features the recordings of microseisms and their interpretation are very complex. This means that their movements are too complex to be of much practical assistance in wave forecasting, although at times they can give warning of approaching storm swells.

7.3 Tsunami

Another type of long-wave activity is the direct result of seismic activity, rather than the cause of it. These are the very long waves generated by a submarine earthquake, called 'tsunami' by the Japanese. At one time they were referred to by the term 'tidal wave', which is misleading because they have no connection with tidal phenomena. It is reasonable to suppose that any sudden movement in the crust of the earth under the sea will affect the water above. The Pacific is particularly liable to tsunami on account of the crustal instability of the land around its edge, but they also occur in the Atlantic. The damage done by the seismically generated waves due to the Lisbon earthquake of 1755 made this apparent. This earthquake affected Loch Lomond to the extent of producing an oscillation of 0·8 m amplitude and 10 minute period.

The waves produced by earthquakes have lengths of about 160 km, although their height in the open ocean may be only 30–60 cm. Because the waves are so long, their velocity depends only on the depth of water, even in the deep ocean, as their length is great compared with the depth of water. If the mean depth of the oceans is taken as 4580 m, the speed of the tsunami would be 755 km/hour. They are normally so long and low that they are not felt by ships in the open sea, but as they approach close to shore their height increases and they may break on the shore with devastating effect as waves 6–9 m high.

The earthquake of 1 April 1946 in the Aleutian Islands generated seismic waves which moved across the ocean to reach Hawaii after 5 hours. They arrived in increasing size, the third wave being the highest (figure 5.11). The time interval between the first two was 12 minutes. Waves from this same earthquake reached as far as the Antarctic, where a hut was washed away (Shepard, 1959).

Still more extensive and widespread damage was done by the tsunami resulting from the earthquake in southern Chile on 22 May 1960. This exceptionally severe shock produced waves which reached and caused damage in New Guinea, on the other side of the Pacific. The velocity with which these waves travelled across the ocean varied between 670 km/hour and 745 km/hour. They reached Wake Island about 19 hours after the shock (Robinson, 1961b). As the water depth decreases close to the shore, the speed of the wave slows down and its height increases. A wave 3·4 m high was recorded on the California coast, while a 10 m wave caused much damage in Hokkaido and Honshu in Japan. Lyttelton Harbour, New Zealand, and even Sydney in Australia were affected by this tsunami. The period between the arrival of the wave crests varied between 10 and 25 minutes. The effects of the tsunami were much more disastrous in Japan than they were at Hawaii, which is very much nearer the source of the disturbance. Partly because of the relatively minor effect of the tsunami in Hawaii, it was not thought likely that it would reach Japan with much energy left. This assumption was, however, very far from justified, as the damage in Japan demonstrated. It also indicates that there is still a lot to be learnt about the way tsunami travel across the ocean. The way in which they are influenced by the shallow water around the coast in relation to their direction of approach also requires elucidation. Thus a tsunami may give waves only 1·5 m high only a few hundred miles from the point of origin, but 10 m high waves at places as much as 16,000 km away. Long waves of this type can travel immense distances without substantial loss of energy.

One of the most recent and severe tsunami was that initiated by the strong earthquake

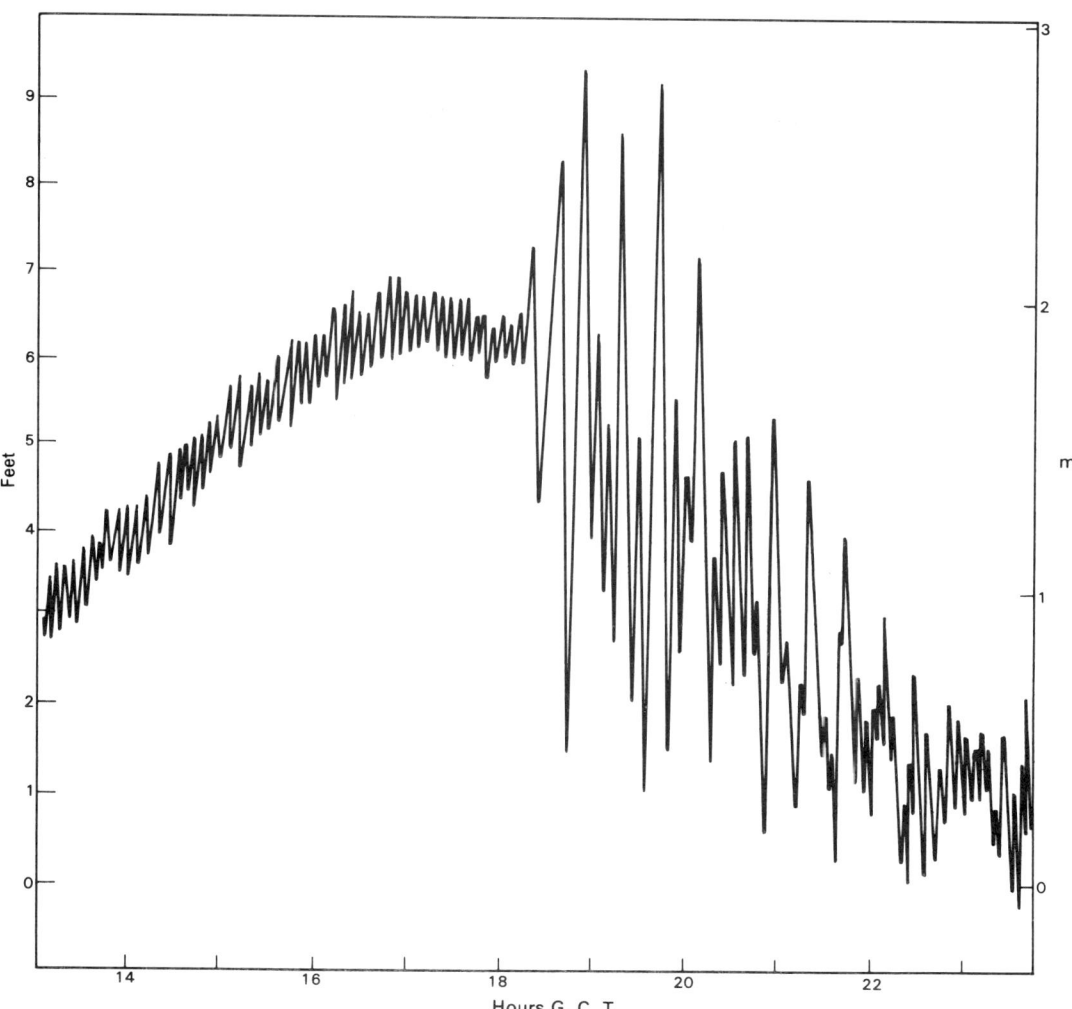

Figure 5.11 Tide-gauge record for San Luis Obispo Bay, California, to show the effect of tsunami generated by the earthquake of 1 April 1946 in the Aleutian Islands. (*After Shepard, 1959.*)

that occurred in Alaska on 27 March 1964. An account of this tsunami has been published by Wilson and Tørum (1968). It had an intensity of 8·5 on the Richter scale—an intensity expected to occur once in 30 years in this area. Extensive earth movement took place on the continental shelf off Prince William Sound at 61·1°N. The main tsunami waves were generated off Alaska and travelled all over the Pacific; there were subsidiary ones caused by submarine slumping of deltaic sediments. Seismic sea waves were generated as far away as the Gulf of Mexico as a result of ground vibration. Damage was reduced due to the coincidence of the main tsunami with low tide on the open coast of Alaska. The tsunami generated by the earthquake had an inferred period of about 1·8–2·5 hours on the coast of Kodiak and Alaska, the first wave being a negative trough caused by subsidence. The waves were overridden by free waves of 1·3–1·8 hours' period, and followed by the free shelf oscillation, which had a period of 5 hours. The waves that spread right across the Pacific were fairly pure in form, having a period of 1·8 hours. They were recorded on tide gauges all round the Pacific. One area where resonance allowed exceptionally high waves to occur was along the indented Canadian coast. At Port Alberni wave height was increased by a factor of 10 between the mouth and the head of the bay. The waves reached Lyttelton in New Zealand in remarkably pure form.

The height of the waves in the west and central Pacific suggest a height decay law of $H = r^{-2/3}$, where H is the height and r is the distance. A relationship between the main period, T (1·8 hours in this instance), and the intensity of the earthquake, M, has been suggested in the form, $\log_{10} T = (5/8)M - 3.31$. Large-intensity earthquakes thus yield long-period tsunami. The study has brought out the importance of resonance in the height of the tsunami at different points along the coast. The probable initial height of the tsunami on the Alaskan shelf was between 9·2 and 18·3 m. It had a continuous front of about 650 km. There were five waves in the first beat.

7.4 Meteorological surges

The North Sea owing to its shape, is particularly liable to the development of surges due to unusual weather conditions. They also affect the English Channel and Irish Sea. They tend, however, to be more disastrous in their effects on the North Sea coast, as the abnormal sea level is liable to flood low-lying ground, as in Lincolnshire, Essex, and Holland.

Although surges have no direct connection with the tide, they behave in some respects in a similar way to the tide, particularly as they progress as a wave in an anti-clockwise direction round the North Sea. Surges are not infrequent in the North Sea, eight having occurred during the present century, following a very severe one in 1897. Farquharson (1954) suggested that the frequency of storm surges has increased from 1850 to the present. The increase may be due to the present rise of sea level, amounting to about 0·23 m in 120 years. A surge in 1894 reached a height of 2·9 m above the predicted tide level at Southend in Essex. This height, however, occurred nearly 6 hours after high water, at which time the surge was only 1·2 m high. The maximum of the highest surges do not coincide with the time of high water. For example, surges over 3 m in height have occurred at Southend and Sheerness in 1905, 1921, and 1943, but they all reached their peak between 4 and 6 hours before high water, being only 0·3, 0·6, and 0·8 m high respectively at high water. Smaller surges may be more dangerous as they sometimes almost coincide with high water. At Sheerness in 1897, a 2·14 m surge arrived 1·5 hours before high water and in 1928 a 1·67 m surge reached Sheerness only 1 hour before high water, while a 1·83 m surge in 1936 hit Southend within half an hour of high water.

The greatest and most disastrous of the recent surges was that of 1953, which caused much damage and loss of life on the east coast of England and in Holland. This surge was exceptional in that the water level was 1·83 m above the predicted tide for 15 hours in some places. At Southend its maximum height was 2·74 m 2·5 hours before high water. It was still 1·67 m at the time of predicted high water. Other places also recorded abnormal heights. It was 1·52 m high in the Tyne, 2·28 m at Immingham, 2·97 m at King's Lynn, where it arrived 3·5–4 hours after high water. The record levels of this surge would have been exceeded by 1·52 m had the maximum of the surge arrived at the time of predicted high water level. In Holland levels up to nearly 3·36 m above predicted were recorded, with very disastrous consequences in such low-lying country.

The meteorological situation is fundamental in the generation of surges. North Sea surges form when a very intense depression to the north of Scotland passes east and south-east into and across the North Sea, often passing onto the land at about south Denmark. The wind direction in front of such a depression is from the southwest, but as the front moves across, there is a sudden veer of the wind from southwest to northwest or north. When the pressure gradient is steep these winds will be very strong, as was the case in January–February, 1953.

The centre of the depression as it passed into the North Sea was 966 mb. At the same time a strong north–south ridge of high pressure, with 1033 mb at its centre, built up behind the depression. This produced very strong northerly winds, blowing over an exceptionally long fetch, because the isobars ran north–south as far as Spitzbergen. The winds themselves did a lot of damage in north Scotland, where they reached velocities of about 160 km/hour on the ground, while the geostrophic wind speed was 280 km/hour. Sudden change of wind direction is important in surge generation. The exceptionally strong winds behind the depression were also important as they drove a considerable amount of water into the North Sea. The Straits of Dover at the southern end act to a certain extent as a safety valve, allowing some of the excess water to escape from the North Sea.

The pressure acts as an inverted barometer: high pressure will lower sea level, while low pressure will raise it by 30 cm for a pressure fall of 34 mb. The total amount of water blown into the North Sea by the extremely powerful winds has been calculated by Rossiter (1954). The speed of movement of pressure change is related to the generation of the surge. When the disturbance moves with a critical velocity it will generate a large surge. The critical velocity is the speed of travel of a free wave in the depth of water available. When the pressure change moves at the same speed as a free wave, a state of resonance will be set up and a large surge will be generated. Slow changes of barometric pressure will not generate surges, which are associated with fast-moving storms. High winds associated with the steep pressure gradients also account for other methods whereby the level of the sea is increased under suitable conditions. In the North Sea surges travel at a rate appropriate to the depth of water as a free wave. They progress in an anti-clockwise direction around the sea, increasing in amplitude as far as the Flemish Bight and thereafter decreasing as they move northwards along the coast of Holland and Denmark; at times they may be traced as far as the Baltic.

Two types of surges occur in the North Sea. External surges are generated outside the sea, and move as free waves anti-clockwise round the sea. Internal surges originate inside the sea itself. In 1953 the surge was of complex origin, being influenced by four factors: 1) low pressure, 2) the external surge caused by the movement of the low pressure centre north of Scotland, 3) the internal surge, developed as the depression passed across the North Sea, and 4) the effect of the wind in blowing extra water into the North Sea from the ocean to the north.

The level of the whole North Sea was increased by a considerable amount as a result of the operation of the last factor. Rossiter has calculated that about 15 × 10^{12} ft^3 (42·45

$\times$ 10^{10} m^3) of water entered the North Sea between Scotland and Norway between 2100 hours on 31 January and 1200 hours on 1 February 1953. This caused an increase in the level of the whole North Sea of over 0·6 m during most of this period. The level increased rapidly during the latter part of 31 January and maintained its high level till well into 1 February. The reason for this influx was the action of the gale-force northerly winds, which transported a great deal of water into the North Sea. Owing to the rotation of the earth the transport of water in the open sea would be at about 45° to the wind direction, but where the influence of the coasts becomes marked this angle is reduced to about 20°. Most of the water entering the North Sea did so on its western side along the coast of Scotland, therefore, as the deflection is to the right in the northern hemisphere. Some of the extra water escaped from the North Sea through the Straits of Dover. The rotation of the earth again affected the water escaping southwards through the Straits of Dover, resulting in higher levels on the coast of England to the right of the flow, rather than on the coast of France. The surge recorded at Dover, in the Straits, was over 1·83 m, but by the time it reached Newhaven it was reduced to 1·22 m; across the channel at Dieppe it was only a little over 0·9 m. The rise in level at Newhaven took place in spite of a very strong offshore wind, which would normally be associated with a fall in level, while the wind was onshore at Dieppe.

It has been estimated that a total of 1·7 $\times$ 10^{11} ft^3 (4·81 $\times$ 10^9 m^3) of water escaped southwards through the Straits of Dover. This is only about 1/100 of the extra water entering the North Sea from the north. It was sufficient, nevertheless, to lower the level approximately 0·27 m in the southern North Sea between Orford Ness in East Anglia and Brouwershaven in Holland. This had a beneficial effect on the water levels in the southern part of the North Sea, in the area where the external surge might be expected to be at its greatest amplitude.

The water level variation during the surge could be obtained at the places where tide gauge data were available. From these data (shown in figure 5.12), it is possible to trace the height of the surge and its passage down the coast of Britain and northwards along the continental coast. The crest of the surge wave was at Aberdeen at 1500 hours on 31 January; by 1800 hours the water level was 1·22 m higher than predicted in the latitude of Northumberland, while by 2100 hours the surge had reached Yorkshire. By 0001 on 1 February, the surge peak had passed between the Humber and the Thames and was approaching the coast of south Holland. Heights above predicted high water levels were 2·37 m in south Lincolnshire and nearly 3 m in the Scheldt estuary. The surge passed north along the coast of Holland and Denmark between 0400 hours at Ijmuiden in Holland and 1000 hours at Esbjerg on the west coast of Denmark, reaching south Norway at 2200 hours on 1 February.

The normal tide and the surge did not travel at exactly the same rate south along the British coast, but in many areas the time of high water of the surge was within two hours of the predicted time of high water of the tide. The actual time of high water was affected by the height of the surge. Record heights were established in many areas. The Thames estuary was severely affected. At Harwich, for example, the predicted high water level was 1·66 m, while the height of the water at the time of predicted high tide was 3·72 m, and the water rose subsequently to 4·0 m. The consequence in the low-lying areas in the vicinity was very serious, and flooding and damage were extensive. The coast of Holland was

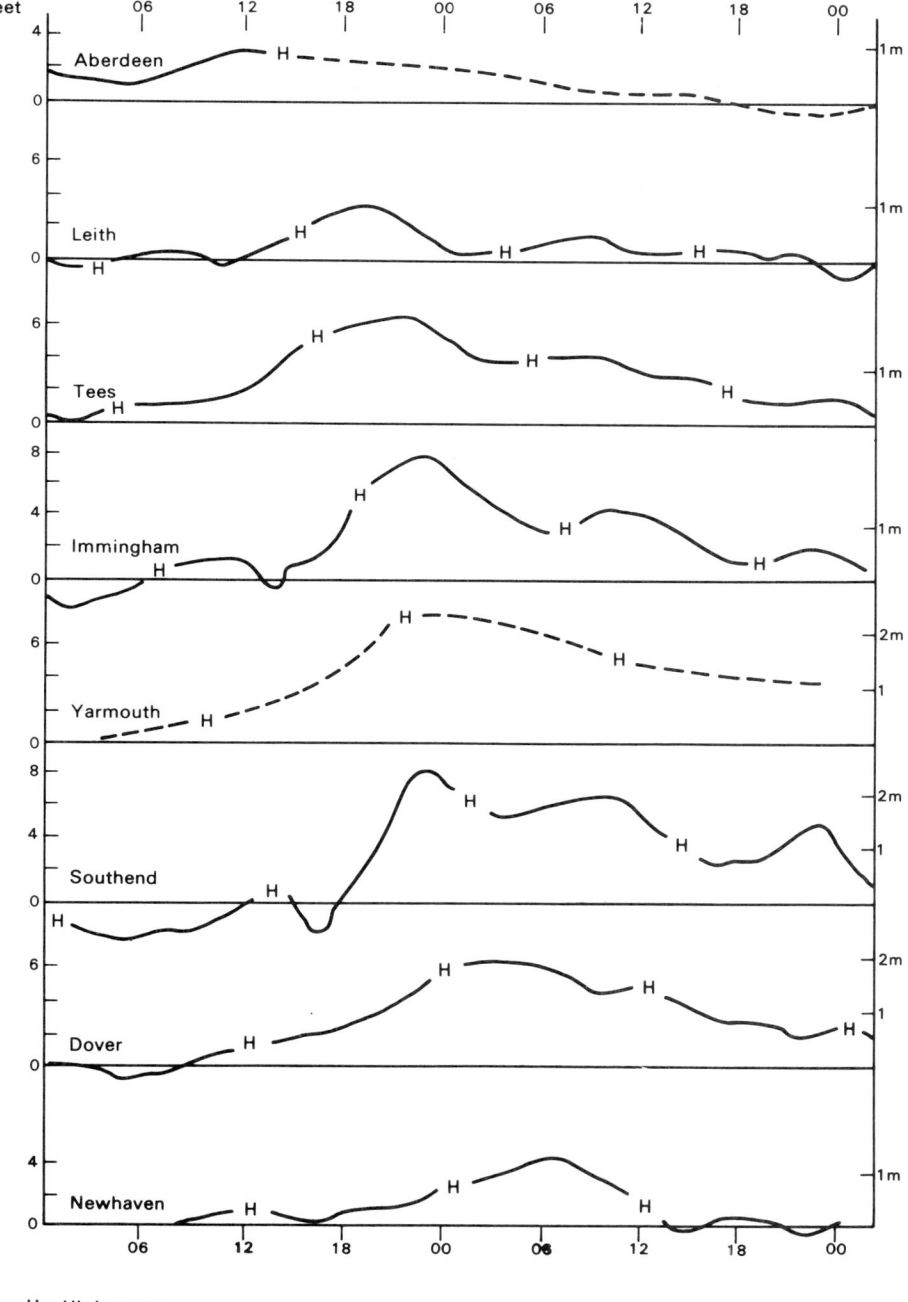

Figure 5.12 The surge of 31 January–1 February 1953, showing the residual values observed after deduction from the predicted tide curves. (*After Robinson, 1953.*)

H – High water

even more severely affected, partly because sea level was also raised by the influence of the wind, which was blowing directly onshore, rather than obliquely offshore as it was in England. At the Hook of Holland the surge height at the predicted time of high water was 3·0 m.

In considering the effects of such a surge on the coastline, particularly where it is vulnerable to erosion, the storm waves which accompanied it must be taken into account. The waves on this occasion were about 6 m high and they did much damage. They were more effective because they were working much higher up the beach owing to the exceptionally high water level. In many areas, therefore, the damage to sea defences was initiated by their over-topping. This allowed the rear to be weakened and they collapsed under the onslaught of the extremely high and destructive waves.

In view of the great damage that surges can inflict their prediction is of considerable value, and various methods have been developed. Heaps (1967) has developed empirical equation for predicting surges at Southend, based on observations at Dunbar, where the surge occurs 9 hours earlier than at Southend. His equation is given by:

$$R_s = R_d + 0.033N - 0.055E - 0.075n - 0.095e$$

R_d is the level of the disturbance at Dunbar at $t - 9$ in feet, N and E are the pressure gradients in the north and east directions at a point between the Hook and Southend at time t, and n and e are the pressure gradients in these directions at a point between north Denmark and Dunbar at $t - 6$. The surge takes 9 hours to travel from Dunbar to Southend. Darbyshire and Darbyshire (1956) have provided equations for estimating the surge height at Lowestoft and Aberdeen:

$$h_L = 0.125(d(p_w - p_b) - p_F)/dt) + 0.125((p_f - p_e) \pm pos (p_a - p_f)) \, t - 6$$

The subscripts refer to the pressure at Wick, Bergen, Aberdeen, Felixstowe, and Esbjerg respectively. The term pos is only included if $p_a > p_f$. This equation has been applied successfully to surges occurring between 1954 and 1957. These methods are empirical, and some partially or entirely theoretical ones have also been developed. Doodson (1956) found that there was no appreciable relation between the tide and the surge, although a study of the Thames estuary showed that the surge is amplified if it occurs on a rising tide.

A numerical method of surge analysis has been carried out by Heaps (1969). He used a computer program to establish sea-level disturbances due to surges in the North Sea. Three models were investigated. In one the Straits of Dover were considered closed, and in the second open. The third model was enlarged to include the deep water area to the northwest of the shelf between Scotland and Norway. Thrée surges were tested on the model. Water depths were averaged for rectangular units of 36·8 km of longitude and 25·8 km of latitude at the northern boundary, and 37·2 km at the southern one. The procedure was iterative at time intervals of 0·1 hours.

The first surge tested was that of 13–15 September 1956. This was an external surge generated by a depression to the north of Scotland, with strong west to northwest winds in its wake. Conditions were fairly calm in the North Sea. The computed sea levels agreed closely with actual ones over most of the area. The external surge was generated mainly on the shelf, as model 3 did not differ greatly from model 2. The surge passed as a free wave south, down the east coast of Britain between 0600 and 1800. It then lost its progressive character, and produced a general rise of sea level, which rotated anti-clockwise around the southern North Sea, decreasing in amplitude as it moved. The water was piled up to the north of the North Sea by strong south to southwest winds off northwest Scotland. This water was then released southwards and travelled along the east coast owing to the Coriolis effect.

The second surge, which occurred on 24–26 February 1958, was an internal one, caused by a depression passing along the English Channel eastwards into the southern North Sea, with strong northeast winds on its northern flank. The surge was generated in the southern North Sea between Holland and Lincolnshire and the Straits of Dover. A negative surge occurred at Esbjerg and Cuxhaven of 1·37 and 1·07 m respectively. Flow through the Straits of Dover significantly lowered the levels at Southend by nearly 1·5 m, as indicated by the results of models 1 and 2 respectively. Model 2 gave a current flowing out of the North Sea at 0·55 m/sec at its maximum. Actual measurements varied from 0·46 to 8·2 m/sec.

The third surge investigated occurred on 15–17 February 1962. It caused serious damage in Hamburg. A depression passed eastwards to the north of Scotland with strong west to northwest winds at its rear, producing a major North Sea surge. The rise in sea level at Southend and Esbjerg was nearly 2·13 m, but at Cuxhaven it was 3·36 m. Part of the surge effect was external, and lowering of pressure was also responsible for a sea-level rise of 30 cm at Esbjerg and 49 cm at Bergen; elevations were smaller elsewhere. West coast surges in Britain are caused by secondary depressions that approach from the southwest or west across Ireland. Winds from the south build up sea level on the eastern side of the Irish Sea. In November 1954 a surge of 1·95 m occurred at Milford Haven and at Avonmouth it was 2·53 m. The critical speed that depressions must move over the continental shelf if a surge is to develop is 74 km/hour.

Surges resulting from tropical hurricanes are among the most intense. They occur especially where wide, shallow shelves are liable to hurricane activity. These conditions occur along the Gulf of Mexico and the Atlantic coasts of the United States. The typical hurricane surge in this area consists of three stages, the forerunner, the hurricane surge, and the resurgences. The forerunner begins several hours before the storm arrives and is a slow change of level over a wide area, as shown by the correspondence between several neighbouring areas. The forerunner occurs as a rise of water level if the longshore movement of the surge is upcoast (to the right along the coast facing towards the land). If the longshore movement is downcoast then there is a fall in water level.

The hurricane surge is a sharp rise in level that occurs when the hurricane centre passes near the point of observation. This phase usually lasts from 2·5–5 hours and water levels can attain heights of 3–4 m above normal. Adjacent areas do not now show similar changes, indicating that the strongest winds in the centre of the hurricane are now responsible. The peak water level occurs to the right of the hurricane track in the northern hemisphere. The resurgences occur after the passage of the hurricane. They may be higher than the main surge if they occur at the time of high water of the normal tide. The resurgences are free waves that are probably generated in the wake of the hurricane and so are somewhat similar to a ship's wake. Their period is that of a free edge wave having a velocity equal to that of the speed of movement of the hurricane. Some of them may be due to the development of an onshore–offshore standing wave on the shelf.

On an irregular coast of bays and estuaries the forecasting of hurricane surges is complicated by the configuration of the coast, as indicated by the work of Wilson (1960), who has studied the prediction of hurricane storm-surges in New York Bay. The normal empirical or semi-empirical methods do not operate successfully in this area because of the local environment and omission of inertial effects. The method he adopted was to correlate observed storm-surges with meteorological parameters of the hurricanes that induced them. The hydrodynamical equations were used, taking into account wind stress and pressure. The Coriolis effect and edge-wave effects were allowed for. The final correlation and prediction equations involved 8 terms. Two hurricanes of 1938 and 1944 and two other storms of 1950 and 1953 were analysed. Data concerning the character of a design hurricane were evaluated for six different storm wind speeds from 37 km/hour to 93 km/hour. The highest surge height was found to occur with the 65 km/hour design storm and this reached a height of 2·7 m. The results confirmed the empirical correlation between height of surge and the central pressure of the hurricane. Predicted storm-surge wave heights are given for a number of places in New York Bay. The maximum surge-height is likely to reach 4·67 m at Sandy Hook and Port Hamilton, exclusive of the astronomical tide.

7.5   Internal waves

Internal waves form at a strong discontinuity between two adjacent but different water-masses. Internal waves can be progressive or standing, and can vary greatly in period, ranging from tidal period to 20 minutes or less. At Mission Beach 50 per cent had a period greater than 7·3 minutes. One of the best-known phenomena associated with internal waves is the so-called dead water that makes movement through the water difficult in some arctic areas. It occurs where cold fresh water overlies dense, warm saline water at a sharp discontinuity. Because the density difference is small, little energy is needed to set up large internal waves. A slow-moving ship can generate large internal waves and hence lose power, but if its speed is high they cannot be generated.

Internal waves can form at a strong thermocline, where a sharp shear zone exists. The waves are not sinusoidal, but they have flat crests and deep narrow troughs. A surface slick is associated with this type of internal wave, due to convergence at the surface as the sharp trough arrives. The slick lies above the steepest gradient, and it moves slowly towards the shore. The relationship is shown in figure 5.13, which indicates the convergence at the slick.

Internal waves seem to be associated with the Atlantic equatorial undercurrent. They are 500 m long and have a period of 17 minutes, moving south south of the equator. Slicks appear when it is calm. Internal waves may be several hundred feet high in deep water. They are normally 6–15 m high at the main thermocline, but the amplitude is negligible at the surface and the bottom. They can be observed with Swallow-type floats, or by measuring changes of density and temperature at depth. Their form reverses near the sea floor. The waves are progressive in the sea, but standing internal waves can occur in lakes. They can be initiated by bottom relief, when this disturbs the stratification of the water. The wave length decreases as they move up a slope. Currents flowing over a ridge cause the thermocline to rise downstream. This process or any one that causes a vertical displacement of the thermocline can create or modify internal waves.

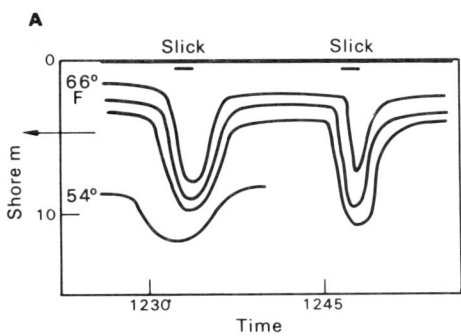
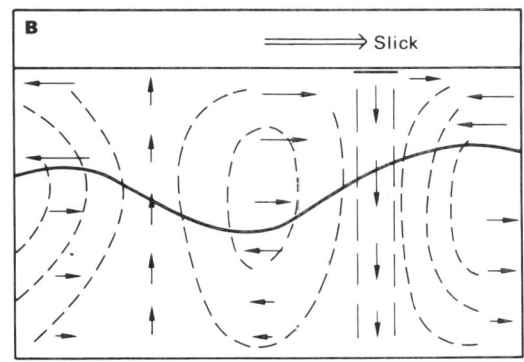

Figure 5.13 A: The isotherms show the pattern of the internal waves and the position of the slicks related to them as a function of time and depth. **B**: The nature of the water movement in internal wave flow is indicated, showing the position of the slick where currents converge. (*After Neumann and Pierson, 1966*.)

Krauss (1969) has discussed current measurements in the Baltic. He pointed out that it is necessary to obtain measurements at several depths in order to elucidate the spectra of internal wave systems. He concluded that in the Baltic inertial waves and internal waves are the most important features with periods between 7 hours and 40 hours.

Further reading

BRETSCHNEIDER, C. L. 1965: The generation of waves by wind. State of the art. *Nat. Eng. Sci. Co. Off. Nav. Res.* **SN** 134–6, 96 pp.

CARTWRIGHT, D. E. 1967: Modern studies of wind generated waves. *Contemp. Phys.* **8**, 171–83.

CARTWRIGHT, D. E. 1968: A unified analysis of tides and surges round north and east Britain. *Phil. Trans. Roy. Soc.* **A 263**, 1–55.

COASTAL ENGINEERING RESEARCH CENTER (CERC). 1966: Shore protection, planning and design. 3rd edition. *Tech. Rep.* **4**, 50–114.

DRAPER, L. 1967b: Wave activity at the sea bed around northwestern Europe. *Mar. Geol.* **5**, 133–40.

HEAPS, N. S. 1967: Storm surges. *Oceanog. and Mar. Biol. Ann. Rev.* **5**, 11–47.

HINDE, B. J. and GAUNT, D. I. 1967: Microseisms. *Contemp. Phys.* **8**, 267–83.

KRAUSS, W. 1969: Typical features of internal wave spectra. In M. Sears (editor), *Progress in oceanography* **V**. Oxford: Pergamon, 95–102.

MILES, J. W. 1967: On the generation of surface waves by shear flows. Part V. *J. Fluid Mech.* **30**, 163–75.

PHILLIPS, O. M. 1966: *The dynamics of the upper ocean*. Cambridge University Press.

RUSSELL, R. C. H. and MACMILLAN, D. H. 1952: *Waves and tides*. London: Hutchinson.

SNODGRASS, F. E., GROVES, G. W., HASSELMAN, K. F., MILLER, G. R., MUNK, W. H., and POWERS, W. H. 1966: Propagation of ocean swell across the Pacific. *Phil. Trans. Roy. Soc.* **A 295**, 431–97.

TRICKER, R. A. R. 1964: *Bores, breakers, waves and wakes.* London: Mills and Boon.

WILLIAMS, L. C. 1969: CERC wave gages. *CERC Tech Memo.* **30.** 117 pp.

WILSON, B. W. and TØRUM, A. 1968: The tsunami of the Alaskan earthquake, 1964: engineering evaluation *CERC Tech. Memo.* **25.** 401 pp.

6 Biological productivity in the oceans

Plants are the essential basis of life both on the earth and in the oceans. They alone can synthesize living matter from the chemical nutrients in sea water. The process is achieved with the aid of light, derived from the sun, by photosynthesis. Animals can then live on the plants, and flourish in the ocean in an infinite variety of forms, each adapted to deal with the special conditions of its environment. Life exists in the ocean from the intertidal foreshore right down the greatest depths of the ocean basins. Each environment poses its own problems to its inhabitants.

The sun provides the energy by which the nutrients are organized into living matter, on which all the other marine organisms depend more or less directly, according to their feeding habits. The most important zone from the point of view of marine plants is, therefore, the layer into which the sun's rays can penetrate. This is very shallow compared with the total depth of the sea. Water absorbs light much more rapidly than air, and at least 10 per cent is lost by reflection from the surface. The clearness of the water greatly affects the penetration of light, resulting in large variation in the depth at which plants can grow. In the clearer seas, such as the Caribbean, it may be 110 m, while on continental shelves in temperate seas it may be reduced to less than 40 m. Near the coast it may fall as low as 15 m or less, depending on the amount of matter in suspension.

The lower limit at which planktonic plants can grow in the ocean is the depth to which only 1 per cent of the light penetrates. Seaweeds growing on the bottom can extend a little further to depths where the light intensity is less than 0·3 per cent of the surface value. In the clear waters of the Mediterranean this depth may be 160 m. Fish cannot see below 500 m and it appears quite dark below 1000 m, although even at this depth and below fish can live. Another essential feature for the production of marine plants is an available supply of nutrient materials in the water. Both light and nutrients are essential to marine life.

1   Fertility of the sea

The fertility of the sea is determined by oceanic circulation, biological processes of uptake and mineralization, settling of organic debris and regeneration of nutrients, migration of animals, and the nutrient supply from the land (Postma, 1971). There is usually an inverse relationship between light and nutrients, except in the upwelling areas of the tropics (for example, along the equator and the eastern side of oceans). Below 1000 m the distribution of dissolved organic matter is uniform. The rate of mineralization of phosphates and nitrates decreases exponentially from the surface to 1000 m. The rate of water renewal is more rapid in the bottom water. These variables lead to layering, with phosphate and nitrate maximums in the intermediate water, where there is an oxygen minimum in excess of the phosphate and nitrate maximums, especially in the tropics and subtropics. The deep water of the Pacific and Indian Oceans is three times higher in nutrients than the Atlantic, where the type of circulation reduces the nutrient concentration. The circulation of the north Pacific Ocean forms a trap for nutrients.

The pattern of nutrient circulation, shown in figure 6.1, is based on a number of considerations. 1) The bottom water is formed in the Antarctic at $0·8 \times 10^{15}$ m³/year, and phosphate in this water is $1·6$ μg-at/l. Bottom water occupies one-third of each ocean. 2) Deep water forms in the north Atlantic at $0·2 \times 10^{15}$ m³/year, with $0·8$ μg-at/l, and is carried through reservoir VI as north Atlantic deep water. 3) Phosphate is carried into the warm surface layers IV, V, VII, and VIII. The amount is calculated by assuming the evaporation rate to be 40 cm/year, and the salinity difference of 35·3‰/34·6‰. The phosphate difference is 1 μg-at/l; net upward transport would work out as 20 μg-at/lm²/year. The resulting total over all the warm layer would be 5×10^{12} g-atP/year. 4) The amount of phosphate liberated from the deep water in III and VI with a volume of 10^{18} m³ can be derived from oxygen consumption in abyssal water. The phosphate supply to abyssal water is $0·5 \times 10^{12}$ g-atP/year; 5×10^{12} g-atP/year return to the warm layer, the difference being liberated from sinking organisms in zones IV and VII is $4·5 \times 10^{12}$ g-atP/year. 5) The assumed phosphate concentration is as follows in the different waters: warm layer 0·5 μg-at/m³, intermediate water at 40°s 1·7 μg-at/m³, bottom water at 40°s 2·25 μg-at/m³, Pacific deep water returning to the Antarctic 2·5 μg-at/m³, and Atlantic deep water in corresponding position 2·0 μg-at/m³.

The warm surface layer receives 5×10^{12} g-at/year phosphate or 20 mg-at/m²/year. The primary production in the tropics and subtropics varies between 150 gC/m²/year in upwelling areas and 30 gC/m²/year in the subtropical gyres, the average being 60. This corresponds with 36 mg-at/m³/year phosphate and is twice the amount supplied from deeper layers. Fifty-five per cent of the phosphate used annually is supplied by vertical

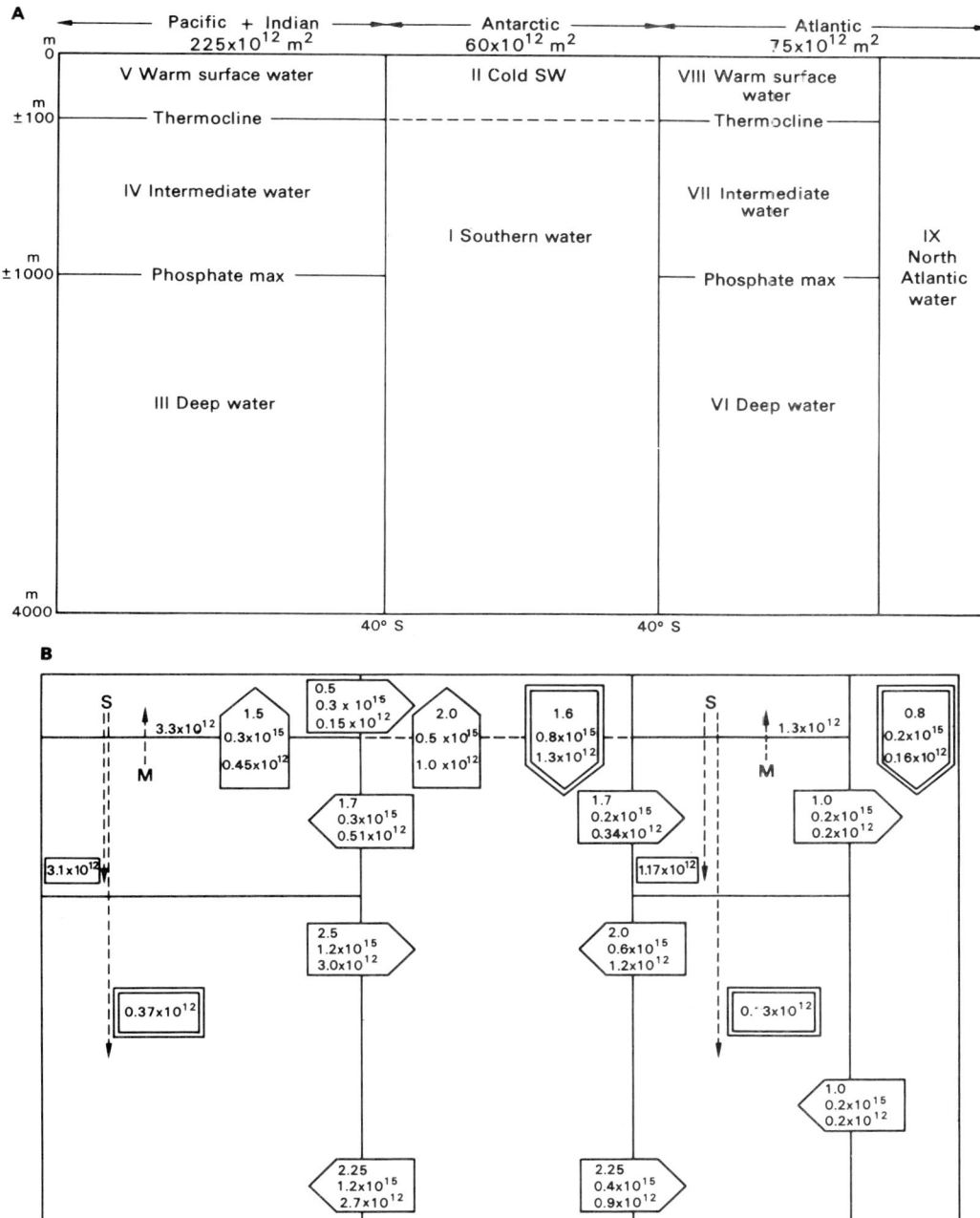

Figure 6.1 A: Simplified model for calculation of net water and phosphate transport in the ocean. **B**: Net transport of water and phosphate between reservoirs. Phosphate concentration is given by the arrows in μg−at/l at the top, centre value is net amount of water in m^3/yr, and lower value is net amount of phosphate transferred in g−at/year. S is transport by sinking of organic matter, M is transport by mixing. A steady state is assumed. (*After Postma, 1971.*)

transport. Each molecule must be used twice. The value is 75 per cent in temperate areas, which are only stable for part of the year. Vertical movement, causing mixing and upwelling, occurs over a length of 19,200 km and width of 50 km at an average velocity of 50 m/month. The volume of ascending water is 0.6×10^{15} m³/year. This would supply phosphate at 0.6×10^{12} g-at/year in the eastern boundary current zone. Twice this is supplied in the equatorial region, amounting to 1.2 g-at/year. The total supply is 5×10^{12} g-at/year, so that 30 per cent is supplied by upwelling. Vertical mixing supplies 3.2×10^{12}

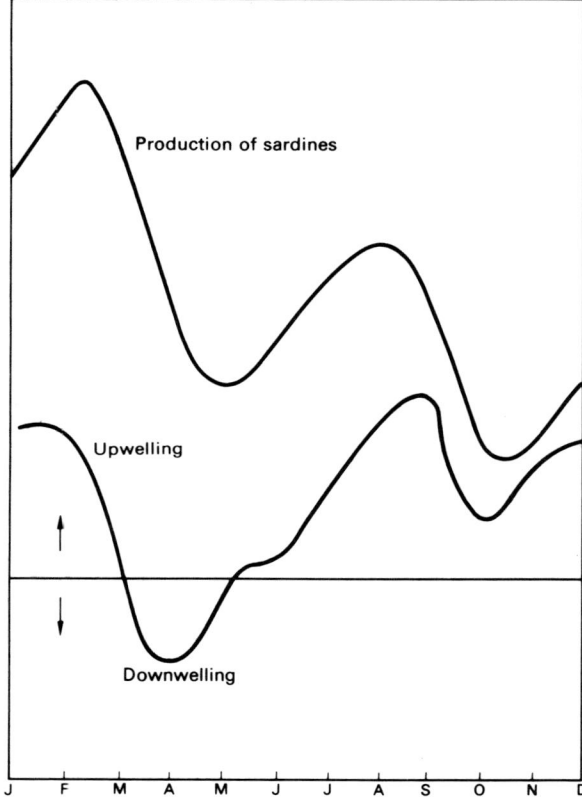

Figure 6.2 Production of sardines related to season and degree of upwelling, off Brazil. (*After da Silva da Castro Moreira, 1971.*)

g-at/year or 14 mg-at/m²/year. Recycling of 50 per cent allows 38 gC/m²/year productivity, corresponding to the production in areas beyond the zones of upwelling.

Ferguson Wood (1971) suggests that phosphate and nitrate are seldom, if ever, the limiting factor in the growth of phytoplankton in the Antarctic. He comes to this conclusion by means of multiple regression and recurrent group analysis, which show that the phytoplankton are closely associated with sigma$_t$ curves, and are thus related to the character of the water-mass.

A detailed study of the upwelling that takes place off south Brazil by Da Silva (1971) shows the variability with the seasons and is illustrated in figure 6.2. The upwelling in this area supports sardines. The pattern of upwelling is closely paralleled by the production of sardines and is related to wind variations. Upwelling and downwelling follow changes in the wind with about a one-week lag. The cycle of wind change is fast in the second half of the year, and slower in the first half. East and northeast winds dominate in the summer and autumn, and southwest and west winds in winter and spring. Upwelling is dominant with winds between northeast and northwest, and downwelling with those with a southerly component. The maximum upwelling occurs in January and February, with a secondary peak in September.

The Peruvian coastal waters are among the most fertile in the world, so that a study

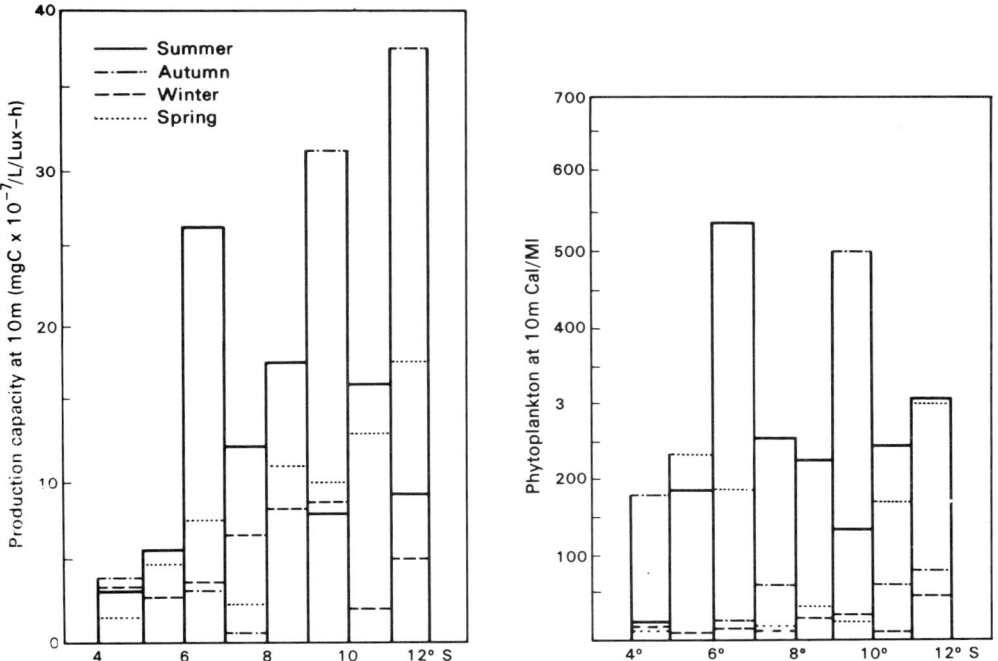

Figure 6.3 Production capacity at different latitudes and in different seasons off Peru. The right-hand graph refers to phytoplankton at 10 m. (*After Guillen* et al.,*1971.*)

of their primary productivity and its variations is of interest and has been reported by Guillen, de Mendiola, and Rouden (1971) and Guillen (1971). There is considerable variation in productivity both with season and with position along the coast (figure 6.3). These variations are related to changes in temperature and salinity, which also vary seasonally as shown for 1964 at a depth of 10 m. Productivity reached very high values close to the coast south of 6°s in summer. An area of great productivity occurred far from the coast owing to an anticyclonic eddy, with low productivity nearer the coast, northwest of 82°w. Productivity was lowest in winter south of 6°s despite intense upwelling due to lack of light. Spring was a more productive season in this area owing to the greater light intensity. A positive correlation, with r = 0·82, was found between production capacity (X) and primary production (Y), the regression equation being Y = 1·9706 + 5·2576X.

The correlation between Z, the total primary production and X was higher, with $r = 0.87$, and an equation $Z = 0.2062 + 0.0043X$. The phytoplankton was most abundant in the Peruvian coastal current with salinities of 35·1–34·8‰. The effect of the 'El Niño' incursion of 1965 is discussed by Guillen (1971). In this year the equatorial water extended as a tongue 30 m deep as far south as Supe at 11°s, with a temperature of 24–27°C and salinity of 33·8–34·8‰. South of 11°s the temperature decreased and salinity increased to 16–22°C and 35·1‰ in the Peru coastal current. Phosphate was low in the equatorial water, and

	Summer	Autumn	Winter	Spring
T°C	26–17	24–15	17–14	22–14
S‰	35·2–34·9	35·3–34·3	35·2–34·7	35·3–33·4

productivity was 0·05 gC/m²/day, compared with 1·56 gC/m²/day in the most productive area off Salavery at 8°s, where the Anchoveta were concentrated. The greatest upwelling took place in Pisco Bay, where the productivity was 0·80 gC/m²/day.

A great variety of oceanic water motions are important in increasing the fertility of the sea. Lafond and Lafond (1971) enumerate some of these and other important variables. Fertility depends on the biological climate, nutritional factors, and physical factors, including light, topography, temperature, pressure, salinity, and water motion. Under the last factor are included river runoff, turbulence and eddies, convection currents, wind and wave mixing, cascading and capsizing, tidal circulation, upwelling, internal waves, and shoal effects. The term 'cascading' refers to the sinking of cold water, and 'capsizing' is the return of lower water to the surface to replace sinking water. The latter process increases fertility. Islands and promontories produce eddies and mixing. Slicks occur above a descending thermocline. The boundary zones of major currents create relative motions and eddies, which are favourable to life. Internal waves cause vertical oscillations that can be beneficial. The microstructure layering of the water provides a variety of conditions, some of which may be optimal.

The flora and fauna of the sea are immensely varied although they represent only 16 per cent of all species in the world. There are over 1 million species in the world, of which about 160,000 live in the sea. The virtual lack of insects, which make up 75 per cent of land species, accounts for the relatively low percentage of marine organisms. Of the non-insect species, 65 per cent live in the sea, and of these 2 per cent (or 3000 species) are free-living in the water, while 98 per cent (157,000) live on the sea bed (Thorson, 1971).

At the lowest level of life in the oceans are the bacteria, which inhabit all depths and latitudes, ranging in density from less than 10 to more than 1 million in each ml. Below a depth of 1000 m there are only a few in each litre, the average for the open ocean being 10/ml, with a total crop of 2700 million metric tons. These could double every 1·25–8·3 days, according to the species. They perform a very important function in the chain of life in the ocean by decomposing dead matter into plant nutrients, and they form bacterial cells that can be assimilated by protozoa upwards.

The phytoplankton consist of flagellates, which are the smallest members and which can live between 500 and 1000 m depth, and the more important members, the diatoms and dinoflagellates. The latter can thrive with less light and nutrients, but they produce cellulose which some animals cannot digest, and some of them make the red tide. These are *Gonyaulax* and *Gymnodinium*, whose concentration can rise to 200,000–500,000/l or even up to 6 million/l at times.

2 The basis of marine life

2.1 Phytoplankton

The term 'plankton', of Greek origin, can best be translated as 'that which is made to wander or drift', according to Hardy (1956). The planktonic plants, or phytoplankton, and the zooplankton, are carried passively by the currents, while the 'nekton', which means 'swimming' in Greek, are those organisms strong enough to swim where they please. Even the free-swimming fish pass their earliest life phases in the plankton in the form of eggs, or larvae too small to move against the currents.

The phytoplankton are minute plants, drifting at the mercy of the surface currents, on which all the other marine creatures depend for their living. Because they have no means of self-propulsion, they must be able to float in the upper layers of the ocean, where they obtain the life-giving light. In order to float, phytoplankton use the principle that the smaller an object, the larger is its surface area in relation to its volume. The volume increases by the cube of the linear measure, while the surface area does so only by the square. The greater surface area will increase the friction against the water and help to keep them afloat. For this reason most of the phytoplankton are extremely small. Their small size has another advantage, as well as helping to keep them afloat. They absorb their mineral nourishment through their surface, so that it is an advantage to have this area large, relative to volume.

Where conditions are favourable the production of phytoplankton is very great, but each unit is very small. Temperature, salinity, and light are all important, and the supply of nutrients is essential. Temperature affects their rate of growth and reproduction, which in general declines as the temperature falls. A higher temperature, in reducing the viscosity and density of the water, makes it more difficult for the plankton to keep afloat in the upper layers. Salinity has the inverse effect, a reduction leading to a decrease in the density of the water.

The reproduction rate of phytoplankton depends to a considerable extent on the availability of nutrients, which in turn depend on a number of different factors. The nutrient elements in sea water are in weak concentration, but nevertheless they must be present to allow plant growth. According to Lee (1958), the main elements are as follows:

Table 6.1 Nutrient elements

Element	Part per million by weight
Phosphorus	0·001–0·10
Nitrogen (dissolved gas not included)	0·01 –0·70
Silicon	0·02 –4·00
Copper	0·001–0·01
Iron	0·002–0·02

In order to be of use to the phytoplankton these nutrients must be in the uppermost layers of water, where there is also light. For this reason the water will be most fertile in those zones where the nutrients are being continuously replenished from the supplies at depth, where they cannot be used directly. Once the surface supply has been used by the plants, the sea will lose its fertility unless the nutrients are replaced.

There are marine deserts and areas of great fertility. The former are areas where the nutrients, once used, are not replaced, while the latter are those areas where processes of heating, cooling, and wind action allow the renewal of water on the surface from below. Nutrients are replaced at depth by the slow decomposition of dead organisms by bacteria.

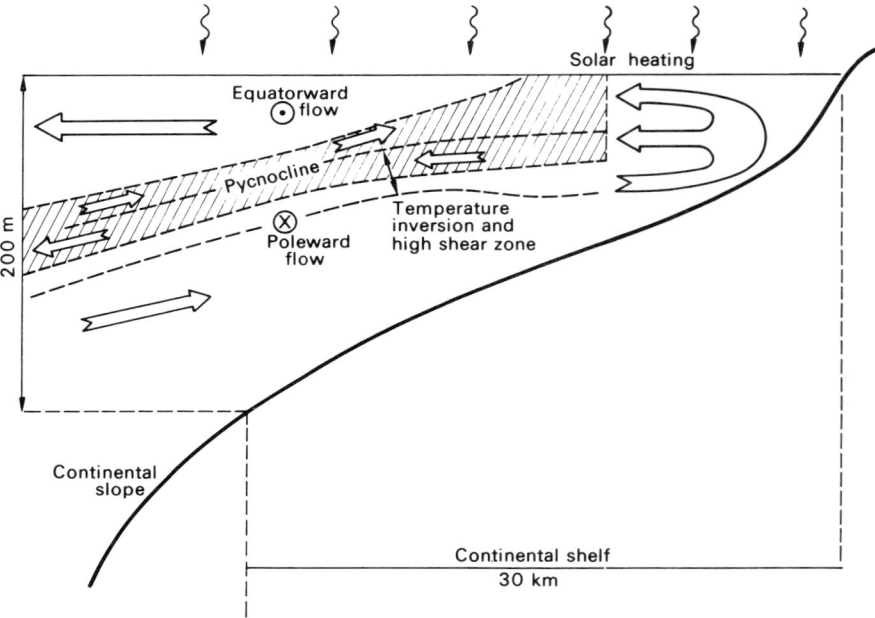

Figure 6.4 Diagram to illustrate upwelling off the coast of Oregon. (*After Smith* et al., *1971.*)

Phosphates and nitrates are essential plant foods. Any process that can stir up the water is important because it brings nutrient-rich water to the surface. Temperature affects the nutrient supply, through warming, rendering the water less dense and making stratification more stable. Cooling, by producing denser water, may allow surface water to sink; it will then be replaced by water from below. Wind can stir up the water and cause nutrients from shallow depths to be brought nearer the surface. This effect can be important in very shallow seas (like the North Sea), where much material is in suspension so that light penetration is very limited.

Some of the most fertile areas of the sea are those where the wind blows alongshore or offshore for most of the year. These zones are regions of upwelling and the water that is brought to the surface from moderate depths is very rich in nutrients. The process is self-generating as the large amount of life in these areas sinks when it dies to make more nutrients, which will again be brought to the surface by upwelling. This process goes on most effectively in those areas in mid-latitudes on the west sides of continents, such as California, the coasts of Chile and Peru, southwest and northwest Africa, parts of north-west Australia, the coast of Somaliland, and southern Arabia (figure 6.4). Water is brought to the surface in zones of divergence, such as the equatorial divergence. Another zone of fertility is associated with the formation of the Antarctic circumpolar water-mass, which is formed by upwelling of large quantities of water in the region to the south of the Antarctic convergence. Zones of continuous upwelling are permanently fertile and produce a fairly steady crop of phytoplankton, variations depending largely on changes of temperature that control the growth rate. In some areas phytoplankton activity shows a strong seasonal rhythm.

The waters of the North Sea and other temperate seas illustrate the pattern. The production of phytoplankton is very low during the winter months from October to February, and the numbers of plants in the water is low. There is, however, normally a very sudden upsurge of activity in March (known as the spring flowering) when the rate of reproduction reaches extremely high values. This phase of vigorous activity does not last for long. The rate of increase of phytoplankton is such that in a week they may increase one hundred-fold, while in two weeks they may have multiplied ten hundred-fold. By April, however, the peak has passed and the numbers decline as rapidly as they increased (due to grazing by the zooplankton), till in May they are down to the low winter level. There is a second smaller upsurge (the autumn flowering), which usually takes place during late August and the first part of September. The cycle and some of the factors on which it depends is shown in figure 6.5. In winter the waters of the shallow North Sea are well stirred by winds and tidal currents and nutrients are well distributed throughout the water. The weak light and dirty water, however, prevent much planktonic activity, while the lower temperature also slows down reproduction. In spring the sea warms up slightly and the sun becomes more powerful and the process of reproduction can start. It progresses rapidly in the nutrient-rich waters, until all the upper layer of nutrients have been used up. By this time in early summer the upper layers of water are getting warmer, and the water stratification is becoming stable, so that the deeper nutrients can no longer reach the surface. At the same time the zooplankton graze down the standing crop of phytoplankton. Towards the end of the summer the surface waters begin to cool, while the strong winds of the equinoctial period stir up the water and allow the nutrients to reach the surface once more. The lesser light intensity of this part of the year means that reproduction does not reach the high level of the spring. As winter approaches the light decreases to a limit below which the activity is again reduced to a low level. The nutrients accumulate ready for the spring flowering of the following season.

The effect of surface heating is even more marked in the tropics, as there it is a permanent state. This has led to the belief that tropical waters are relatively poor in phytoplankton compared with those of higher latitudes (except in the zones of upwelling). It has been suggested, however, that the greater depth of light penetration and higher temperatures of the tropics may render production there almost as efficient as some areas in high latitudes (Graham, 1956).

One of the areas where the nutrients are particularly rich is the southern North Sea off the Thames estuary, where the sewage of London reaches the sea. These very rich waters provide an extra 2900 tons of phosphorus a year, leading to a fish catch 25 times

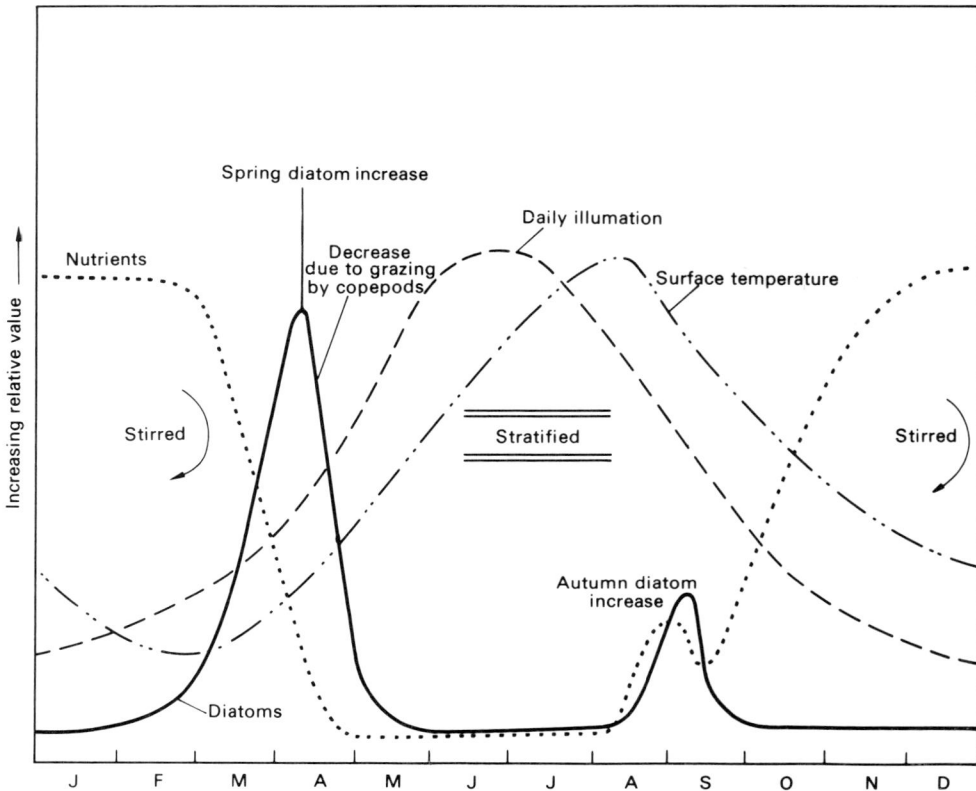

Figure 6.5 The yearly cycle of diatom production in the North Sea and some factors to which it is related. (*After Lee, 1958.*)

that in the Baltic and double that of the rest of the North Sea. This region may be contrasted with the oceanic desert of the Sargasso Sea, where the water remains stratified throughout the year and nutrients cannot be replaced.

Sometimes phytoplankton is so dense in the water that it appears slimy, or takes on a colour related to that of the organisms within it. The Red Sea for instance, gets its name from the colour of the water when a particular species is exceptionally abundant. The very rapid reproduction characteristic of phytoplankton under favourable conditions can be important environmentally. Increase can go on as long as nutrients are available but stops suddenly when they are exhausted. Such sudden increases in numbers are called 'blooms' and give rise to the so-called red tides that gave the Red Sea its name, due to great increases of *Trichodesmium erythraeum*. The red tides often have toxic effects. They are associated usually with dinoflagellates and the toxic effect may kill fish because of depletion of the oxygen supplies.

Many of the phytoplankton are diatoms, and these sometimes impart a brown-green colour to the water. Some of the smaller members of the phytoplankton are the coccolithophores, whose protective plates (coccoliths) form a constituent of some deep sea oozes.

Others are called dinoflagellates because they have tiny whip-like organs to help them keep afloat. These are very small; when magnified 1500 times they are still less than 1·25 cm long. Sometimes the coccolithophores are so numerous that they give a milky appearance to the sea. Such 'white water' is said to be an indication of the presence of herrings.

It is extremely difficult to get an accurate idea of the number of phytoplankton in the water, but an estimate by Johnstone, Scott, and Chadwick (1924) suggests that in an average April in the Irish Sea there are about 727,000/m³; this may be compared with the number of zooplankton in the same volume, which was 4500. The figures show how much smaller the phytoplankton are than the zooplankton. The total amount of primary food produced by phytoplankton might be expected to exceed that produced on land by plants. In fact, estimates suggest that the total production of all the sea amounts to about

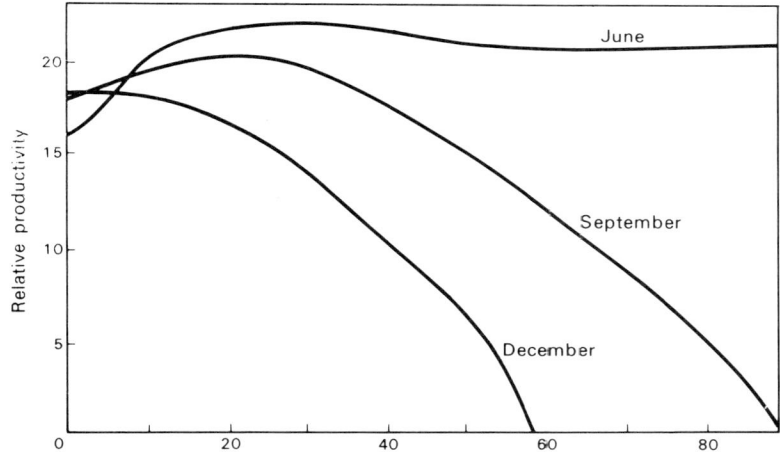

Figure 6.6 Relative photosynthesis as a function of mean radiation. Relative productivity is shown against north latitude. (*After Ryther, 1969.*)

1·5 × 10¹⁰ tons of carbon annually; other values given are 1·9 × 10¹⁰ tons and up to 15 × 10¹⁰, although the lower values appear more reasonable. The estimate for land production is about 2·5 × 10¹⁰ tons annually. Daily production off Plymouth has been estimated at 0·4–0·5 g/m² for phytoplankton and 0·15 for zooplankton, with a standing crop below each m² of 0·4 for phytoplankton and 0·5 g/m² for zooplankton. On the whole, the larger sea area and greater depth of production in the ocean does not compensate for the more efficient processes on land. The phytoplankton, in using the radiant energy of sunlight, does not do so very efficiently. It has been calculated that much less than 1 per cent of the sun's energy is used for plant production in the sea. Carbon is produced by photosynthesis from the carbon dioxide obtained by solution from the air, from carbon salts in solution, and from respiration. Geographical variations in the environment control the efficiency of photosynthesis and plant growth. Production is affected by the incident radiation derived from the sun. The variation is great in high latitudes and has little limiting capacity in low latitudes (see figure 6.6). Transparency of the water affects the depth at which photosynthesis can take place. This depth is called the compensation depth, and it occurs where photosynthesis equals respiration. The zone above it is called the euphotic zone. This zone is typically 50–25 m deep in the more turbid middle and high latitudes, but it is 100–200 m in the clearer tropical and subtropical seas. The relation between productivity and the depth of the euphotic layer is shown in figure 6.7. It applies where the euphotic layer is deep owing to the paucity of organisms. Nutrients are another essential requirement and are much less abundant in the sea than on land. A good, rich soil 1 m deep can produce 50 kg of dry organic matter. The richest sea can only support 5 g of dry organic matter in 1 m³ of water. Figure 6.8 shows the phytoplankton that can

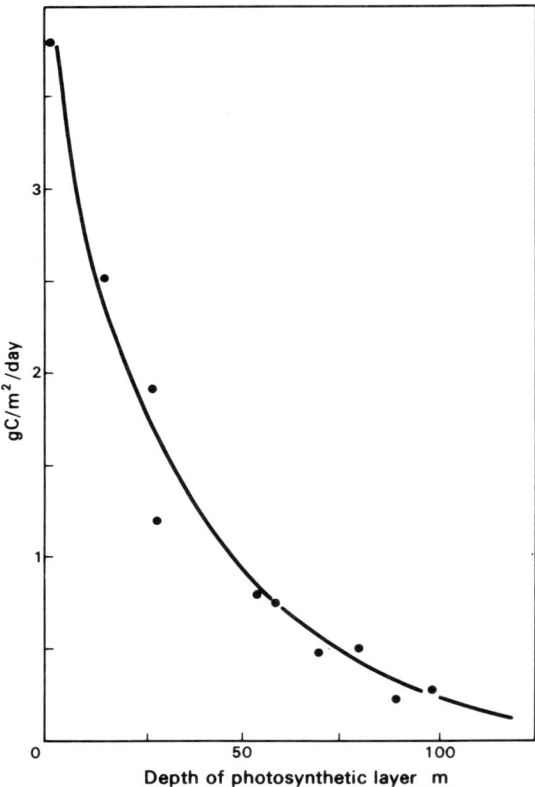

Figure 6.7 Productivity in relation to the depth of the photosynthetic layer. (*After Ryther, 1969.*)

develop at different depths under favourable conditions. The total is less than 1/1000 of the standing crop on land. Temporary rates of production may equal those on land, but they only occur for a day or two, rather than for years as on land. Local turbulence can provide nutrients in some areas for considerable periods, but at the same time this cuts down the available light by rendering the water turbid. Fertility based only on nutrient supply would be expected to be low and constant in the tropics, low in summer but higher in winter at 40° and considerably higher all through the year at 60°. Some measured values of productivity range from 0·89 in April to 0·10–0·20 gC/m²/day when thermal stability returns in the Sargasso Sea. Elsewhere values were as low as 0·05 in the Sargasso Sea. An annual rate of 72 gC/m² was measured—twice that expected in the

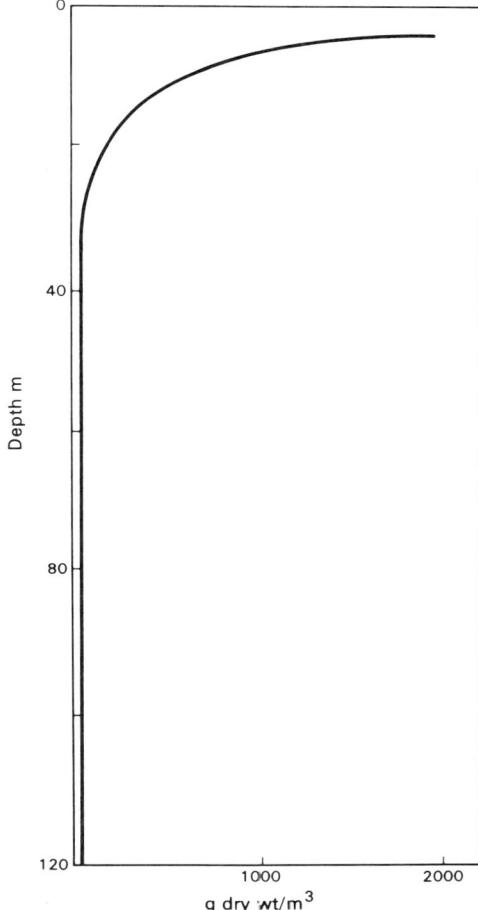

Figure 6.8 Phytoplankton population in relation to depth.

tropics. In temperate and subpolar areas an annual production of 120 gC/m² was recorded, twice the Sargasso Sea and four times the tropical production. In the Arctic annual values less than 1 gC/m² were found, while in the Antarctic the value is around 100 gC/m² and more in certain fertile areas. The annual production in the Fladen Ground on the North Sea was 68 gC/m², with a total range from 45 to 110 in the North Sea. Water circulation is responsible for the contrasts, and circulation has a very important effect on productivity.

2.2   Algae; benthic marine plants; seaweeds

One of the areas of densest benthic plant life in the oceans is the kelp beds off California. They probably provide several kg/m² compared with the maximum of 25 g for phytoplankton. They grow at a rate of 33 cG/m²/day, producing a crop of *Laminaria* of 4400 g/m². This is 10 times the phytoplankton annual production and is produced in two months. The algal members of coral reefs in the Pacific produce 12 gC/m²/day, and about the same is produced by turtle grass (*Thalassia*) in Florida. The net coral reef production by plants is 2000 gC/m²/year. Coral algae can achieve a high rate of production because they are not confined to one patch of water as are phytoplankton. Benthic plants produce a large proportion of marine plant production considering their small areal extent. It has been calculated that the large kelps off California produce 1/10 of the whole phytoplankton production in only 0·1 per cent or 1/1000 of the ocean area.

Off Nova Scotia 1000 ha of sea contains 80,000 tons of seaweed at 23 tons/ha in places. Several crops could be grown each year. Japan has experimental seaweed cultivation projects which produce up to 6 crops a year. Commercially important seaweeds include *Laminaria* and *Ascophyllum nodosum*, a fucoid rock weed. The former grows up to 3·7 m long, consisting of a round stipe that branches into a large deeply divided frond. The latter is brown-yellow and grows up to 1·2 m long with a short stipe, which expands into a cylindrical frond branching irregularly and containing air vesicles. These all belong to the brown seaweeds of algal class *Phaeophyceae*. The fucoid weeds grow around mid-tide level and the *Laminaria* from low water neap-tide to a depth of 18 m or 27 m according to

Table 6 Seaweed around Scotland

	Fucoid algae			Laminaria
	Tons	Hectares	Density, tons/ha	Tons
Outer Hebrides	125,136	1415	88·5	700,000
Inner Hebrides	8,263	113	73·0	—
Orkney	38,774	735	52·7	1,200,000
Mainland, W and NW	8,540	144	58·5	930,000

species, and they require rocky shores. They are only of harvestable density where they grow at more than 4 tons/ha for the fucoids and 2 tons/ha for the *Laminaria*, which must be collected from the beach, while the fucoids can be collected *in situ*. Around Britain they are mainly restricted in workable densities to western Ireland and Scotland. The standing crop in Scotland is given in table 6.2 (Boney, 1965). Seaweed forms a valuable fertilizer, adding organic matter, minerals, and calcium to the soil, although frequent applications are required because it decays quickly.

2.3 Zooplankton
Zooplankton depends for its food on the phytoplankton. The little creatures, although they are small and still float passively with the currents, are very varied in character (Plate 6). There are tiny jelly-fish, arrow worms, small crustacea or copepods (*Calanus*) which belongs to the latter group, being particularly important as it forms the main food of herring. Some members of plankton are only temporary inhabitants, they are the young and larval stages of various other creatures which will eventually be able to live a more independent life in their adult form.

Different planktonic animals require different conditions of temperature and salinity. At times, therefore, they are indicative of the water-mass in which conditions are optimum for their life. The water-masses of the western approaches to Britain can be differentiated

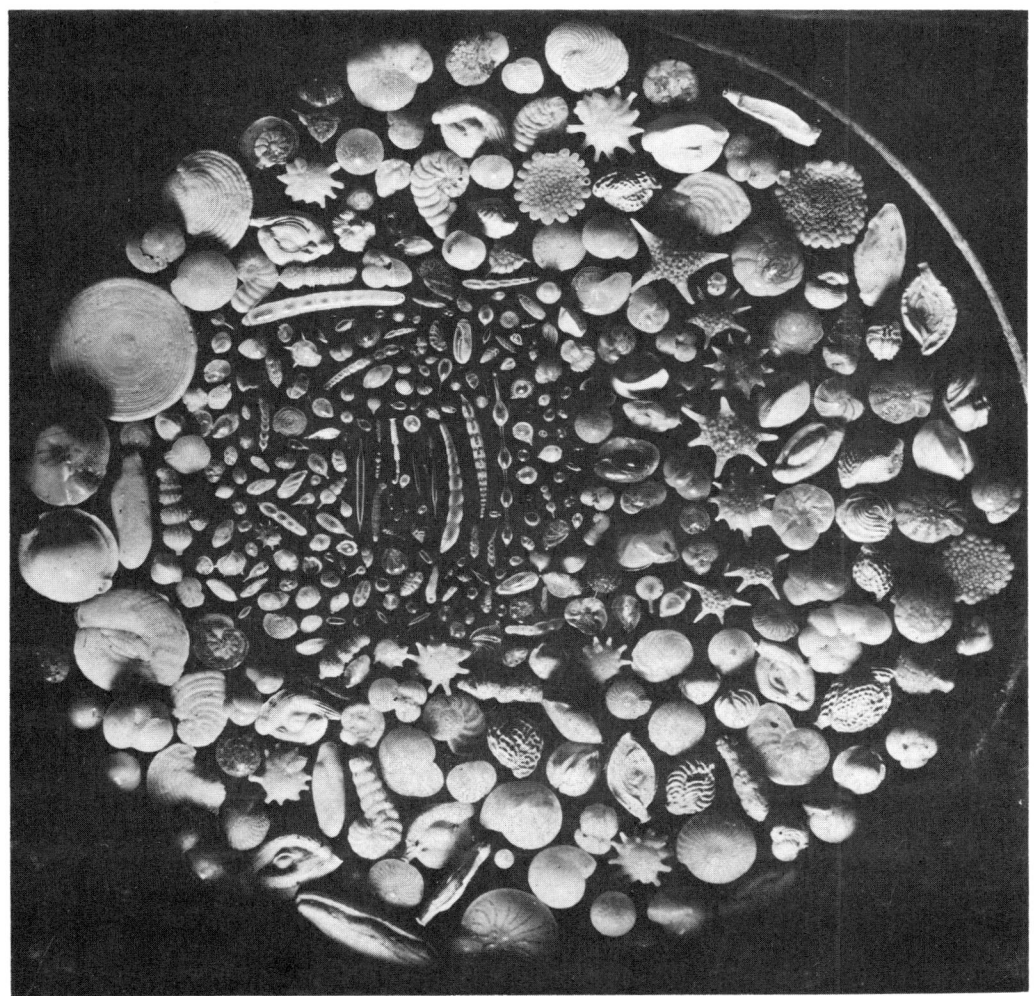

Plate 6 Zooplankton: a wide variety of different types are illustrated (*Natural History Museum*).

by their planktonic fauna. Arms of Atlantic water extend eastwards into the English Channel and their extent at any time can be determined both by the colour of the water, and by the characteristic arrow worm which lives in the planktonic community. The channel water and the North Sea water is green and is characterized by *Sagitta setosa*. Atlantic water to the west and north is deep blue and is the home of *Sagitta elegans*. The boundary between these two water-masses is sometimes very sharp.

Other interesting water movements have been associated with planktonic fauna of a special type. Thus Mediterranean water, flowing out at the bottom of the Straits of Gibraltar, turns in part north and gradually approaches the surface, coming to the upper levels along the western continental shelf off western Britain. This water brings with it planktonic fauna typical of far more southern latitudes. The northward extent of this southern water varies from year to year, but it may stretch as far as the Shetland Islands, where it reaches the surface. The distribution of these water types is shown in figure 6.9.

Some members of the zooplankton have developed a habit of vertical migration. They sink down to the lower layers during the day and climb to the surface during the night, sometimes migrating through a depth of more than 100 m. This movement is illustrated in figure 6.10. One reason for the movement might be related to the behaviour of the species with regard to light. Each species may have an optimum light intensity, which it follows

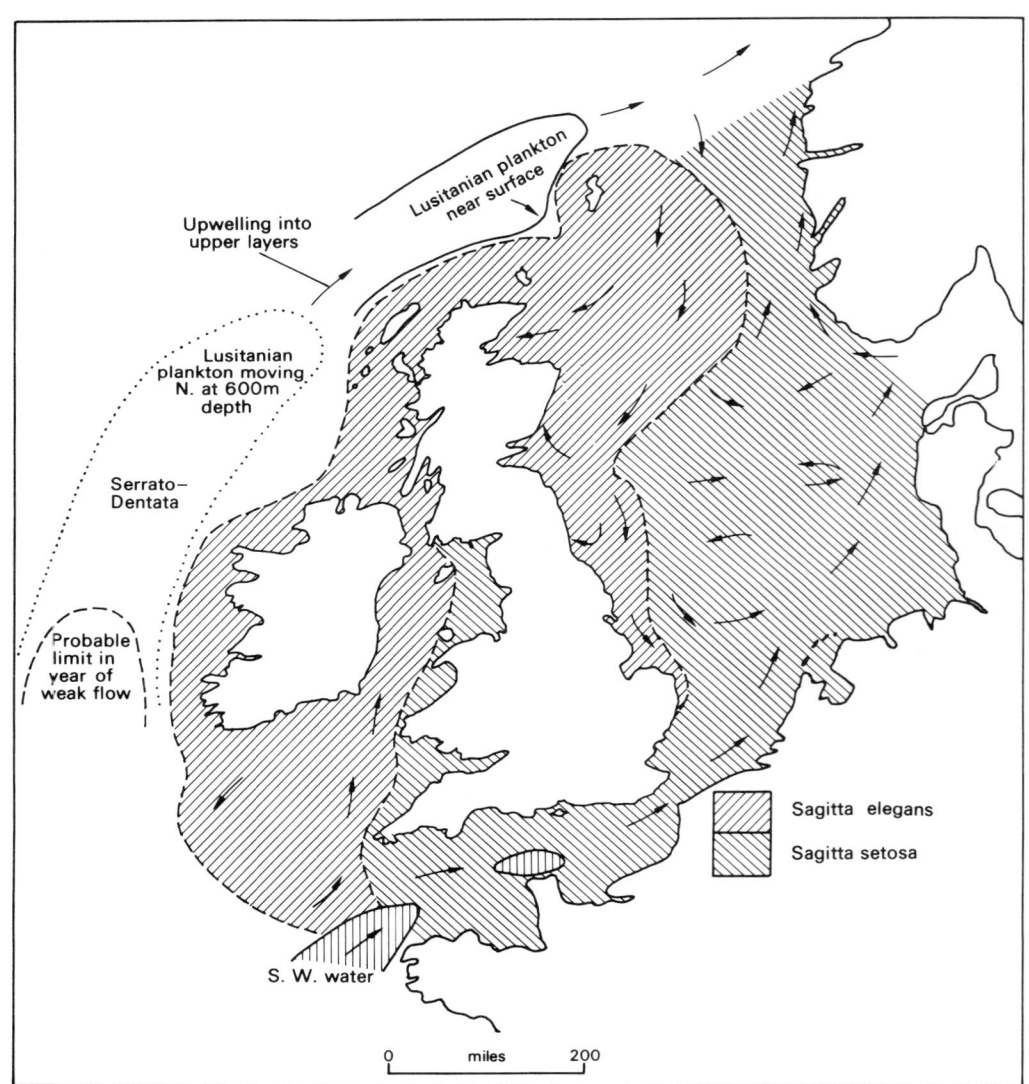

Figure 6.9 Water types around the British Isles defined by species of *Sagitta* and the northward extension of Lusitanian plankton to the west of Britain. (*Modified after Hardy, 1956.*)

up and down, during the day and night. Experiments have shown that this factor does play a considerable part. *Calanus*, which is only the size of a grain of rice, can swim upwards at a rate of nearly 15 m/hour, while larger creatures can achieve rates of 93 m/hour or more over shorter periods. Vertical migration gives an animal, which cannot move about by its own powers to a new environment laterally, the possibility of changing its surroundings by moving into different layers of water vertically. Other factors probably also affect vertical migration and it is a complex process, not yet fully understood.

The distribution of both phytoplankton and zooplankton is patchy and they often seem to be mutually exclusive. Alternatively some product of the processes of plant production may lead to antibiotic conditions, which excludes many of the animal species. Some species are, however, positively correlated with dense phytoplankton. The extremely varied life of the sea is founded on the abundance of plankton. The movement of some important food fish can be related more or less directly to the distribution and type of plankton. Some fish and other marine creatures feed directly on the plankton, and they are naturally more directly affected by it; herring and whalebone whales are important examples.

The plankton sometimes form an essential link in complicated ecological chains. One example is the relationship between herring and cod through the ctenophore *Boliopsis* and

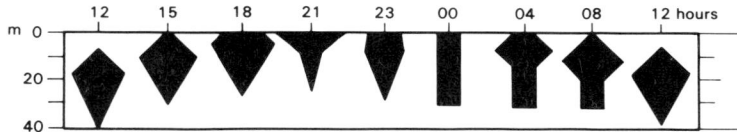

Figure 6.10 Vertical migration of the planktonic Copepod *Calanus*. (*After Russell, 1926–7.*)

the herring's main food, the copepod *Calanus*. One ctenophore 2·5–3 cm long can eat 10–11 copepods in 2 hours. Where *Calanus* occur at 300/m³ two *Boliopsis* could eat all the *Calanus* in one month. There is a larger stenophore *Beroe*, which eats *Boliopsis*, and which is itself eaten by cod. When cod are abundant they eat the *Beroe*, which allows *Boliopsis* to increase, and their consumption of *Calanus* has an adverse effect on the herring stocks. Thus a good herring fishery may be associated, through this relationship, with a poor cod fishery. The distribution of some plankton species gives useful information concerning fish, such as the relationship established between *Calanus* and herring, but this is not statistically significant for some of the data available.

A particularly important branch of the zooplankton are the *Euphausiids*, which are common in the Pacific. In this ocean it is thought that about 28 million tons of *Euphausiids* exist, or about 0·08 g/m² of ocean. This is probably a conservative estimate, as 100–500 probably exist in 1000 m³ in the upper 100 m, giving a likely concentration of 0·5–2·5 g/m², a value which is also probably an underestimate. The ratio of the standing crop of this creature in the tropic : subtropic : subantarctic : Antarctic is about 1 : 1·3 : 2·7 : 3·3. The standing crop in the Antarctic of *Euphausia superba* is probably nearer 29·28 g/m². These creatures are a major element of the total plankton biomass. Some direct use of the *Euphausiids* is made in Japan, for example, where they are used to feed rainbow trout and as a fertilizer. *Euphausiids* are attracted to light and can be caught at a rate of 500 kg in one night under optimum conditions, such as obtain in some Norwegian fjords. They are used as bait and food for trout and salmon. The densest concentration occurs in the upper

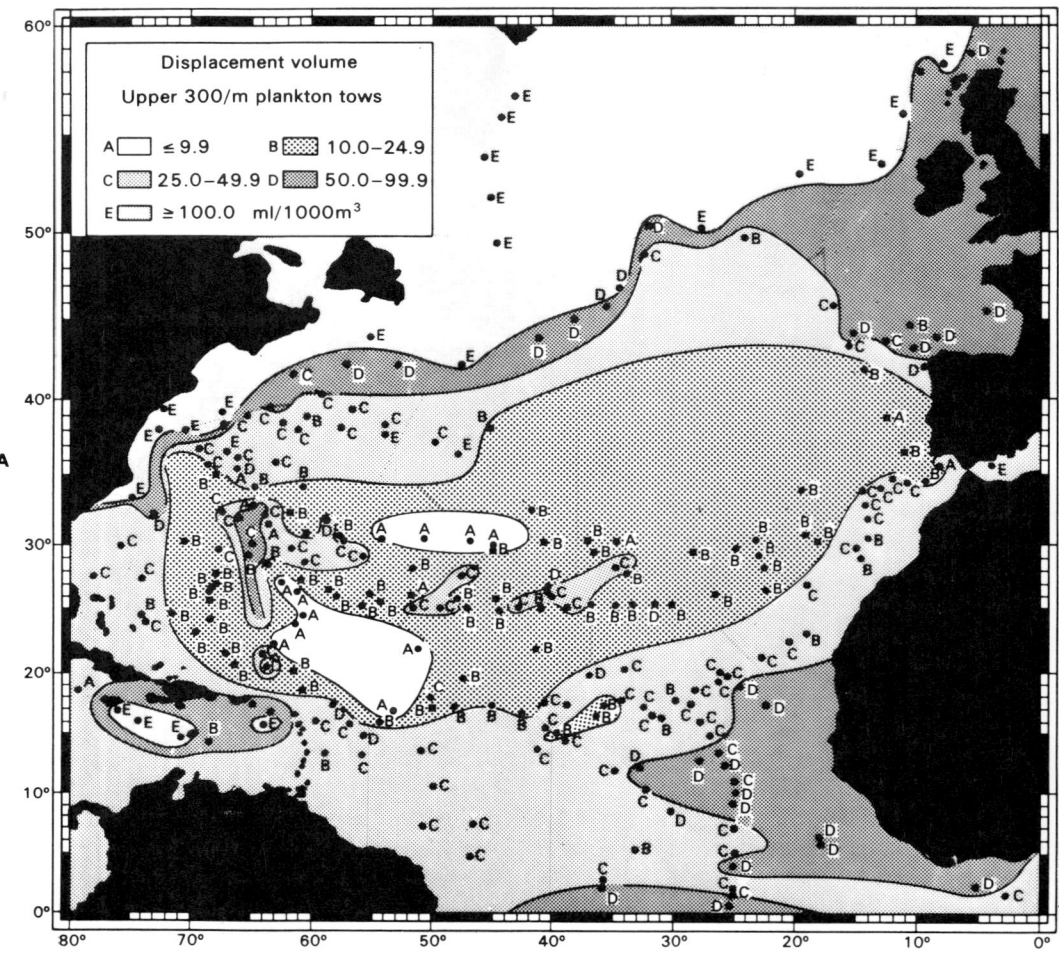

Figure 6.11 A: Geographical variation in zooplankton displacement volumes in the north Atlantic Ocean based on plankton tows in the upper 300 m. **B** (*opposite*): Geographical variation based on ash weight. (*After Bé et al., 1971.*)

10 m in the Antarctic, where they extend down to 50 m. In this area they swarm in late summer and early autumn, when the temperature is 1·5–1·9°C. It is possible that in this area the decline of whale numbers could lead to an increase in the *Euphausiid* population.

The most important species of zooplankton are the copepods and *Euphausiids*. The density of copepods can reach 15,000/m³ on the surface. A record of 28,000/m² has been observed off Murmansk. Even as deep as 500 m they are found at 1500/m³. In the North Sea the zooplankton has been estimated to weigh 10 million tons net, of which two-thirds are copepods. These animals have several generations each year. The smallest copepods are about 1 mm, and the largest about 5 mm in length, for example *Calanus finmarchicus*, which is the most important single food animal in the sea. Their reproduction depends on

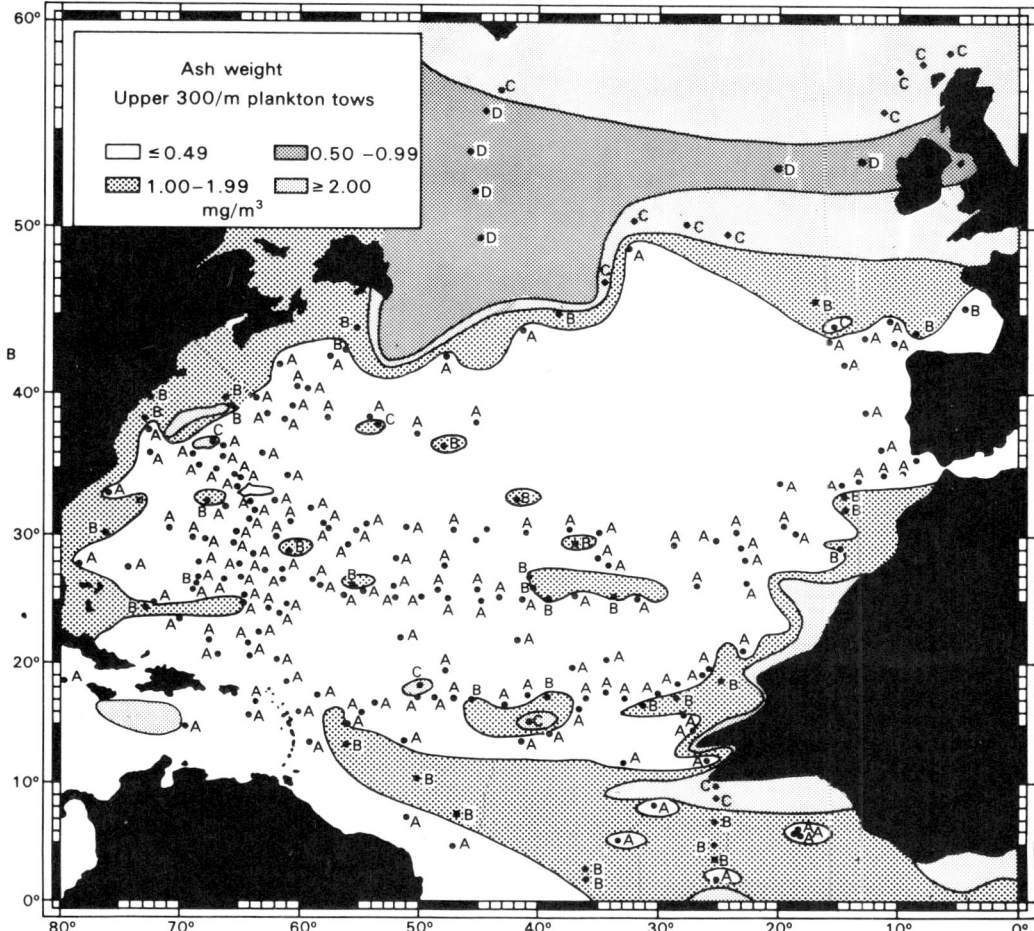

conditions, and takes one year off east Greenland, 3 months off west Norway, and 2 months in the English Channel. They eat 50–70 per cent of their body weight in diatoms each day. *Euphausiids* are 2–5 cm long and grow by moulting as shrimps and lobsters do. They extend from the surface to depths of 5000–6000 m, and they eat algae or copepods. In the cold conditions of the Antarctic they take two years to mature and breed, but in the tropics they only take one year, though they are not so plentiful here. *Euphausia superba* could yield 100 million tons from Antarctic waters. Their fertility is fairly low, each female producing less than 10,000 eggs. The term 'neuston' has been introduced to include all the organisms that live in the top film of water where the nutrient supply is greater. The neuston include bacteria, which are eaten in this layer by protozoa; and the very few marine insects also live in this zone. The zooplankton include many larval forms (who are only temporary members) which feed on the phytoplankton. The fact that nourishment is available all around as phytoplankton means that larval forms need not be highly developed when they hatch from the egg; thus marine eggs mostly have only a small yoke.

One problem concerning plankton is the difficulty of assessing its variability and density. Colebrook (1969) has discussed the variability of plankton in terms of the value of statistical analysis. He suggests that plankton sampling is more efficient if a large number of samples containing few organisms are taken. The concept of patchiness that comes from a stochastic model is probably not a useful one, and although patchiness can occur it does so at very different scales, varying between a few centimetres to tens of kilometres. Many variables affect the distribution of plankton, including water movements on all scales from small-scale turbulence to major current movement, locomotory behaviour of the plankton, social behaviour, food supply, diurnal migration vertically, and reproduction processes, as well as factors concerned with the environment, temperature, salinity, and nutrient supply. Mortality and predation must also be considered. The system is complex and dynamic, and standard statistical measures are not applicable as the system is not random. It is possible to recognize systematic variations, which are non-statistical. Serial correlation is involved in these—for example, changes in numbers over a seasonal cycle, and changes in area. Statistics could be used to study these systematic variations. The method has been applied to results obtained with the continuous plankton recorder. This technique was used in 1965 in crossings of the north Atlantic and North Sea. A correlation matrix of the annual fluctuations of the abundance of *Temora longicornis* for 1948 and 1963 provide three groups in which high correlations occur between the abundance for the two years. These groups have a meaningful geographical distribution— group B, for example, covering the central and southern North Sea. There appear to be real year-to-year changes in abundance as indicated by the patterns of relationships within and between the different species. Principal component analysis, parametric analysis, and analysis of co-variance can all be used to bring out aspects of the pattern of plankton distribution and variability.

An extensive study of plankton abundance has been carried out by Bé, Forns, and Roels (1971) in the north Atlantic. They collected 342 plankton samples between 1958 and 1968. The zooplankton abundance was highest in the subarctic and cold temperate waters, along oceanic margins, in regions of upwelling, and in active current systems. Low plankton was found in central water-masses, such as the Sargasso Sea, where the minimum was in the southwest sector. Seasonal variations in high latitudes are very large. Standing stock and not the productivity was measured. Five trophic levels were established. Sampling covered a range of 300 m vertically, thus exploring the depth of vertical migration. The biomass was recorded in different units, including wet weight, displacement volume, dry weight, and ash weight. The mean ratios for these measures are displacement volume : wet weight : dry weight : ash-free dry weight 15·9 : 13·0 : 1·1 : 1·0. Distribution in terms of some of these is shown in figure 6.11. The ratio of displacement volume between coastal areas, slope waters, and the Sargasso Sea are 16 : 4 : 1. The plankton of coastal waters is 20–40 times greater in summer than in winter.

Parker and Berger (1971) have used statistical classification methods to study the distribution patterns of planktonic foraminifera in the surface sediments of the south Pacific. The sediment patterns reflect conditions both in the upper ocean, where the animals live, and solutional processes in deep water as they sink to the bottom. The similarity value used for cluster analysis was found by $S = (\Sigma\, p_i - q_i)/\Sigma\, (\frac{1}{2}p_i + \frac{1}{2}q_i)^2$ where p_i is the fraction of the ith species in sample 1 and q_i is that in sample 2 of a pair.

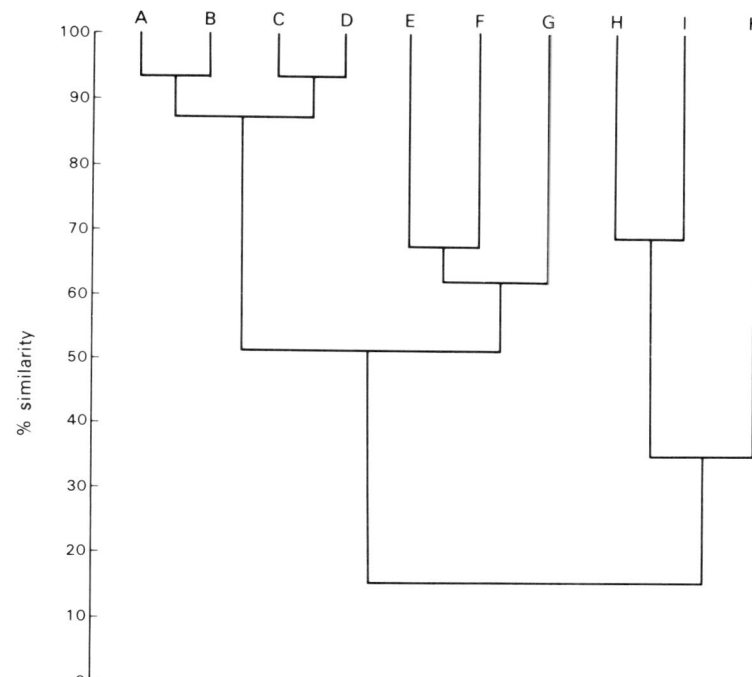

Figure 6.12 Classification of planktonic foraminifera showing percentage similarity in linkage tree form. (*After Parker and Berger, 1971.*)

Ten clusters were formed by successively reducing the similarity percentage as shown in figure 6.12. The clusters are partly related to the latitude, as a result of temperature patterns and ocean currents, and to the oceanic circulation. The boundaries between clusters A–G and H–I, and between H–I and K are major faunal changes and relate to oceanic boundaries: 1) the subtropical band of minimum fertility and the subtropical convergence, and 2) the Antarctic polar front. The three main areas of clustering (shown in figure 6.13) correspond to the three major fertile areas, the equatorial, subantarctic, and Antarctic zones. The large number of tropical clusters are due to a large number of species, giving an east to west pattern, which is probably associated with the intensity of the equatorial current system.

The clusters correlate with depth due to differential solution. There is a discontinuity at 4 km in low and middle latitudes, but not in high latitudes. The lysocline is a plane at depth separating well-preserved from dissolved foraminifera. The calcium carbonate compensation depth is where the rate of supply and rate of solution balance. The lysocline is about 4 km depth over much of the south Pacific, rising towards the east to about 3 km, with marginal zones of intense dissolution to the south and east along the south American shore and the Antarctic shore. This zone is related to circulation and fertility. Circulation imposes variations on the general depth pattern by mass balance, thermodynamic, and deep circulation. The variation between the level of the lysocline and the compensation depth suggests that the supply of calcareous shells and their dissolution vary together in a non-linear fashion. Many of the species belong to the *Globigerina* genus; 37 different species were used for the analysis.

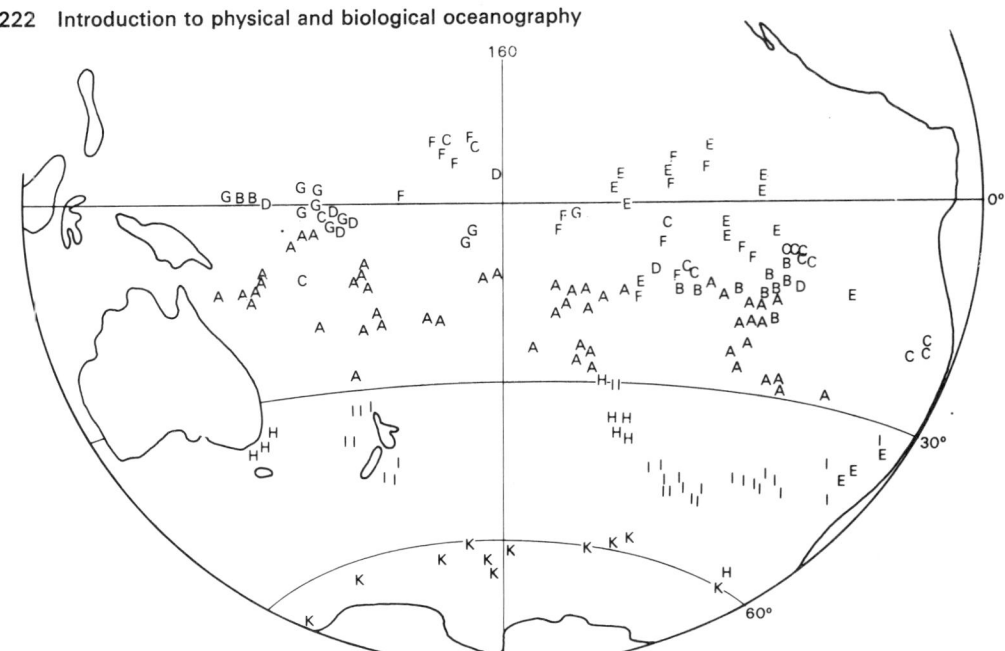

Figure 6.13 Map of the Pacific Ocean to show clustering zones of the groups of planktonic foraminifera shown in figure 6.12. (*After Parker and Berger, 1971.*)

Beers and Stewart (1971) report observations of micro-zooplankton in the eastern tropical Pacific, where samples were taken between latitudes 10·5°N and 12·5°S on longitude 105°W. Three sites were in the north equatorial current and 7 were in the south equatorial current, while two were in the equatorial counter-current. The highest concentrations were not found in the uppermost 10 m except in a few samples. The total abundance varied with a factor of three between the lowest in 12°S and the highest just south of the equator, the actual values being 15·0 mm³/m³ and 47·1 mm³/m³. The smallest organisms, less than 35μ, were most abundant at the northern sampling sites in depths of 20–50 m. The phytoplankton support the micro-plankton directly and primary production passes largely through this stage to the larger organisms. It has been estimated that the micro-zooplankton consume about 70 per cent of the daily production of the phytoplankton, which is considerably higher than the percentage calculated for the area off the Californian coast. In the equatorial Pacific a strong positive correlation was established between the chlorophyll α level and the total micro-zooplankton volume. The percentage of micro-plankton to the larger zooplankton in the euphotic zone is calculated to be about 24, although different sampling methods make comparison difficult. The standing stock biomass of micro-zooplankton averaged 34 per cent of that of the phytoplankton in terms of estimates of dry weight within the euphotic zone. Again this percentage was considerably higher than that calculated for stations in the Californian current.

It has been estimated that 1 gC/m² of zooplankton would ingest 0·75 per cent of a phytoplankton crop. Various assumptions suggest a total primary production of 800 mgC/m²/day, by an average phytoplankton crop of about 10,000 mgC/m². The mean biomass of zooplankton was about 6400 mgC/m², and the mean production was 200 mgC/m² or 3 per cent of the standing crop. Annual production of zooplankton was about 25 per cent of the net primary production. In temperate latitudes prcduction varies with the seasons, and there is a seasonal imbalance between phytoplankton and zooplankton, while in the tropics annual values are steadier. In the Sargasso Sea the standing crop of zooplankton was stable at 600–1000 mgC/m² in a 2000 m water column. A major peak occurred in the spring. Other estimates were three times lower than this. Harvey's estimates for the English Channel give an annual average of 250 mgC/m²/day for phyto-plankton production in a 50 m column, from a mean crop of 2000 mgC/m². The phyto-plankton bloom in the North Sea gives values increasing from 52 to 2330 mgC/m² and then declining to 10 between March and early June, giving an average net production of 230 mgC/m²/day. The biomass of zooplankton increased from 36 to 1090 mgC/m². The best method of calculating production is probably to deal with one species that is abundant and to extrapolate from this to the total population.

3   Zoogeography

The distribution of marine animals is of particular geographical significance. One important aspect of the biological provinces of zoogeographical regions is the degree of endemism that they have. Both the number and proportion of endemic species are important, the significance growing with the level of the taxomomic group. The higher the group (i.e., genus as opposed to species) is, the greater the importance of endemism. A study of zoogeography gives insight into evolution and geographical changes, for which purpose a historical study is essential. In the oceans the depth factor adds a complication by adding another dimension. Two broad faunal divisions are recognizable: the epipelagic fauna of the continental shelf, and the bathypelagic fauna of the deep ocean. The two are separated by the mud line, which lies near the outer edge of the continental shelf (McConnaughey, 1970).

The epipelagic fauna are influenced crucially by temperature; most animals live nearer the upper thermal limit than the lower one. It is easier for an organism to survive in colder conditions, however, than in conditions warmer than its optimum. In general there is more movement from warmer to colder conditions, and evolution seems to have developed in this direction. Once animals live in colder water they tend to remain, because adaptations render them more specialized, as they adapt to the more rigorous conditions. The greater range of conditions, as seasonal effects strengthen, also require greater specialization. The higher latitudes are characterized by many more individuals of fewer species, with a more marked seasonal variation in growth and numbers. In lower latitudes there is greater diversity, but less tendency to produce a very large biomass of particular species. The tropics show a high degree of endemism, some species being circumtropical, while others' are more localized.

Neritic tropical waters have two great ecological groups that are endemic. These are the coral reef and mangrove swamp communities. The coral reefs in particular consist of a very rich fauna. There are three main types because land barriers have created three distinct tropical zones: the Indo-west Pacific, the tropical Atlantic, and the tropical east

Pacific. The first region extends from the Red Sea in the west to the Malay Archipelago off southeast Asia in the east, reaching as far as Hawaii. This zone contains a rich fish fauna, and sea snakes are characteristic. A very rich fauna in all groups has developed in the extensive shallow water area of the Malay Archipelago. A similar, but not quite so rich, fauna exists in the tropical west Atlantic (in the West Indies), where shallow water also occurs around islands. This rich fauna contrasts with a poor one in the eastern tropical Atlantic. Communications between the two groups is restricted at present.

The west Atlantic fauna has more affinities with the Indo-west Pacific than with the eastern Atlantic, indicating a former connection between the west Atlantic and the Pacific through the Panama Isthmus. There is also evidence of past closer communications between the west and east Atlantic than at present. Indeed in the Mesozoic there was open communication all around the tropical latitudes, with land-masses to north and south. The Cretaceous and early Tertiary fauna of the east Atlantic was rich, and similar to that of the Indo-Pacific realm, with which it had contact. It has since become much poorer in its faunal types, while the West Indian fauna is now relict. The tropical eastern Pacific fauna shows a closer relationship to that of the West Indies than the western Pacific, despite the barrier of Panama. The last direct connection in this area across the isthmus was in the lower Pliocene, the straits having been open during the early Tertiary. There has been an effective deep water barrier between the east and west Pacific for a longer period than this.

North temperate and cold regions contrast strongly with the warm tropical and sub-tropical regions, in that only 38 per cent of the families and 8 per cent of the genera are common in these two major divisions. On the other hand, all families and between 50 and 75 per cent of the genera are found in Arctic, boreal and temperate regions. In the north Atlantic faunal boundaries are not clear-cut. There are many endemic species on both sides of the Atlantic, but more on the east than the west, as currents facilitate the transference of species from west to east. One important feature of the boreal Atlantic is the massive seasonal growth of copepod *Calanus*, the basis of the herring's food. In the Arctic Ocean there are a high number of endemic species, most of which are circumpolar. Many invertebrates and 60 per cent of the 80 species of fish are endemic. The fauna can be divided into high and low arctic groups and in general the number of species is lower than in warmer regions.

In the north Pacific there is a high degree of endemism, with about half the species endemic, compared with a quarter in the Atlantic. Starfish are varied and numerous and crustaceans well represented. The giant kelps are also mainly endemic. The area can, however, be divided into a number of different provinces, including the Panamic, the Californian, the Oregonian, and the Bering.

Southern temperate and cold regions lack the wide and continuous continental shelves of the northern hemisphere, but on the other hand the southern ocean is continuous. In general the dispersal of faunas has been from west to east with the major current system. There are many separate faunal groups in the temperate southern hemisphere. Sixteen separate provinces have been recognized in the Australasian area. Each has recognizably different fauna, indicating considerable isolation between different habitats in the southern area. The Antarctic Ocean fauna, south of the Antarctic convergence, is related to the high fertility of the water. The zooplankton is very rich owing to a spectacular seasonal outburst of phytoplankton, there being few species but immense numbers, especially of *Euphausia superba*. These feed the whales, penguins, and other endemic species. There is a high degree of endemism in the Antarctic, including 73 per cent of species of echinoderms and 90 per cent of 78 species of fish. The fish include one species, the ice-fish, which has no red blood cells. This is an adaptation that enables it to live in cold, oxygen-rich water. The Antarctic is much richer in fauna than the Arctic, particularly in benthic species and there is a higher degree of endemism, probably because of an older origin of the fauna. It has apparently had an uninterrupted cold climate during most of the Tertiary period.

The Atlantic abyssal fauna is quite distinct from that of the Pacific and Indian Oceans, especially at species level. The polar seas also have a distinct abyssal fauna. The deeper benthic communities have probably adapted from those of shallower water in the higher latitudes, after their adaptation to colder conditions from their original tropical homes. Owing to the relative youthfulness in geological time of the ocean floors, because of sea floor spreading, the ocean floors do not provide good evidence of the early development of marine fauna. Few if any of the ocean sediments date back beyond the Mesozoic, so that early growth and evolution of marine organisms must still be deciphered from the records of early rocks preserved on land.

It was thought that the stable conditions of the abyssal zone would provide the best conditions for the preservation of geologically old species, so that it was a surprise to find the *Coelacanth latimeria* in the zone between 200 and 1000 m, the bathyal zone. This is a fish that belongs to a group that is supposed to have died out 200 million years ago. Subsequent catches show that this fish lives in comparatively shallow water. One interesting discovery from the abyssal zone in a depth of 3590 m was caught in a trawl off the Pacific coast of Mexico. This was a mollusc, similar to a limpet, called *Neopilina*—a very primitive animal belonging to the group *Monoplacophora*, of which further species have been found at depth subsequently. One feature of abyssal fauna is its widespread distribution in the uniform conditions that exist at this depth.

Fish of different types occupy all depth zones, each zone requiring special adaptations. In the twilight zone from 200 m to 1000 m there are many fish, including schools of luminiferous fish of genus *Cyclothone*; *Cyclothone signata* is possibly the commonest fish in the sea. The advantages of this zone include warmth compared to the deeper layers. Metabolism is slower in this layer than nearer the surface, thus less food is required than surface fish need. The temperature is more stable, but the fish are still close enough to benefit from the food-producing surface layer. Viscosity is higher, which means easier floating and slower sinking of detritus, so carrion eaters are common in this zone. Many fish from this zone migrate vertically as copepods do. A feature of this zone is the scattering layer at 500–900 m depth in the day and 100–200 m at night. The layer is formed by plankton, the density of which is sometimes so high as to give a false bottom. This occurs in the Pacific Ocean from the Aleutians to Tahiti, and San Francisco to Japan, the Mediterranean, the Red Sea, the Indian Ocean, and the tropical Atlantic Ocean. The echo-layer is poor in the Antarctic. Fish follow the plankton and their swim-bladders, which make a big echo, could account for the effect. Large shoals of fish may be involved. The dry weight of organic matter in this zone is probably between 5 and 10 g/m^2, in places increasing up to 44 g/m^2. Four-fifths of the animals have light organs at this level. The light they produce is a cold light that causes little energy loss as heat. The deep zone, which occupies 90 per cent of the ocean living space, is inhabited largely by predators. It is a very uniform environment throughout the world ocean, forming one connected zone, apart from the Arctic Ocean to the north of the north Atlantic Ridge.

4 Life on the ocean floor: benthos

The creatures that live on the bottom of the sea are called the benthos. They are quite varied and play an important part in the food chain, both as prey and predators. The character of the bottom and the depth influence the type of animals found in each habitat. The variety of creatures is greatest on the continental shelves, in the zone where light can reach to the bottom and the benthos can feed on living phytoplankton as well as zooplankton and other food. Benthos tend to be more plentiful in regions where the plankton production is high, for example in the Kattegat, where it has been estimated that 230,000 tons of first-class food and 380,000 tons of second-class food for bottom-living fish (such as plaice) is produced in an area of 9070 km². This amount of benthos has been estimated to produce 13,000 tons of plaice annually.

There are three main methods whereby the benthos feed. Some filter the finest food particles from suspension. Others live on the detritus which is deposited on the bottom. Yet others are carnivorous, and compete with man and other predators for bottom-living organisms. The greatest variety of forms is associated with a rocky bottom, but the largest area of the sea floor is covered with sediment, including mud, sand, and gravel, each of which has its own typical fauna. Some animals live in, rather than on the bottom. Most of the animals of the rocky floors eat particles suspended in the water, which they filter out. The fine particles are mainly planktonic in character. A lot of these creatures, existing on such small food particles, are polyzoa, consisting of many small individuals living together in a colony as one unit, such as the corals and *flustra*.

Some of the bivalve molluscs are also suspension feeders. Their gills are enlarged to enable them to sieve large amounts of sea water. Mussels, which live on rocks, and oysters and scallops, living on gravelly or sandy bottoms, belong to this group. Other molluscs, such as the razor shell, *Ensis*, the cockle, *Cardium*, and clam, *Mya*, bury themselves in the bottom sediment, but feed through a siphon which projects above the surface. They can then remain safely buried in the sand while feeding.

Deposit feeders make use of the detritus and remains which fall to the bottom. Worms exemplify this group. Some of them form a U-shaped burrow in the sand. They pass the material through their bodies, extracting nourishment from it as it passes through. Some worms bury themselves in the bottom, while exploring the surface for food with their tentacles. Heart-urchins (*Echinocardium*) and brittle stars (*Amphiura filiformia*) bury themselves in the bottom and feel for their food, the former eating sand grains, covered with organic matter, while the latter use their flexible arms to search for food. Some foraminifera are benthic in type. Unlike the planktonic foraminifera, which secrete limy shells, such as *Globigerina*, they may make themselves houses. Some of these houses built by these tiny animals, without sense organs such as eyes, are quite elaborate. For example *Psammosphaera rustica* builds a polyhedral or nearly spherical chamber, using minute sponge spicules, the longest of which is only 2–3 mm in length for its construction. The spicules are built into a framework and spaces are filled with carefully fitted fragments of sponge spicule of just the correct length. The longer 'tent poles' project from the structure and help to prevent it sinking into the ooze on the sea floor. That such small and primitive animals, without sense organs, can select their building material and fashion it into a structure (rightly termed a marvel of constructional skill) shows that there are still mysteries to be solved concerning the nature of simple organisms (Hardy, 1956).

Benthos are an important means whereby the finely divided food on the ocean floor is made available for larger creatures who are not adapted to filter the finest particles of nourishment for themselves. Benthos also prey on some of the bottom-living fish. There is, therefore, an intimate prey–predator relationship between different types of marine creatures. Petersen (1918) developed a grab in which he obtained samples of the sea bed and the creatures living in it to further his study of benthic ecology. He divided the bottom into zones according to characteristic faunal assemblage, the type of bottom, and availability of food. The results also give useful information concerning the availability of food for bottom-living fish.

A survey of the Dogger Bank has shown that the distribution of different species is patchy. The bivalve, *Spisula subtruncata*, which is important for plaice food, was surveyed in detail. Very young specimens were found in one area, where their density increased to over 2590/km² in one small patch. The one-year old specimens were concentrated some distance away, where the maximum density (again is a small patch) exceeded 386/km². The area of two-year olds was even more restricted. This indicated the large mortality this species suffers. Fish were not the only predators, some of them being eaten by a carnivorous gastropod.

The bottom-living invertebrates also pass through a larval planktonic form, when they drift about with the currents. Some, however, have the ability to postpone their metamorphosis into the adult form, for days or even weeks, until they reach a favourable type of bottom. Some larvae are probably lost when the currents are abnormal and they do not reach a suitable ground before they change form; and some may be carried beyond the continental shelf to water too deep for their survival on the bottom. There are two opposing tendencies at work during the planktonic larval phase: 1) the need to mature quickly to propagate the species and 2) the need to remain afloat long enough to spread the species widely.

The third category of benthic animals is carnivorous. They compete with man and other predators for the demersal or bottom-living fish. Some of the bristle worms or *polychaetes* are voracious carnivores. One of these is known as the sea mouse, *Aphrodite*; it burrows into the substratum and feeds on the worms. Some molluscs eat others, which in turn form the food of plaice. The beautiful sea slugs or *nudibranchs* browse on sea anemones, hydroids, or even seaweed; they are usually coloured in such a way that they blend with their main food. Starfish are one of the most voracious carnivores, eating bivalve molluscs by opening their shells with their own muscular arms (Plate 7). In fact they are perhaps the most serious rivals of the demersal fish in obtaining food from the bottom.

Work by Thorson (1971) has shown that plaice take about 3–5 per cent of their weight in food each day in the warmer part of the year, but only one-tenth of this during the colder season. Considering many bottom fish, he reached the conclusion that they rarely took more than 5–6 per cent of their body weight in food per day. Most invertebrates,

Plate 7 Starfish: a plague of common starfish (*Asturias*) feeding on mussels at Sandgate, Kent (*G. W. Potts*).

however, take a much higher proportion. For example, a gastropod, *Conus mediterraneus*, took 25–50 per cent of its weight of worms, *Nereis*, at one meal, while newly settled starfish ate up to 3 times their own volume of clams in 6 days. In youth the invertebrate predators take up to 25 per cent of their own weight of food a day, while active adults take about 15 per cent a day. On the whole they take about four times the food taken by bottom-living fish each day.

About one million tons of food is available in the Kattegat, consisting of gastropods, lamellibranchs, polychaetes, and crustacea, and it must feed 5000 tons of fish and 75,000 tons of invertebrate predators. If they both consumed food at the same rate, the fish would have 6–7 per cent of the supply, but since the invertebrate eat four times as much as the fish, this percentage falls to only 1–2 per cent. Even if the food supply were increased the quicker developing invertebrates would take advantage of it long before the fish were able to. This shows what a serious menace to fish stocks invertebrate predators can be.

Predators usually have a longer life than the prey, thus it would seem likely that when the new generation of prey arrives, the longer-lived predators would consume them before they had a chance to grow and reproduce. For example, the newly settled bivalves must escape the searching arms of the brittle stars during the period when they are small enough to be eaten easily by them. This is achieved by the brittle stars entering on a passive period, during which they reproduce but do not feed, for two months. By this time most of the bivalves have grown large enough to escape being eaten when the brittle stars begin to feed again. The same sort of relationship between predator and prey applies to other benthic species and seems to be the method of survival developed by natural selection, for the survival of each species.

Another important element of benthic fauna are the crustacea, which include crabs, shrimps, and lobsters (Plate 8). They live mainly on rocky or stony ground, where they can be caught in pots, particularly off the west coast of Britain. Only one of the many types of crab is eaten by man; the others include many species, such as the hermit crab, which lives in empty gastropod shells. Recent work has shown that benthic animals are not entirely confined to the shallower water of the continental shelf. Even in the greatest depths at which photographs have been taken, there are signs of benthic creatures (Laughton, 1959a).

Benthos can be divided into epifauna and flora and infauna; the latter live buried in the sediment of the bottom, while the former live on the bottom. The epifauna includes four-fifths of all benthic animals, the shallow tropical seas being especially rich in species. The infauna are much poorer in species, but are often very rich in numbers. The fauna are closely controlled by depth. In the intertidal zone the *Macoma* or mussel dominates, while the *Amphiura* or brittle stars are dominant in mud at depths of 15–20 m in colder waters. Benthic fauna can use quite large territories, for example *Macoma secta* is only 6–7 cm long, but it has a siphon 1 m long. Many benthic animals are also very voracious, eating a large proportion of their body weight each day.

Benthic creatures in particular have developed habits that fit them to different depth zones. The highest is the supra-littoral zone, followed by the littoral zone. One of the

Plate 8 Crab: note unusual formation of right front claw (*Natural History Museum*).

most important aspects of the littoral zone is the tide, and many animals have developed spawning habits associated with the tidal cycle, including the grunion. Off Samoa and Fiji in October and November and the day before the moon is in its last quarter, the palolo worm, *Eunice viridis*, spawns in shallow water near the surface in great profusion. The hind part of the worm breaks off, in which the eggs and sperm are situated, while the other half of the worm continues its normal existence. The polychaete bristle worm, *Nereis virens*, spawns at full moon in spring. These worms are up to 70 cm long. The reason for the moon-associated spawning is not known, nor is it known how the animals that spawn at such specific times know when this time comes. This applies to fish, bristle worms, oysters, mussels, winkles, sea-urchins, and medusae. Some bivalves, such as *Donax*, move up and down the beach with the daily tidal cycle in the intertidal zone, always maintaining a position near the swash.

The supra-littoral and littoral fauna, especially the sessile forms, must be able to survive great extremes, and are often adapted to breathe both air and water (barnacles, *Balanus balanoides* are an example) (Plate 9). Some fauna can withstand large temperature ranges and even freezing. The littoral fauna on exposed rocky coasts need hard shells to withstand waves. Barnacles also have another problem as they must settle close together in order for their eggs to be successfully fertilized; this can lead to overcrowding as they grow. Limpets (*Patella*) are another species found on rocky coasts. They occupy one place on the rock

Plate 9 Barnacles: examples of large barnacles with smaller ones settled on them (*Natural History Museum*).

that they have adjusted to exactly fit their shell. Although they wander around at high tide, they always return to the same spot to maintain their airtight seal during exposure to the air at low tide as the tide falls again. Mussels are the most numerous species in some areas in the littoral and supra-littoral zones. On suitable rocks they occur at 30–50 kg/m², even reaching a density of 140,000–170,000/m².

Coral reefs are among the most prolific areas for both abundance and variety of fauna. Some corals live in symbiosis with *zooxanthellae*, the vegetative stage of the dinoflagellate *Gymnodinium*, the density of which can be 30,000/mm³ of coral tissue. Plants also play a useful part in the reef communities as they use the carbon dioxide and other waste products of the coral, in turn supplying oxygen to the coral through photosynthesis. They also stimulate calcium production in the coral. The Indo-Pacific reefs have 700 species of coral fauna, but there are only 35 in the Atlantic. *Acropora* and *Porites* are the most important. Large creatures also live in the coral community, an example being the giant clam, *Tridacna*. It is a giant mussel and can reach 1·35 m in length and weigh 200 kg. They can kill unwary people by trapping them in their giant shells.

The infauna of the littoral zone depend on the nature of the sediment. On a sandy bottom the common cockle (*Cardium edule*) can occur at densities of 350/m², and in the Dutch Waddenzee they have been recorded up to 2000/m², each being 2·5 cm in diameter. They are eaten by oyster catchers at the rate of 214–315/day in cold weather. There are about 30,000 oyster catchers in southern England; they eat about 642 million cockles in 100 days. Bristle worms also occupy sandy bottoms and filter sand at the rate of 210 kg or 2100 × their weight each year. Mud infauna lack oxygen below the top 10 cm, but they have the advantage of very low gradients that mean that the mud does not dry out at low tide. Mud contains more food as a rule than sand, and it is this that feeds the mud dwellers, which are mainly detritus feeders. The fauna include mud-burrowing amphipods, *Corophium*, and mud snails, *Hydrobia*, who eat the bottom layer and its included bacteria. Most of these animals are small. Another specialized mud community consists of the tropical mangrove swamps, the community of which includes many crabs, such as the fiddler crab, *Uca*.

The sublittoral zone extends from low tide level to a depth of 200 m, approximately the edge of the continental shelf. Of significance to man in this zone are the borers and encrusters, by their habit of infesting ship keels and boring into marine structures. One of the borers is *Pholas*, the piddock, which can bore into limestone and effect considerable erosion. *Teredo* bore through wood and are worm-like and 15 cm long. They bore into wood when small and gradually grow inside the hole so that they cannot escape, although they can feed by siphon through the opening as well as ingesting the wood through which they bore. They can do much damage to wooden structures and ships.

The ecology of the sublittoral zone is complex. For example, limpets can feed on seaweed, which is also eaten by sea urchins. Mussels are preyed on by sea stars and crabs, which also eat other molluscs. The brittle stars are voracious predators except when they are about to spawn. In the Kattegat they occur at densities of 400–500/m², with a total weight of 300,000 tons. They eat mussels, snails, worms, and other animals. There are also many bacteria in the microfauna, estimated at 160 million/cm³ in the West Indies. They can double in 1·24–8·3 days. Diatoms can raise 150–200 generations/year, and there are 4–11 million in each cm³ in the Danish shallow waters. Slightly larger are the meiofauna, which are 0·5–2 mm in length.

The abyssal zone has mud eaters and predators in a uniform environment. Holothurians become dominant increasingly as the depth increases in the first group, and include echinoderms and sea spiders. The predators include sea stars, crustaceans, and fish. The live weight at 4000–5000 m averages 2–5 g/m² or less, the dry weight being 2–3 per cent of the live weight. At 9000–10,000 m only 20–30 mg/m² occur and large predators cannot live on this. Estimates of faunal densities are probably on the low side, but life does exist right to the bottom of the greatest depths. The abyssal fauna is relatively recent on the geological time scale. Below 6000 m depth in the hadal zone, 366 species have been found; they are dominated by the mud-eating holothurian, which can live without danger of predators at this depth. The environment in the deep trenches is uniform, with the temperature at about 1·2°c–3·6°c. Because of their proximity to land their fauna is rather more varied than that in the open ocean, but is limited compared to the abyssal zone and the continental slope area. Despite the paucity of species in the greatest hadal depths, numbers are sometimes large; for example, a single trawl in the Sunda Trench at about 7000 m produced 3000 holothurians (sea cucumbers). Sea anemones have also been found at the greatest depths, together with bristle worms and crustaceans, and thread-like worms, *Pogonophora*, which lacks a gut. The availability of food sets a limit to the depth to which different species can penetrate. Predators like fish, for example, have not been found below 7580 m, crabs have not been caught below 5200 m, brittle stars below 8006 m, and sea stars stop at 7630 m. Omnivorous species who do not depend on living prey can go right down to depths between 8210 and 10,700 m. Large predators cannot exist at depth as more energy is needed to catch prey where it is sparse than the prey provides; thus the minimum life density that will support large prey probably lies between 20 and 100 mg/m².

The differences in epifauna in canyon and non-canyon areas have been studied by Rowe (1971a) by means of photographs, which revealed considerable differences in the two environments. Brittle stars (*Ophiomusium lymani*) are most common in the Hatteras area non-canyon environment, but were lacking from the canyon where sea cucumber (*Peniagone willemoesia*) was very common and sea pen (*Kophobelemnon*) also occurred. The canyon current readings gave an average of 38 cm/sec, with a range of 36–42 cm/sec. The flow was predominantly up-canyon. The current was presumed to be the western boundary undercurrent, and it had a westward trend. The sediment forms in the canyon, however, suggested easterly or southeasterly flow, probably caused by intermittent stronger down-canyon flows. The study showed that submarine canyons do have a distinct fauna. Movement of water outside the canyons is predominantly along the contours within the undercurrent, but within the canyon the flow is mainly up or down. This canyon movement would disrupt the dispersal pattern of benthic species. Differences in the character of the sediment could also help to account for the distinction of the fauna, especially the degree of induration. The sediments of the canyon axis on the continental rise and slope are strongly indurated, but they are soft on the continental rise adjacent to the canyon axis. In some canyons benthic life is sparse, for example the Corsair Canyon off Nova Scotia, a characteristic that could be due to intermittent flushing by currents flowing down the canyon.

5 Demersal fish

The fish that live near the bottom of the sea are called demersal fish. They can be divided into two broad types, the round fish (such as cod, haddock, and hake) and the flat fish (such as plaice, sole, and halibut).

The second group, which live right on the bottom, are well adapted to this environment, having undergone structural modification. Their bodies are flattened and both eyes have moved over to one side of their head. The plaice, *Pleuronectes platessa*, is a good example of this type (Plate 10). It has been intensively studied and is also a valuable commercial species, particularly in the North Sea. It has been modified in such a way that it lies on the bottom with its right side uppermost. Its colouring is such that it blends perfectly with the surroundings, and it has developed a habit of moving its fins in such a way that sand stirred up as it settles falls on the edge of its body to break the hard outline.

This bottom-living fish lays floating eggs. A large female plaice may lay up to 0·5

Plate 10 Plaice: a common flat demersal fish (*Natural History Museum*)

million eggs at each spawning. The main area where plaice spawn is the Flemish Bight at the southern end of the North Sea, where the eggs are laid in mid-winter. In this zone a tongue of rather more saline Atlantic water penetrates through into the North Sea from the English Channel. It has been estimated that 60 million plaice come to this area to spawn each winter. Other smaller spawning grounds are off Flamborough Head, off northeast Scotland and in the Irish Sea and northwest Heligoland. The eggs deposited in the Flemish Bight, drift towards the coast of north Holland, a movement that has been confirmed by the study of drift bottles as shown in figure 6.14. The eggs drift northeast at about 1·5–3 miles/day (2·4–4·8 km/day), as members of the plankton (on which they feed at first when they hatch) to their nursery grounds. Until it is a month old, the little fish has the normal shape of any round fish, but at 4–6·5 weeks old it changes into a tiny flat fish, which is only 2 cm long at this stage. During this critical period the young fish are very dependent on the wind and currents. If the winds should prevent the normal northeasterly movement, or carry the little fish too far, they will fail to reach their most favourable nursery grounds. The amount of planktonic food available at this early stage is also an important factor. These factors lead to great variation in the success of broods in different years. Those that spawn earliest may not have so much planktonic food available as those spawned later, when the spring flowering is available for food.

Plaice stay in their nursery grounds in the coastal waters off Holland and south Denmark for two years. As they grow they gradually move northwards into deeper water, increasing in size till at 4–6 years old, they are 40–44 cm long. By this time they are found in the western North Sea off north England and south Scotland. Their movements are shown in figure 6.14. One-year old plaice are only 7·5 cm long. Their diet varies seasonally, both in quantity and type of food. In winter they live on polychaete worms and in summer on bivalve molluscs, although in winter only 10 per cent of the fish feed.

Experiments made with marked plaice give details of the growth rate and the percentage of fish caught. The capture rate is sometimes very large. Up to 70 per cent of the marked fish were recaught in one year. This is a minimum number as some of the marks may never have been recovered. Plaice travel to and fro between feeding grounds and the spawning area. Large catches are made in the Flemish Bight in January when spawning is taking place, while in July the main catches are further north, although still concentrated in the southern North Sea. Similar movements take place around the other spawning grounds.

Other experiments have shown the relation between growth rate and food supply. The main nursery ground for young plaice is off Holland. There are so many fish here competing for food that the growth rate is slow. Some marked fish have been taken to the Dogger Bank, which is not a normal plaice nursery, but where food supply is rich. The growth of these fish has been compared with similar ones, which were marked and left in the original nursery. The transported plaice had grown about 13–14 cm during the year after their move, while the fish remaining behind had only grown about 4 cm. This seems to be one practical method whereby the wealth of the sea may be increased for the benefit of commercial fishing and human food supply.

Plaice are one of the prolific breeds of fish and it has been estimated that the southern

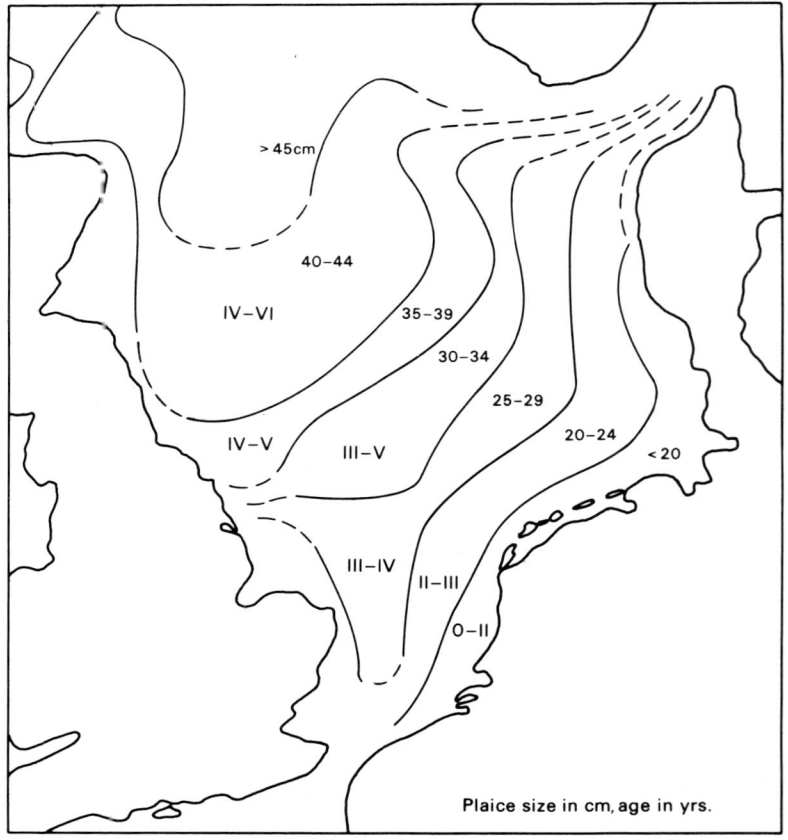

> 45cm

40–44

IV–VI

35–39

30–34

IV–V

III–V

25–29

20–24

< 20

III–IV

II–III

0–II

Plaice size in cm, age in yrs.

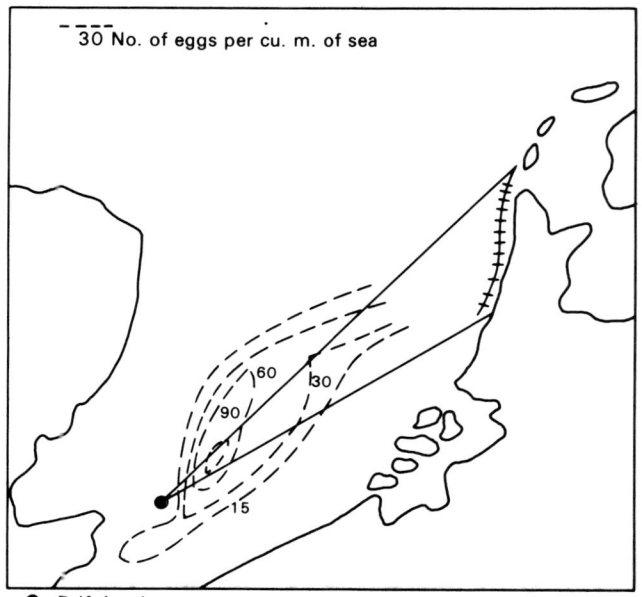

‒ ‒ ‒
30 No. of eggs per cu. m. of sea

60

30

90

15

● Drift bottle source
✝✝✝✝ Recovery area

Figure 6.14 The lower figure shows the spawning area of plaice in the southern North Sea in relation to the drift of bottles released in the same region and their recovery zone on the Dutch coast. The upper figure shows the northward spread of plaice from their nursery grounds off the Dutch and Danish coasts. (*After Wimpenny, 1953.*)

North Sea stock contains about 246 million mature plaice (Wimpenny, 1953). The whole of the North Sea fishing grounds probably contains about 2000 million plaice at or near catchable size, although the figures can only be approximate. Plaice are not limited to the North Sea, but are distributed widely around the north Atlantic. The southern North Sea, however, probably contains the major concentration. They live on a sandy bottom, extending to a depth of 70 m or even to 180 m, in an area stretching from the western Mediterranean as far as the extreme north of Norway and into the Murman coast off north Finland. Plaice are also widely distributed around Iceland, where their main

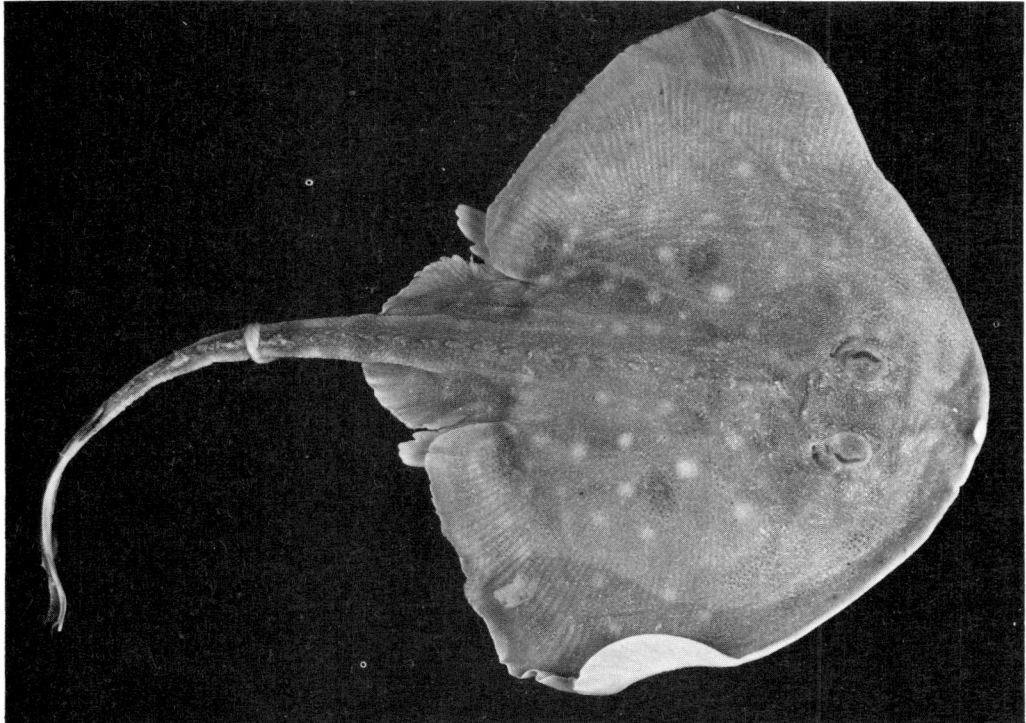

Plate 11 Ray: a flat fish that is flattened symmetrically (*Natural History Museum*).

spawning ground is situated off the west coast. These widely distributed plaice probably belong to different races which can be differentiated by certain physical characteristics, such as the number of vertebrae, although these variations may be more closely related to their different environments than to separate races.

Other creatures also adapted to life on the bottom by flattening of the body include such fish as skates and rays (Plate 11). In these fish the flattening is in the opposite direction, and their bodies are more symmetrical as a result. Their eggs are laid in cases and when young they feed on small crustacea. Later their diet is mainly fish, mostly herrings, which they hunt by smell. They are found to the west of Britain, in the water of the north Atlantic drift.

Of the round demersal fish, cod, haddock, and hake are the most important commercially. Cod is a good example of this group of fish, because of its great commercial significance and its wide distribution (Plate 12). Cod are found around nearly all the coasts of the north Atlantic and penetrate into the Arctic waters to the north of Norway. Southwards they extend to Cape Hatteras on the west of the Atlantic and to the Bay of Biscay on the east. They are particularly concentrated in the Barents and White Sea in the north, on the Faroe Plateau, round Iceland, off western Norway, in the Baltic and North Seas, and around the western coasts of Britain. On the western side of the Atlantic, cod are particularly prolific on the Grand Banks off Newfoundland and around the coasts of Greenland and the eastern coasts of America. Somewhat similar fish, also called cod, are found in the north Pacific, around Japan and off North America. They are cold water

Plate 12 Cod : a large, round demersal fish (*Natural History Museum*).

fish, who require a temperature lower than 10°c, and can live in temperatures of 0°c. A temperature between 1 and 5°c is probably the optimum value.

The cod that inhabit this wide area belong to many groups. They move around a lot in these northern waters, as indicated by tagging experiments. The fish round Newfoundland move to Iceland, while the latter may link up with the stock around north Norway, but the North Sea cod seem to be a distinct stock. They are limited in their distribution by the fact that they live near the bottom and therefore are restricted to shallower water. Their depth limit seems to be about 366 m, but they must be able to cross deeper water to account for the results of the tagging experiments. They are rarely caught in depths exceeding 370 m, however.

In their earliest phases of life the cod are classed with the plankton, as they are incapable of independent movement. The eggs and larvae drift from the spawning grounds to the nursery grounds at the mercy of the currents. Thus the Barents Sea cod spawn near the Lofoten Islands in March and April. They then drift in the west Spitzbergen current to their nursery feeding grounds on the Spitzbergen shelf, while others drift with the North Cape current to the southeastern part of the Barents Sea. They are very dependent on the currents to carry them to the nursery grounds at the right time, when the plankton is rich. If the currents differ from normal, the brood may fail. This is not uncommon owing to the variability of the currents.

The North Sea cod are an important group of these fish. The spawning period is from February to April on the banks in the North Sea, particularly on a small area off Flamborough Head, the Great Fisher Bank, the Forties, and Ling Banks (figure 6.15). A large female cod may spawn up to 4 million eggs, which float with the plankton, at one spawning. The small fry when hatched spend 2·5 months in the plankton, living on small copepods, and drifting with the currents, which do not take them out of the North Sea. When they are 2 cm long, they leave the plankton and take to the bottom. They stay on the bottom in rough ground and are very difficult to catch, so that it is not easy to trace

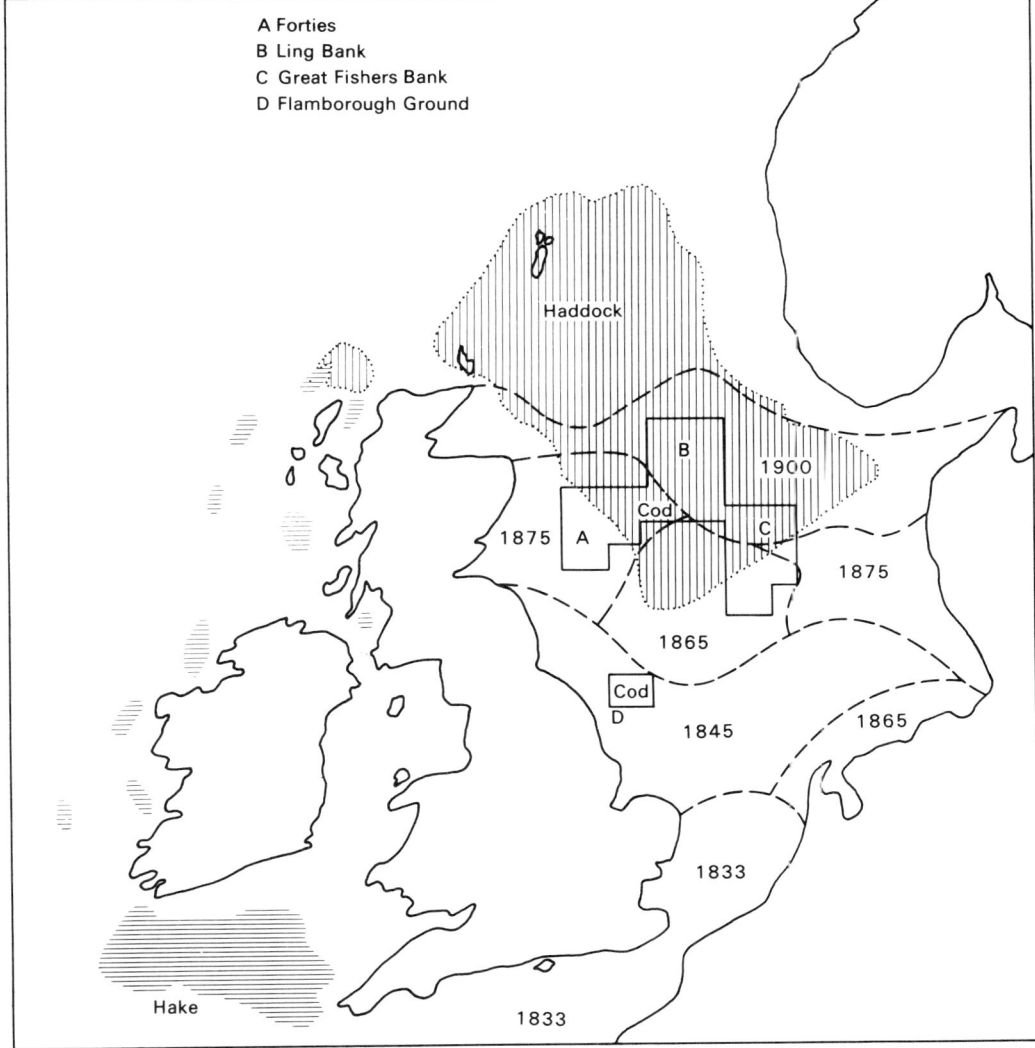

A Forties
B Ling Bank
C Great Fishers Bank
D Flamborough Ground

Figure 6.15 Map to show the spawning grounds of cod, haddock and hake around Britain. The dates indicate the spread of trawling in the North Sea. (*Modified after Hardy, 1959.*)

their subsequent development. In 1·5–2 years the cod are 30–35 cm long, but they do not breed until they are approximately 5 years old, when they are about 70 cm long. Their age can be estimated by studying their scales, which show annual growth rings. Adult cod feed on other fish, particularly herrings, mackerel, small haddock, and sand-eels. They differ in their food from their near relative, the haddock, enabling the two species to live together on the same ground. The haddock live mainly on invertebrates, such as crustacea, molluscs, and brittle stars, and they also feed on herring spawn when this is available. Cod are caught over wide areas in the North Sea, the heaviest catches coming from off Denmark in winter, while in summer the greatest catches are made on the western side, off northern England. Throughout the year the northern part of the sea is more prolific than the Flemish Bight at the southern end.

The haddock is another important fish in this group. It is very abundant in the northern North Sea, extending up to the Faroes, Iceland, and Barents Sea areas, and is also prolific on the western side of the Atlantic, in the coastal waters from Cape Cod to Cabot Strait, and on the southern part of the Grand Banks of Newfoundland. Like the cod, the haddock spawns from February to April. It is smaller and matures earlier, at 2 years, when it is only 25 cm long. Female haddock lay about 30,000 eggs at this stage, while a 6-year-old fish, about 40 cm long, will produce 280,000 eggs. The fish in the northern North Sea grow rather less fast than those in the south. At 5 years old the northern ones are about 32 cm, while the southern ones are about 39 cm. This is due to the extra warmth of the shallower sea bed of the southern part of the sea.

Haddock spawn mainly in the North Sea between Scotland and south Norway, and in small areas to the west of northwest Scotland (figure 6.15). Hake, on the other hand, spawn along the edge of the western shelf of Britain, between the coast and the 183 m line. Hake have a rather more southerly habitat than the cod and haddock. Like the cod, haddock remain in the plankton for their early life, staying rather longer till they are 5 cm long; during this time the circulation keeps them within the North Sea. The success of the brood depends very much on their survival in the plankton stage. There is great fluctuation in the success of broods of different years; for example, very poor broods survived in 1922, 1937, and 1946. Work by Rae (1957) has suggested that the fate of the haddock brood can be correlated with the extension of Atlantic water (as indicated by plankton species) down the coast of the North Sea. The movement of Atlantic water can in turn be related to the wind pattern over the area. The distance which this water penetrates off the east coast of Britain in late autumn depends on the strength and direction of the wind.

6 Pelagic fish

In contrast to the demersal fish, who spend most of their life near the bottom of the sea, the pelagic fish swim for part of the time at least near the surface. One of the most numerous types, which is of great importance commercially, is the herring family, *Clupea* (Plate 13). The family also includes the sprat, pilchard, sardines, and other near relatives. One important characteristic of this group of fish is their habit of swimming in shoals (schools) (Plate 14), which renders their capture easier and more rewarding. The herring itself is the most important of the family. It feeds entirely on zooplankton, particularly the copepod *Calanus.* Its movements and distribution are closely associated with those of its favourite food. Although members of the *Clupea* family are widely distributed around the world, the herring (*Clupea harengus*) is a northern fish. It extends from latitude 30°N to the Arctic. It is found on both sides of the Atlantic, on the west in the St Lawrence estuary and off the coasts of Labrador and Greenland. It lives around Iceland and in the open sea between Iceland and Norway, extending north of Novaya Zemlya and Spitzbergen, and along the coast of Norway from the White Sea southwards. It also is found all around Britain and the coast of Brittany as far as 47°N.

The most intensive fishing for herring takes place in the North Sea and off the coast

Plate 13 Herring: a common pelagic fish (*Natural History Museum*).

of Norway. Another branch of the herring family which lives in the Pacific is *Clupea pallasi*, which is found on both the Asiatic and American coasts. The herring around the North Sea can be differentiated by their characteristics into three stocks—the Baltic herrings, the Norwegian herrings, and the North Sea herrings. There is some intermixing of the different stocks. They also move and decline from time to time. The Baltic herring were particularly prolific during the Middle Ages. From the thirteenth to the fifteenth centuries they formed the basis of the wealth of the Hanseatic League, who fished for the herring in the Baltic. It has been suggested by Pettersson (1926) that their presence in the Baltic during this period was due to particular tidal conditions there which may have been responsible for the movement of the herring into the area. Between 1416 and 1425 the herring disappeared from the Baltic, followed by the decline and ruin of the Hanseatic League. The North Sea fishery for herring by the Dutch meanwhile increased in importance. It is likely that two different stocks of herring were involved, rather than that the Baltic herring moved to the North Sea. A more recent sudden change in the habitat of herring is indicated by the sudden disappearance of herring off Plymouth (Hodgson, 1957).

The herring that occupy Norwegian coastal waters are considerably larger than those in the southern North Sea. The former live to about 20 years and the latter only about 11 years. The age of a herring can be determined by counting the rings on its scales, which show annual growth rings, rather like those of a tree. The growth is quicker in summer, when the herring feed well, and slow in winter when food is scarce. The herrings of the North Sea and those of the Norwegian coast can also be differentiated by the character of their scale rings.

The herrings of northwest Europe can be divided into groups according to their period of spawning. Some spawn in the spring, another group spawn in late summer and autumn, while a third group spawns in winter. The positions and periods of spawning are shown in figure 6.16. The position of spawning influences the place of fishing. Thus the fish congregate in large spawning shoals off Shetland and northeast Scotland in summer, off Yorkshire in early autumn, and East Anglia in October and November, while they used to be caught off Plymouth in December and January. The North Sea herrings are divided into two groups, one of which spawns in the spring and the others in autumn and winter. The latter group itself seems to be composite, one part seems to move round the North Sea with the main current feeding in the north and moving south to spawn near the Dogger Bank, then migrating back to the northeast as spent fish. The second group feeds in the summer off the east coast further south, moving south in the autumn to make what used to be the most important herring fishery almost anywhere in the world off the East Anglian coast. They finally spawn at the eastern end of the English Channel in winter. They then drift northwards via the Dogger Bank to their more northerly feeding grounds.

The eggs of the spawning herrings are laid on the sea bed, the fish congregating in large shoals to spawn. The spawning shoals may be 14 km long and about 4 km wide. Each square kilometre probably contains about 200 million herrings, and each female lays about 10,000 eggs. This vast number of eggs attracts bottom-living predators, such as haddock and cod. The eggs are mainly deposited on the western side of the North Sea in autumn and winter, with spring spawning taking place in the north around Shetland and off southwest Norway.

When the herrings hatch they drift upwards to join the plankton for a time, feeding first on the minute phytoplankton and then changing to a diet almost exclusively of *Pseudocalanus*. If this particular species is in short supply there may be a heavy mortality among that particular year class of young herrings. At this stage they are very vulnerable to a large number of predators. As they grow they move up the estuaries, where they may be caught as whitebait. They live in this form for about 6 months and at this stage can eat almost the same diet as the adults. Thereafter they disperse for a number of years over the North Sea, only joining into shoals when they are sexually mature at 3–5 years old. Their nursery ground for this period seems to be situated east and southeast of the Dogger Bank.

A herring shoal usually contains a considerable variety of year classes. It was found that in some particular year the fish spawned were extremely successful. That particular year class could be followed for a considerable number of years, during which they made up a significant proportion of the whole stock. A study of Norwegian herrings over the period from 1904 to 1921 showed that the fish spawned in 1904 survived as a successful

Plate 14 (*opposite*) Bass: these fish illustrate the schooling habit (*G. W. Potts*).

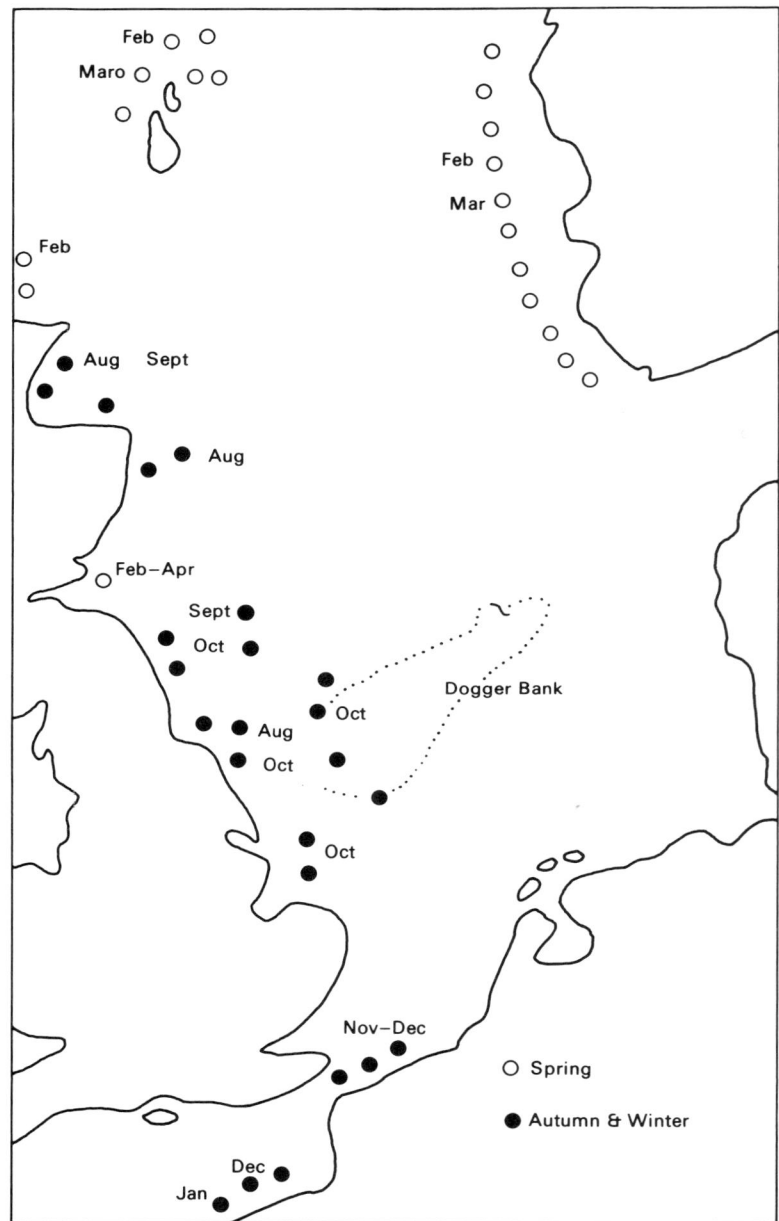

Figure 6.16 Herring spawning grounds in the North Sea. (*After Hodgson, 1957.*)

year class right through until 1922, when they would be 15 years old (figure 6.17). In 1910 when they were 6 years old they formed 80 per cent of the shoal. In 1917 another successful year class was spawned. The early period of their lives is vital to the success of the year class.

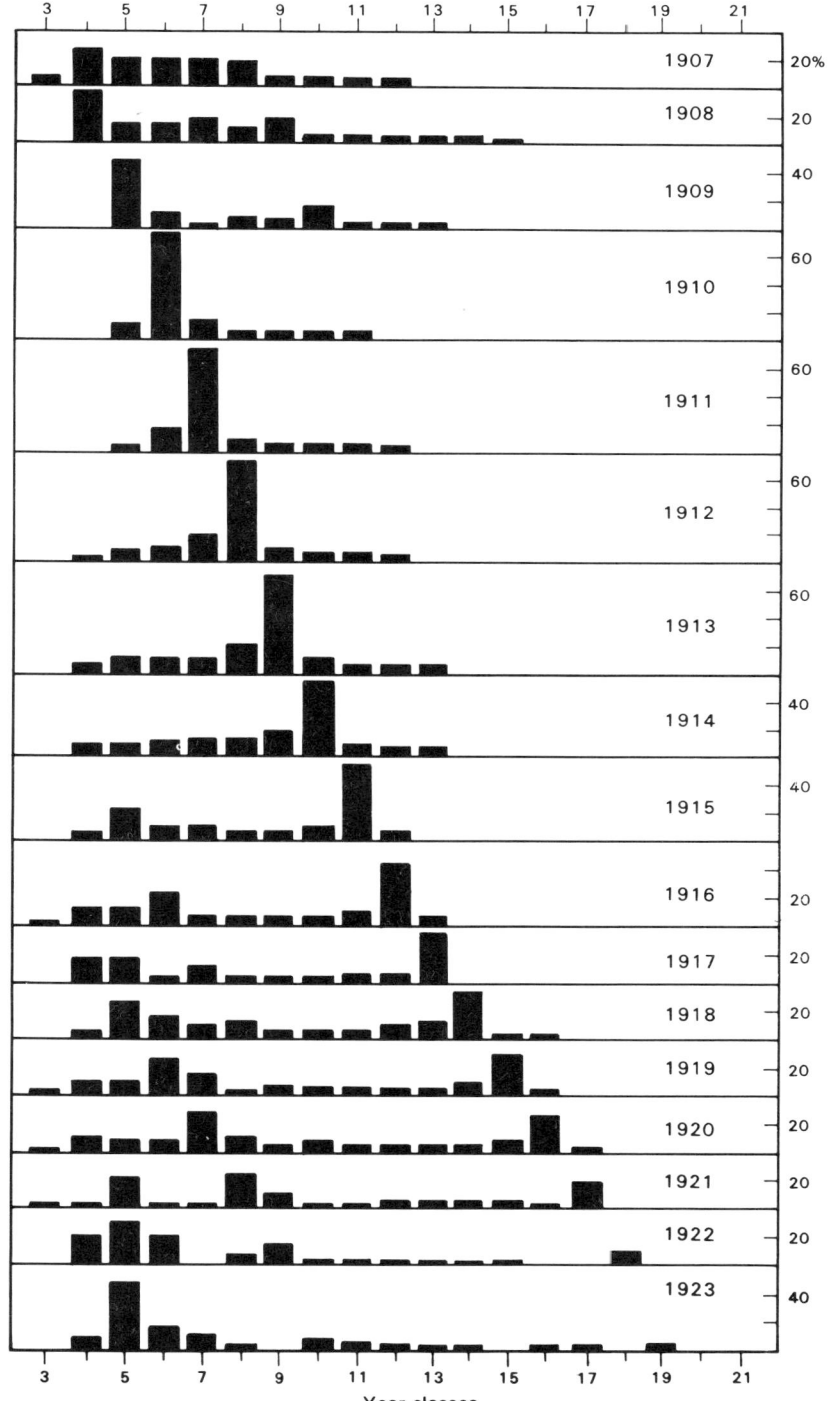

Figure 6.17 Age groups of Norwegian herrings in the North Sea. (*After Hogdson, 1957.*)

The adult herring feeds on zooplankton, but it is itself the prey of larger fishes and other creatures. Haddock eat its spawn, and its vulnerability at the planktonic stage has already been mentioned. The adult herring is preyed upon by killer whales and other whales, porpoises, sea-gulls, and particularly cod. It is attacked by the cod when it goes down the lower levels at night as a result of its diurnal migration.

Although herrings are usually caught in the upper layers of water, they are known to spend part of each day on the bottom or in deeper water. Like zooplankton, the herrings seem to follow the light. They go down during the day and rise to the surface at night. The Norwegian herrings have been observed to descend to 137 m during the day. They do not go to the bottom when the water is deeper than this. In the North Sea the situation is different, as light penetrates to the bottom. The fish, therefore, spread out on the bottom. Another characteristic of the herring is its shoaling instinct, which normally keeps the fish in well-defined groups, a tendency which starts when the fish are small whitebait. Each herring is trying to get into the safest place, which is the centre of a large shoal. This habit is of considerable importance when methods of fishing for herring are considered. The shoals are compact and rounded, but they vary with the life cycle of the fish. When feeding they spread out somewhat, but while spawning they are very compact. In shallow water they spread out on the bottom during the day. At dusk they rise in a dense shoal in what is called the 'swim', and spread out on the surface during the night. Herrings tend to break shoal at night when they stop feeding. Their habit of shoaling may be related to their pectoral fins. At times very large shoals occur, containing 50–200 tons of fish, while one exceptionally large one contained 22,000 metric tons. Shoals at times reach 1500 m in length and are cigar-shaped, with the long axis parallel to the tidal streams. The density of east American shoals is about 2 kg/m³. Large shoals off Iceland correlate with the numbers of *Calanus*, with mixed water, and with shallow irregular bottoms. Migrations of herrings are associated with the movement of their most favourable environment, so patterns vary from year to year, and are related to kineses. The fish may drift with the residual current after spawning to over-wintering or feeding areas, but a more directed movement is needed to get to the spawning grounds, and this has yet to be explained. It may be a reaction to temperature gradients or chemical factors, or possibly a result of orientation by the sun. There is no evidence of learning. Vertical migration is associated with avoidance of light, or following the preferred light intensity and temperature. They tend to remain at the base of the thermocline.

Herring are an important commercial species and work done on improvement of stock is important. Blaxter and Holliday (1963) described the successful moving of 19 million herring eggs from the Baltic to the Aral Sea between 1954 and 1956. The growth rate in length was 1·5 times and weight 4 times the normal. Other experiments of this type in Japan were less successful. A positive relationship has been established between growth rate and salinity. In the North Sea and English Channel herrings showed an increase in growth rate during and after 1950 due to an increase in the abundance of *Calanus*. The North Sea herrings do not mature unless they reach a minimum length of 21·6 cm in the southern part of the sea. Metamorphosis occurs when they are 25–45 mm long, depending on the species, the process taking 10 days where the temperature is 15°C.

Marine animals on the whole have high reproductive rates, and most of them have a larval phase, in which they are highly vulnerable, hence the need for large numbers of eggs. There is a relationship between larval mortality and spawning rate, and the losses are not mainly due to lack of fertilization, but to loss by being eaten. There is more than a 90 per cent chance that any marine animal will be eaten. One way some fish overcome this problem to a certain extent is to develop the habit of schooling, a habit that is adopted particularly by prey fish, such as herring (Plate 15). Schooling or shoaling gives the fish a better chance of escaping being eaten by predatory fish, which only have a certain

Plate 15 Mackerel: a school of pelagic fish (*G. W. Potts*).

capacity, than if they were to swim singly. The predators can catch fewer schooling fish, although as far as catching by man is concerned such fish are more vulnerable because of this habit.

7 Fish migration

The movement of fish and other marine animals is of interest from several points of view. It is closely associated with the life cycle and abundance of the fish. It is also closely related to the environment in which the fish live, and because of its spatial aspects, it is of particular geographical significance. The movement of fish can be divided into three classes: 1) local and seasonal movement, 2) dispersals, and 3) true migration. The second can be differentiated from the first by the fact that in dispersals the numbers remain most concentrated in the breeding area from which dispersal takes place. In a true migration the whole population moves from one area to another. Fish movements can be distinguished as denatant and contranatant movements. The former are those that follow the currents, either by swimming or by passive drifting, as in the plankton, while the latter involves moving (by swimming) against the current. In general young fish and eggs follow denatant movements, while mature fish follow contranatant movements (in reaching their spawning grounds, for example). Spent fish may use either type, but often the first. The return movement against the current by the adult is often necessary to bring the eggs and young fry into a region where denatant movements will carry them to suitable nurseries for the critical early stages of their life cycle.

There is a close connection between the migratory habits of fish and their commercial importance because the migratory habit is associated with abundance. When a fish population is very abundant the food may not suffice for both adult and young in the same area, so that a migratory habit that spreads the stock of different ages over a wider range of territory is valuable in maintaining abundance. Commercially important species are important because they are abundant, and they are abundant because they are migratory. The use of the same spawning areas by migratory species year after year ensures that favourable conditions enjoyed by the parents are also available for the young of the next generation. Commercially important species that are migratory include herring, salmon, cod, plaice, and eels.

In considering the migration of fish it is always necessary to take currents into consideration. These include ocean currents, tidal and also more local currents, and for some species, river currents. The speed of the current is important in relation to the speed that a fish can swim in contranatant movements. In general a reasonable estimate of the cruising speed of fish is that they swim at a speed about three times their own length per second. Thus a cod of 80 cm length could travel at 240 cm/sec or nearly 5 knots, while a herring of 25 cm would travel at 75 cm/sec or 1·5 knots. Even a powerful swimmer, such as a large cod, could not swim against the stronger ocean currents, such as the Gulf Stream, which at times reaches 350 cm/sec.

Evidence for fish migration has been obtained from tagging experiments, and from the study of otoliths, or ear bones, which at times give evidence of the origin of the fish, because different patterns develop under different environmental conditions. Fishery statistics can provide evidence of fish movement if they are used with care. Blood type can also be used to assess movements because each stock has a characteristic blood group. Direct observations are not often possible at sea, although some data are available. Fish shoals can sometimes be observed from the air and some useful evidence of movement has been obtained in this way. Success is only possible with calm surface conditions, clear water, strong sunlight, and fish. Sea planes and helicopters are used for tuna fishing in the Pacific and in locating sardines off California. Fish can also be identified on echo-ranging instruments and this technique is being increasingly used in fishing. The identification of fish echoes requires experience. Individual fish can be traced as they move by this method, or the method can be used to show changes in the distribution of fish over time.

The eel is one of the best known of migrant fish. They spawn in deep water, which is warm and saline, at a depth of 400–700 m, producing pelagic eggs that float. Freshwater eels are found in the Indo-Pacific region and on both sides of the north Atlantic. The American and European eels can be distinguished by their vertebral count, the American eel (*Anguilla rostrata*) has a 107·2 mean value and the European eel (*Anguilla anguilla*) has a mean of 114·7 and a range of 110–119. There is very little overlap in the counts. In 1965 the total catch of European eels amounted to 17,000 metric tons. There are two main theories of eel migration and its relation to their spawning habits. The first, according to Schmidt (1932), considers that the mature European eels return to the Sargasso Sea to breed, the elvers then moving with the currents to the rivers of western Europe, where they spend much of their life, only returning to their birthplace to spawn and then die. The second hypothesis, according to Tucker (1959), suggests that there is only one species of eel and that two types become distinct as they separate from one breeding area in the Sargasso Sea. This second view holds that mature eels do not return from Europe to the Sargasso Sea to breed, and in fact no mature eel has been caught making the journey. One important point is to determine whether the difference in vertebral count is environmentally determined or not. So far the question has not yet been solved.

The common European eel, *Anguilla anguilla*, lives much of its life in freshwater rivers, around the coasts of Europe, arriving as young elvers, 5–7 cm long, and eventually going to sea again years later as mature eels, ready to spawn, but never to return. The eels are about 5–7 years old when they leave the rivers, possibly to travel 3200 km across the ocean to spawn in the Sargasso Sea area. The eggs float in very deep water and as the young gradually emerge as larvae they are swept along by the north Atlantic drift, with the plankton of which they form part, until they eventually reach the coast. It is not yet known how the young elvers find their way back to the coastal waters and rivers. It has, however, been established that the young elvers do in fact move out from their spawning ground. Isopleths, linking points where the average size of elvers is the same, show that they increase gradually in size as they move away from their small spawning area in the Sargasso Sea, in latitude 25–30°N and around 60°w. In their first year they are found in the west Atlantic and average 25 mm in length; in the second year they are found in the central Atlantic with an average length of 52 mm; while in their third year they have reached the seas off Europe and average 75 mm in length. The American eels at least must return to the Sargasso Sea to breed, and it is likely that the European ones do also.

The herring is a very widespread pelagic fish, occurring both in the north Pacific and Atlantic Oceans, in the latter from Cape Hatteras to northern Labrador on the west and from the Bay of Biscay to Spitzbergen and the White Sea on the east. The northeast Atlantic herring can be divided into seven groups: 1) Atlantic–Scandian, 2) Hiberno–Caledonian (Atlantic), 3) Channel, 4) North Sea, 5) Skagerrak, Kattegat, Sound, and Belts, 6) Baltic, and 7) estuarine, fjord, and local stocks. Most of the groups remain separate and different groups spawn at different times, getting later in the year further south. The Norwegian spring-spawning herring are the best-known group from the point of view of migration. The young fish spend their first year in fjord or coastal waters between Bergen and Finmark, a northern and southern type being identifiable. The fish move north in the summer and south in late autumn, and they come closer to the coast in summer, moving offshore as they get older and in winter, until as they mature they stay offshore until they spawn. The larger fish spawn first, and then move west and northwest into the Norwegian Sea, reaching as far as Spitzbergen, the Mohn Ridge and Jan Mayen, to the north, and the east Iceland current in the south. Smaller, later spawners do not move so far north and west. In autumn their feeding area contracts and they are in the southwest Norwegian Sea area, wintering in the area north of the Faroes, and moving towards the coast for spawning in early spring off Norway. They move 2880 km during the annual migration. They appear to keep within the cyclonic system of currents in the Norwegian Sea, and it has been suggested that currents determine their movements. Seasonal changes in their food supply may also play a part, as they move to the northwest with the biological spring after spawning. Evidence for these movements has been derived from tagging and sonar observations. The herring must be able to return to the right spawning area, and within this to the specific ground and bed within the ground, as each ground has localized beds. Evidence for the return of herring to their own spawning ground to spawn in their turn is provided by vertebral counts. The more southerly the ground, the higher the vertebral count, suggesting that herrings do return to the same ground to spawn. There is no evidence that herring return to spawn on the same bed, however, on which they were spawned. The data concerning the return of the North Sea herring to their own spawning ground are not conclusive. These herrings have well-defined and relatively small spawning areas.

The cod of the north Atlantic are the most valuable of the commercial fish, and like the herring they form separate stocks, although there is some movement between the stocks that occupy different parts of the area. For example, cod tagged on the spawning grounds off southwest Iceland have been recovered off Newfoundland, the Faroes, North Cape, and in the North Sea, although this is rather exceptional.

Cod spawn in spring and the eggs float, thus drifting with the currents, hatching in two weeks, and remaining floating or in shallow depths until they are 3–4 months old, when they take to the bottom. Some may be even older so that during this period they can drift considerable distances; moving with a current of 10 cm/sec they would cover over 1000 km in 5 months. Movements of immature cod are difficult to trace as they are too small to tag. They appear to move about more as they become older, probably moving in search of food. Barents Sea cod go to deeper water in winter. They make a spawning migration as they become mature, swimming near the sea bottom. They have been trawled from depths of 460 m. The Arcto-Norwegian cod mature at 8–10 years and after spawning they return to the feeding grounds, moving into shallower water for the summer, and often leaving the bottom for considerable periods.

The Iceland cod spawn off the southwest coast in the warm water of the Irminger current, which carries some of the eggs and larvae round to the north coast of the island, while other eggs are carried by the branch of the current that sets towards Greenland across the Denmark Strait. It would take larvae about 3–4 months to reach the Cape Farewell area from Iceland, a distance of 1600 km. Young cod probably also reach the west coast of Greenland. The spent fish disperse round Iceland and some have been recovered on the Norwegian coast, the Faroes, or Newfoundland.

Cod also spawn along the coast of Norway and round to the east of the North Cape to the Murman coast. Three-quarters of these fish come from the Barents Sea, the rest being of local origin. Tagging experiments have shown the migration from the Barents Sea area. There is a regular movement of cod with the seasons. In late summer the mature and immature fish are at the northern limit of their feeding migration around Spitzbergen and the northeast Barents Sea. In October they move south and west towards Bear Island and the Norwegian and Murman coasts, retreating into a much smaller area for the winter period. In winter they concentrate along the zone where the warm Atlantic water meets the cold polar water, but are nearer the Norwegian coast in cold winters. The fish move at about 5 km/day during the winter migration. In December the cod are in deep water at the edge of the Bear Island Bank, but the large, mature fish soon move onto the spawning grounds. They move at about 11 km/day towards the Norwegian coast spawning areas from the southeast Barents Sea. They leave the coastal spawning banks as spent fish in April to move rapidly up the Norwegian coast at about 29 km/day, reaching Bear Island banks about May and June. The eggs and larvae meanwhile drift with the north-going current along the Norwegian coast, the young fish taking to the bottom in the Barents Sea. The numbers of cod involved in these movements has been estimated at about 424 million fish, on the basis of the heavy fishing which probably catches about half the exploitable stock of Arcto-Norwegian cod. About 80 million of these fish will be mature.

The movements of mature fish consist of three types: 1) feeding area to wintering area, 2) wintering area to spawning area, and 3) spawning area to feeding area. These movements can also take place in the reverse direction. Winter and spring spawners tend to move from the feeding area to the wintering area and then to the spawning area. The summer and autumn spawners, on the other hand, move from the feeding area to the spawning area, and then on to the wintering area.

One species may have several spawning areas, one of which tends to produce the major group of progeny, according to the environmental conditions, in any one year. Each group would grow up in different nursery areas and have different characteristics as a result. A few survive from each spawning area to which they return, thus developing the separate stocks that are characteristic of so many species. In order to maintain the total numbers some flexibility of movement between the stocks is advantageous, so that areas that become unfavourable may be avoided. Thus, too rigid a migration is not desirable from the stock point of view. This is particularly true in areas where some grounds have been contaminated by pollution. There has, for example, been some change in the balance between the plaice spawning grounds of the German Bight and Southern Bight of the North Sea, while the success of the herring may be partly due to the wide range of spawning times and places. Capacity to meet change is based largely on the number of spawning units, and not on the flexibility of each unit, which is controlled by the migratory habits of returning to the same place to spawn each season. The demersal eggs of the herring facilitate the use of the same spawning ground by each unit, as landmarks are available to assist accurate location, thus putting the larvae in a more favourable position.

Migratory movements by fish require an ability to home to a particular region, which could be reached by several possible methods, including 1) passive drift, 2) search, either random or directed, 3) undirected movement or kineses in relation to environmental stimuli, and 4) directed movements. Poor swimmers must drift with the currents, but powerful swimmers may also do so if their course is not directed by some means. Eggs and larvae drift, and spent fish may also drift with the currents. The Arcto-Norwegian spent cod provide an example of movement by drift. Little detailed work has yet been done on the search movements of fish, but work on salmon has shown that random search is unlikely to be effective. One difficulty is that fish in open water must be able to move in a straight line, and there is little evidence that they can do so. Kineses, or movement related to external stimuli, are unlikely to be effective in enabling fish to home, as most of the environmental gradients in the sea are very slight. Directed movements must depend on some mechanism enabling the fish to react with the environment, to give clues concerning the required direction to be taken.

8 Marine mammals: whales and whaling

Marine mammals belong to the order *Cetacea*. They are warm-blooded and feed their young on milk as do other mammals, yet the whales in particular spend their whole life in the sea and are perfectly adapted to this environment. Seals, on the other hand, still come to land to breed. Whales include the largest animals ever known. They are larger than land animals ever could be, as their great weight can be supported in the water. The largest whale measured reached about 30 m in length, while a whale 27 m long weighed 120 tons. The advantage a whale derives from its great size is its ability to swim faster. It can develop greater muscle power, owing to its large volume, while its relatively smaller surface area creates less friction. It is the reverse of the argument used to account for the small size of the phytoplankton. Whales have been recorded as travelling at 20 knots (36 km/hour) for a short time.

Whales cannot be very small, because being warm-blooded, they would lose too much heat if their volume were small. Volume decreases more rapidly than surface area with diminishing size. The smallest marine mammal is a dolphin, which is still about a metre long. There are a wide variety of types of whales, each with its own feeding habits. They can be broadly divided into whalebone whales and toothed whales.

The whalebone whales have enormous mouths, in which there is an elaborate filtering device, as they are filter feeders. Water is taken in and forced through the filters by the tongue, which then scoops the trapped planktonic animals into the gullet. The large whales eat the larger animals of the zooplankton, particularly the shrimp-like krill (*Euphausiacea*). The food chain of these largest creatures is remarkably short. The whalebone whales belong to two main families, the first of which is now very rare due to overcatching. These are the right whales and the rorquals. A method of telling their age has been discovered by studying their ears, which has shown that whales can live to at least 50 years of age.

The right whales were so called because they were the correct ones for catching. Too many have been caught and the Atlantic right whale (*Balaena glacialis*) is now very rare, having first been hunted by the Basques in the tenth and eleventh centuries and also by the Norwegians. Whaling was mentioned in a report made to King Alfred about a voyage to the White Sea in about 890. At one time it was thought that this whale and the more northern one, the Greenland right whale (*B. mysticetus*) were extinct, but a few have been seen subsequently. Whales were caught for their baleen plate of whalebone, which were up to 2·7 m long. A single whale might produce 1·5 tons of it at a price of £2000 per ton, as well as about 30 tons of oil, so it is not surprising that they were fished nearly to extinction in the seventeenth and eighteenth centuries. The last northern whaling ship sailed in 1868.

The right whales had been hunted up to this time because they were more sluggish than the rorquals, which could not be hunted with the equipment available at the time. The development of the explosive harpoon in 1865, however, allowed these more powerful and often larger whales to be hunted successfully, from the period when the northern catching declined. The rorquals of the north suffered the same fate as the right whales before them. Hunting then moved to the recently discovered wealth of rorquals in the Antarctic seas, which were exploited from the beginning of the twentieth century. The first southern whaling station was set up in South Georgia in 1904 (Brown, 1955). Whale catching in this area rapidly spread and became very extensive, there being eight whaling stations by 1911. By 1953–54 there were 227 catchers, working from 17 stations, who together caught 34,869 whales, providing 2,285,730 barrels of oil. From the commercial point of view, the most important whales are now the rorquals and the humpback and Sei, and to a growing extent the toothed or sperm whales.

The whalebone whales include the blue whale (*Balaenoptera musculus*) which is the largest, reaching 30 m in length. In 1955–56, 1987 of these were caught. The fin whale (*B. physalus*) is about 26 m long, and 31,496 of these were caught in the same season, amounting to 54·2 per cent of the total catch, as against only 3·4 per cent for the blue whale. Then the hunters turned to the smaller Sei whale (*B. borealis*), 15 m long, of which 2076 or 3·5 per cent were caught, and finally to the humpback (*Megaptera novae-angliae*), also 15 m long, of which 3880 or 6·6 per cent were caught. The sperm whale is 18 m long and 18,590 of these were caught, amounting to 31·9 per cent of the total whales caught. During the first part of this century the great toll on whales was reaching such proportions that steps were taken to stop it. A tax was levied on the industry, some of which was used to collect scientific data on whales, and this led to the *Discovery* expedition between 1925 and 1939.

The humpback whales live mainly in the southern hemisphere, and it is only there that they are caught. There seems to be five different stocks of these whales, which live more or less independently. In winter they travel to tropical coastal waters, where they are sometimes hunted, and return to the Antarctic for the southern summer. The whales are rather slower, smaller and fatter than the others so they are more easily caught. They are taken in regulated numbers in winter in Australian and southwest Pacific waters, and also in summer further south. The true rorquals are all world-wide in distribution and also travel towards the equator in the winter where they breed. In the southern population of whales there seems to be about one blue whale to every five fin whales, and it has been estimated that between 1933 and 1939 there were about 250,000 fin whales in the southern population. A limit on the number of these whales that can be caught was imposed by international agreement in 1937 (Brown, 1955). The post-war catch was limited to about two-thirds of the pre-war numbers. Blue whales were protected until late in summer season, which halted their deline. It is extremely difficult to get precise information concerning the state of the fin whale stock, but many biologists consider that the stocks are still declining. The four species of rorquals may move along the western coast of Britain to breed in the warmer seas. There does not seem to be much evidence of an interchange between the northern and southern stocks, as the northern ones have never been recovered despite the great activity in the south over a long period.

Blue whales feed entirely on krill, but the fin and humpback whale may also take herring and other shoaling fish in their diet. Whales in the south remain in the areas where the krill is abundant. Nutrient-rich water to the south of the Antarctic convergence supports a very rich planktonic life, which helps to account for the large number of whales in this zone. The Sei whales have a finer seive than the others and often live on *Calanus*. Blue whale calves are born about every two years (usually in May), one at a time, but occasionally twins are born. They are about 7 m long at birth. They remain with their mother till they are weaned at about 6 months, when they are 16 m long and have reached the areas where the krill is plentiful. At a year old the calf is 11 m long and by the time it is 2 it becomes a mature adult of about 23 m length, although it may grow a further 8 m.

The sperm whale (*Physeter catodon*) has a large blunt head with a narrow underslung mouth, in which there are many teeth (Plate 16). The sperm consists of liquid wax, which fills the broad head. It is thought to be a store of fatty emulsion which can absorb the nitrogen released as the whale dives and surfaces rapidly to and from considerable depths.

Plate 16 Sperm whale, showing broad head and narrow underslung mouth (*G. W. Potts*).

They must go to great depths as they have been found entangled in submarine cables at a depth of 1130 m. They go considerably deeper than rorquals. They do not suffer from over-catching to the same extent as the rorquals because their oil is less valuable. Sperm whales live in warmer waters, and only the males, which are twice as big as the females, go to colder latitudes. They are polygamous, so that they can be relatively easily controlled by limiting the size caught ensuring that only males are taken. An important sperm whale industry is developing in the north Pacific, and there is also a sperm whale fishing industry in the Azores.

The favourite food of the sperm whale is large squid. Occasionally very large ones are eaten, although the average size from the contents of 112 whale's stomachs was just over 1 m, excluding their long arms, but one squid taken intact from a whale was altogether 10 m long (Clarke, 1955). This is, however, not their only food, which may include basking shark, benthic animals such as skate and crabs, and other animals. Recoveries of marked whales numbered 68 in the 1957–58 season, of which 10 were 20–25 years old, all of them being fin whales. One blue whale had moved from 62°47′s, 60°e to 67°24′s, 127°25′w between 17 December 1955 and 22 February 1958, a total of 170 degrees of longitude. It is not known whether the whale travelled eastwards or westwards. By contrast a fin whale was recovered after 22 years within 240 km of the place where it was marked in 1936. The movement of humpback whales from the Antarctic area north of the Ross Sea, in 65°49′s, 179°45′e, to east Australian water has been confirmed by the recovery of a marked whale on 27 June 1958 at 28°38′s, 153°43′e, which was marked on 27 December 1957. These experiments are giving more precise information on the movement of the whales. During the succeeding season, 1958–59, 102 marks were recovered. Some of these whales had been marked 24 years previously, and further evidence confirmed the movement of the humpback whales between the Antarctic and lower latitudes (Brown, 1958 and 1959).

9 The total organic production of the sea and the food pyramid

All the organic production of the sea is derived ultimately from the phytoplankton. These minute plants feed the zooplankton and other fish, and so through the food web till the main commercial fish are reached. The relationship between the different species, or the ecology of the seas is an enormous topic. Some of the interrelationships, however, have already been mentioned. Both the standing crop and the annual production must be considered and proportions of these will vary from species to species. Thus the phytoplankton, which increases with extreme rapidity, has a much larger annual production than standing crop. At the other extreme, the large whales show just the reverse, as they only reproduce biennially, and then usually only one offspring at a time is born.

An interesting attempt to arrive at an estimate of the organic budget of the seas off southern California has been made by Emery (1960, p. 175). The annual production of phytoplankton is estimated at about 42 million tons dry weight, while 1·4 million tons of seaweed is produced, making a total plant production of about 44 million tons. This, however, only represents about 0·18 per cent efficiency of conversion of solar energy into plant tissue. The zooplankton then feeds on the phytoplankton, much of the material eaten being turned again into chemical nutrients to re-enter the cycle. The annual production of zooplankton is estimated at about 3·4 million tons, 7·5 per cent of the phytoplankton production. The ratio between the standing crop and annual productivity is estimated at 35.

The commercial catch of fish over 5 years average was 160,000 tons, plus 7000 tons of fish caught for sport. This averages about 0·03 million tons by weight of dry substance. Tagging experiments with sardines suggest that the fisheries are catching about 28 per cent of the total. Taking 5 per cent as the average catch for all fish, and a life span of about 6 years, the annual production of fish can be estimated as 0·1 million tons dry weight. This is only 3 per cent of the zooplankton production and 0·2 per cent of the total phytoplankton production. The annual production falls as low as 300 tons for the sea mammals, dolphins, and porpoises, a negligible amount in comparison with the fish. A standing crop of 5·5 million tons wet weight of benthic invertebrates is a reasonable estimate. The annual production of this group is fairly high, being about twice the standing crop in shallow water. The total is 7·4 million tons wet weight, or 1·5 million tons dry weight, as their water content is 80 per cent. This is about 3·4 per cent of the annual plant production. Of the total organic production, about 7 per cent reaches the bottom, but only 0·6 per cent of the total production is permanently lost in the bottom.

This type of analysis, although at times the data on which it rests are by no means exact, does give some idea of the nature of the food pyramid in this part of the ocean. There are vast numbers of tiny plankton plants at the base, supporting the rather larger planktonic animals. Varying food chains link these with the benthic invertebrates, the fish, and finally the marine mammals, with their very great size but relatively few numbers. It is more economical to catch and use animals or even plants which have a fairly short

Table 6.3 Ocean productivity

Area	Carbon fixation gC/m²/day	gC/m²/year
Tropical areas		
Open ocean waters	0·05–0·15	18–55
Equatorial Pacific	0·50	180
Equatorial Indian	0·20–0·25	73–90
Upwelling area	0·50–1·00	180–360
Sargasso Sea	0·10–0·89	72
Temperate areas		
Continental shelf off New York	0·33 mean	120
Fladen ground North Sea		57–82
Kuroshio current	0·05–0·10	18–36
Oyashio current	0·25–0·024	?
Arctic Ocean	0·005–0·024	1
All oceans estimated mean	0·137	50 (361·1 × 10⁶ km²) area

Dry weight organic matter below 1 m² in g/m² daily production			
Phytoplankton	0·4	0·4–0·5	The first set of values
Zooplankton	0·5	0·15	gives the standing crop
Pelagic fish	1·8	0·0016	and the second the
Bacteria	0·04	—	daily production
Demersal fish	1–1·25	0·001	
Epifauna and infauna	17	0·03	

food chain. Thus the whales and the herrings, both of which are plankton feeders, have a shorter food chain than the cod, which feeds on herrings and other fish, or the killer whales, which eat the seals, which themselves eat fish and so on down the chain to the phytoplankton, the basis of marine life. Much more nourishment for the growing human population could be obtained from the sea if there were some means by which the phytoplankton could be used directly, but this has not yet been achieved economically.

The limit to the harvest of the sea is set by the primary productivity of the ocean. Some comments have already been made on this aspect in the consideration of plankton at the beginning of the chapter. Table 6.3 summarizes some data on oceanic productivity. The most fertile areas are the upwelling zones, and the general fertility of the ocean is indicated in figure 7.1, in which the fertile areas are shown to be the major zones of upwelling. The reason for this is usually considered to be the increase of nutrients brought to the surface in the rising water. The view has also been expressed, however, that the blooming of the phytoplankton in these areas could be the result of cooler water reaching the surface. In a cool water ·region there are greater seasonal charges in temperature. This concentrates the phytoplankton bloom into a shorter period of time and the greater density during the bloom supports a large zooplankton population at this time. This in turn allows for concentration of larger numbers of fish in the area and hence a more productive fishery. Fishing can only be productive where the fish are concentrated. The seasonal phytoplankton regime assists the concentration of fish in larger shoals and hence in more catchable quantities. This pattern is also characteristic of the higher latitudes, which are generally more fertile than the lower latitudes.

If this concept is correct then the factors on which the phytoplankton bloom depends are very important. The variability of the success of different year classes lends some support to the view, and it seems likely that local weather conditions are important in

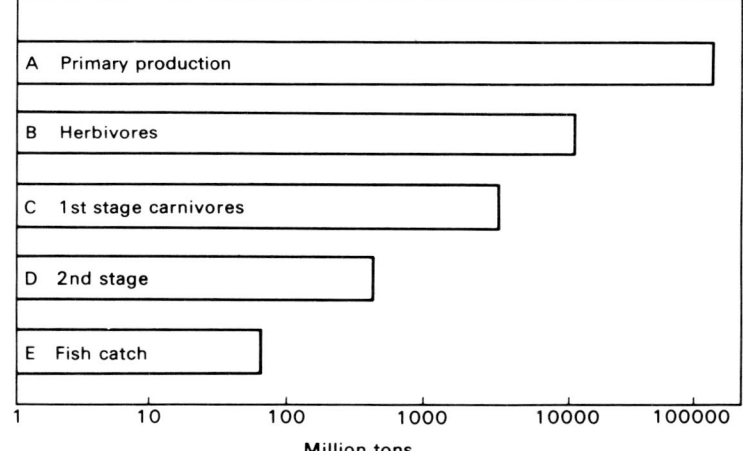

Figure 6.18 The loss in the food web between primary producers at the top and fish catch at the bottom. The horizontal scale is logarithmic. (*After Morris, 1970.*)

determining the exact place and time of the phytoplankton bloom in any one area. A close watch on world weather may, therefore, help to predict the place and abundance of the fish that depend on the primary production of phytoplankton through the food chain.

One of the problems of making better use of the primary production of the sea is the loss in the food chain. Bardach (1968) estimates that there is a loss of 80–90 per cent at each step in the food chain. The weights estimated for different types of marine organisms related to the phytoplankton give the following results: phytoplankton—100, zooplankton —70, bottom invertebrates—11, small plankton feeding fish—5·5, demersal fish—1, and large predatory fish—0·3. It has been estimated that the total fish catch of the world in 1960 used about 15 per cent of the organic matter synthesized by photosynthesis by phytoplankton. The fish catch of 1960 required as its plant base between 60 and 70 times the weight of the world's wheat crop. Ocean fertility in comparison to the land can only produce about one-tenth of land food even without loss in the food chain. Figure 6.18 shows an estimate of such loss.

Suggestions for harvesting plankton directly have been made, but are generally thought not to be feasible. As far as phytoplankton is concerned it would require 1 million gallons of water to obtain 1 lb (0·45 kg) of plankton. The most expensive fish would be 20 times cheaper than the plankton. Cost of fertilizing the sea would also be prohibitive, as $40,000 would be required per square mile, while the fish return would be worth only $4000. The total annual production of phytoplankton is estimated at 150 billion tons in the 2–3 per cent of water-mass that is in the euphotic zone. Because even now a relatively large proportion of this total is used by the fish caught commercially, the total fish harvest from the sea cannot be increased indefinitely. The best estimates indicate that the potential yield from the sea is about three times the present catch of about 64 million metric tons in 1968. The maximum sustainable world catch is of the order of 100–200 million tons. This volume will probably have been reached by 1985, or at least by the end of the century (Holt, 1969).

The relationship between benthic biomass and surface productivity is considered by Rowe (1971b). He compares animal densities and biomass under different ecological conditions. The relationship between the depth in m (X) and the biomass in $\log_{10}$ no./m^2(Y) is given by $Y = 3\cdot4 - 0\cdot00035(X - \bar{X})$ from 0 to 6000 m. The average biomass is higher in fertile regions where primary productivity is higher. Rates of decrease (the b value) were greatest where the productivity varied markedly in an offshore direction. The major control is the depth and productivity plays a secondary role. Where oxygen is at a minimum off Peru the benthic fauna is numerically dense, but low in biomass and diversity. Diverse communities with higher biomass occur in deeper water offshore. In this situation productivity plays a greater part than depth in the biomass variation. The regression equations for depth (X) and $\log_{10}$ mg wet weight/m^2(Y) was $Y = 3\cdot86 - 0\cdot00030(X - \bar{X})$ for Peru in >800 m, and $Y = 3\cdot60 - 0\cdot00054(X - \bar{X})$ for the Gulf of Mexico.

Further reading

BEERS, J. R. and STEWART, G. L. 1971: Micro-zooplankters in the plankton communities of the upper waters of the eastern tropical Pacific. *Deep Sea Res.* **18** (9), 861–83.

COLEBROOK, J. M. 1969: Variability in the plankton. Part II. In M. Sears (editor). *Progress in oceanography.* **V**. Oxford: Pergamon Press, 115–25.

COSTLOW, J. D. (editor). 1971: *Fertility of the sea.* Vols **I** and **II**. New York: Gordon and Breach Sci. Pub.

FRASER, J. 1962: *Nature adrift. The story of marine plankton.* London: Foulis.

HARDEN JONES, F. R. 1968: *Fish migration.* London: Edward Arnold.

HARDY, A. 1956: *The open sea* **I**: *the world of plankton.* 1959: *The open sea* **II**: *fish and fisheries.* London: Collins.

MCCONNAUGHEY, B. H. 1970: *Introduction to marine biology.* St Louis: Mosby.

MAUCHLINE, J. 1969: The biology of euphausiids. *Adv. in Mar. Biol.* **7**, 1–454.

RYTHER, J. H. 1969: Photosynthesis and fish production in the sea. *Science* **166,** 72–6.

THORSON, G. 1971: *Life in the sea.* London: WUL Weidenfeld and Nicolson, 256 pp.

WIMPENNY, R. S. 1953: *The plaice.* Buckland lecture for 1949. London: Edward Arnold.

7 Biological exploitation of the oceans

One of the major problems confronting the world is the continued increase of population, in relation to the resources available to produce food for the growing numbers. There are varying estimates concerning the relationship between these two factors. It is uncertain how soon, if at all, a shortage of land-derived food will be really serious. Given more efficient means of distribution that prevaile at the moment the situation might be contained. At present, however, some parts of the world population are seriously underfed, while others have a superfluity of food.

In comparision with farming on land, the resources of the sea for food are very inefficiently used. Obtaining food from the sea has not yet advanced beyond the stage of hunting. In the sea even the basic knowledge on which marine husbandry could be based is rarely yet available. The problems of the marine environment make research into the ecology of the sea a slow and difficult process. However, study of this type is essential to determine what can best be done to husband marine resources. The difficulty of implementing the findings, particularly as they must depend on international co-operation in an uncontrollable environment like the sea, will be immense.

The food production of the oceans is involved with political and commercial factors which determine the size and character of the so-called territorial waters. Some countries such as Iceland depend mainly on their fishing industry for their trade balance and exports. It is, therefore, natural that they should wish to protect their main fishing grounds, spawning and nursery areas from over-fishing by other fleets by extending their territorial waters and excluding other fishing fleets. It is equally reasonable that trawlers from Britain and other European countries should wish to fish in the waters that they know, especially when the fish near at hand are being heavily over-fished by boats of many nations, as is happening in the North Sea. Thus quite apart from the difficulty of reaping a controlled harvest from the sea due to lack of knowledge of the ecology of the sea, there are the human problems as well as rival political and commercial claims.

One of the problems is to relate the major areas of need with those in which the natural production of fish is greatest, that is, to relate the supply to the need. It is difficult to determine how closely the pattern of primary production can be related most profitably with fishing activity. The bulk of fish are caught in coastal upwelling zones and in favoured shallow seas such as the North Sea, Newfoundland Banks area, and other wide shelf zones; nutrient supplies are rich and fish grow abundantly. The rate of fishing, however, exceeds natural replacement in some areas, resulting in the problems of over-fishing.

The high southern latitudes are very rich in nutrient supplies, but fishing in this region is limited for economic and physical reasons. It is far from the main centres of population; physical difficulties include the lack of suitable shallow shelves which provide the best fishing in the northern hemisphere. In the far southern latitudes, however, there used to be large whales which provided some of the best value in marine food because their food chain is short. This valuable source of food has been seriously over-exploited, and international agreement is essential if this source of supply can be revived and used efficiently. There is some room for expansion in fishing in the equatorial and tropical waters, although there are problems of fertility in parts of this area. Growth rates in the warmer seas could partially compensate for the smaller supply of nutrients in parts of the area.

The oceans are probably not so efficient as the land at converting the energy of light into

Table 7.1 Areas proportional to sea food in percentage terms

Japan	185	Malaya and New Zealand	18
Vietnam	120	Burma	18
Korea	100	Philippines	17
Taiwan	69	Indonesia	16
Norway	64·5	Sweden	12
UK	38	Chile	11
Peru	34	Ecuador	8
Portugal	29	Honduras	5·5
China	24	Costa Rica	3·6
Ceylon	23	USA	3·3
Thailand	20	Panama	2·4

organic matter, and certainly the use of fish is much less economic of organic matter than the use of plant food on land. As yet there is no acceptable means whereby the plant production of the sea can be harvested directly for human consumption. Earlier predictions of the potential harvest of the sea have proved to be underestimates. In 1962 an estimate of 55 million tons was made, an amount exceeded in 1965. If the total reaches the value of 200 million tons, this would supply the minimum protein needs of 5000 million people if it were evenly distributed, which would seem unlikely.

The extent to which different countries depend on the sea for food can be assessed by comparison of their fish catches with the arable land that would be needed to equal the sea food obtained (table 7.1). The world fish catch declined for the first time since 1960 in 1968, notably off Peru and in the Arctic. This is despite the probability that the sea may be becoming richer in its total productive capacity, due to the return of sewage detritus and nutrients from land fertilizers, even though much fish and other organic matter is extracted. Plankton in the high latitudes has a higher food value than in the tropics, owing to a higher fat content.

1 Exploitation: general

1.1 Methods of fishing

Modern fishing methods, although they are much more advanced than those used by primitive peoples, can still be classed as hunting rather than husbandry. The seas are not farmed as is the land: fish are caught in their natural state, and man must compete with all the other predators. Methods of fishing are numerous, but they can be classified into a number of types: 1) Fish may be snared. They catch themselves in a net or device into which they swim and from which they cannot escape. 2) They may be lured into capture by some bait, such as fly fishing, or the lobster pot technique. 3) They may be hunted, with nets of various types, which are dragged through the water (Plate 17).

The different methods are used to catch different fish according to their behaviour. Successful fishing, therefore, depends on a sound knowledge of the behaviour of fish, and a successful skipper must be a good naturalist. A good example of the first method of catching fish is that used for many generations to catch the herring. This is the drift-net method. The drifters depend for their success on the shoaling habit of the herring and on their diurnal movements through the water. The drift-net method is only suited for pelagic shoaling fish, which swim near the surface for part of the day. The nets consist of a single wall of netting that hangs down vertically in the water and is supported by floats. A large number of nets may be joined to form a continuous wall of netting up to 3·2 or even nearly 5 km long. They are made of a series of nets each 15·2 m deep and 45·6 m long. Each boat has about 80 nets of diagonal mesh, into which the fish swim and are held by the gills. As the term implies, the drifters merely lie in the water, waiting for the fish to come to them. They go out in the evening, and return in the morning, hauling in the nets in the early hours of the morning. This timing coincides with the upward movement or 'swim' of the herring shoals towards the surface during the night. Once caught the herrings are hauled on board and shaken out of the net by extending the diagonal pattern of the meshes. The skill of fishing depends on shooting the nets in the area and at the depth where the fish are. A good catch of herrings can amount to between 50 and 100 crans. This measure is a volume of fish, which on average includes about 1000 herrings, varying between 900 and 1300 according to their size.

The movement of the herrings is dependent on the wind and tide. A strong offshore wind concentrates them close to the shore and results in heavy catches. This is due to the development of an undercurrent in the opposite direction to the wind-generated surface drift. East Anglian fishing appears to be influenced by the full moon, which is related to the period of spring tide. The heaviest catches take place when the full moon occurs in the middle of the season, during the second week of October (Hodgson, 1957).

Fishing by drifters has been going on for a very long time. There is evidence to suggest that the Yarmouth herring fishery started about AD 495, after the arrival of the Saxons. Herring fishing was important by Domesday when the port of Dunwich, now lost beneath the sea by erosion, paid 60,000 herrings annually to the king. The wealth of the Hanseatic League has already been mentioned in connection with the Baltic herring fishery, until the fish deserted the area. After this, from the fifteenth to the eighteenth centuries, Holland was the leading herring fishing nation and became very wealthy on the proceeds of the

Plate 17 Icelandic fishing boats in Reykjavik harbour (*Icelandic Photo & Press Service*).

fishery. It is said that Amsterdam is founded on herring bones. The industry caused disputes between the Dutch and the English, the latter demanding tribute for fishing in English waters. This led to the 1652–54 war, from which British naval supremacy dates, and which led to the decline of that of Holland. It was, however, not until the early nineteenth century that the British herring fishery became important.

This development started first in Scotland. More than a thousand sailing herring drifters worked in Scottish waters all the summer and then moved south, with the herring. Towards autumn the Yorkshire coastal waters were fished, and then the climax of the herring season took place off East Anglia, based on Yarmouth and Lowestoft, in October and November. The boats followed the herring on their spawning migration to the southern North Sea, so that they were caught in good, full condition, just before spawning. Sailing drifters have now given place to steam drifters, and their numbers have declined rapidly during the twentieth century. The peak of the herring fishing was reached in 1913, when England and Scotland together produced 11,762,748 cwt of herring valued at £4,412,838 out of a total European catch of 22,018,130 cwt. Most of these herrings were exported salted. There was a very rapid decline in fishing in the 1930s when the markets disappeared. Of the 1000 boats of 1913 which frequented the East Anglian fishing ports, only 106 remained in 1959, while the whole season's catch has been reduced to about 50,000 crans, the amount caught in one day in the past.

Successful herring fishing depends on the location of the shoals. Various methods can be used, such as sampling the water, to ascertain whether it contains *Calanus* in the plankton. Echo-sounders are now widely used to locate shoals. The sound wave reflects from a tightly packed shoal of fish, giving information about their position and depth. All modern drifters, now normally diesel-driven, are fitted with echo-sounding devices. The fish can only be caught by the drift nets when they rise to the surface in the swim at night. They cannot be registered by the echo-sounder when they are on the bottom, but only when they are on their way up. One of the problems of catching herring is their continual movement to and fro from the spawning grounds and in search of food. At times they move to avoid water that is distasteful to them on account of noxious plankton. There is evidence to suggest that shoals of herring can move up to 16 km in 24 hours, when they are not spawning. Such movement makes it difficult to track them, and the secret of successful drifting is to know enough about the behaviour of the fish to anticipate where they will rise for the swim.

Line fishing depends on attracting the fish by a bait, fixed to a hook on the line. As a method of catching fish it was used before the much more efficient trawling became widespread, as it could be done from small boats with few men and no power. In the fifteenth century boats with lines were already going as far afield as Iceland to fish for cod. In the eighteenth century line fishing for cod was going on around the Dogger Bank. It is still carried on off western Scotland, where cod and halibut are caught by line in areas where the bottom is too rough for trawling. Cod are also caught by line on the rich Newfoundland Banks.

Some of the line fishing boats from Aberdeen use lines up to 24 km long, with a hook every 5·5 m. The bait used is usually herring, which may have to be caught before the fishing can start. Tuna are fished by American boats in the Pacific, using lines and live bait. They may travel great distances across the ocean in search of this active fish.

Much of modern fishing is done with nets, in which the fish are pursued and trapped. There are two major types of net, the seine net and the trawl. The former is of very ancient origin, having been used in prehistoric times. The principle of the seine net is to surround the fish by netting. This method of fishing is being increasingly used now by Dutch and Danish fishermen in the Central North Sea. The modern seiner starts operations by putting down a marker buoy and steaming about 100 m letting out a cable, then an abrupt turn is made to the right for 55 m, the net fastened to the cable being lowered across the tidal flow. Another cable is lowered as the boat again turns abruptly towards the starting point. By the time the buoy is regained the boat has completed a triangular course. The ends of the two cables are then hauled in, drawing the net towards the ship, gradually closing it and trapping all the fish that were in the triangle on the bottom. The gear is relatively light and the boats can be small, but the method can only be used in fairly shallow water. Some seiners now work from Grimsby and Lowestoft, and the Moray Firth in Scotland is fished by this method. This technique seems to take the maximum number of demersal fish, such as plaice, with the minimum amount of power and effort.

In some types of seine fishing (for example, the purse-seine), two boats are used and the net is drawn together by them. This type of fishing can catch surface shoaling varieties, such as herring. Large catches, up to 1000 crans, can be caught in one haul by this method if the shoal is dense. The South African pilchard fishery also makes large catches with the purse-seine net. The purse-seine net differs from the first type in that it takes fish near the surface rather than from the bottom. The large Peruvian anchovy fishery is also carried out by purse-seiners, of which 1300 of 320–350 tons are used.

Trawling is now one of the most important methods of fishing. It is fairly recent because it depends on power to pull the trawl across the bottom. The trawl came into general use about 200 years ago. It has, however, become an important fishing method and the basis of the industry only in the last hundred years. As early as 1377 the Commons complained to King Edward III about a fishing device called the 'wondyrchoun', which seems to have been the forerunner of the beam trawl. This instrument was said to have been banned, as no more was heard of it or this method of fishing for another 200 years. If this ancient instrument could endanger the fisheries, it is not surprising that the modern trawls do so. There is still a difference of opinion on this point among fishing authorities. The trawl was banned in the sixteenth and seventeenth centuries. This was done partly to encourage the herring drifters, who were setting up in opposition to the Dutch. Trawling, however, still went on despite the bans, and the great destruction caused by this method of fishing was commented on by King Charles I in 1635.

As a method of taking bottom fish, for which purpose it is designed, there is no doubt that the trawl is by far the most efficient technique. The larger the trawl, the larger is its catch, if other things are equal. The decreasing number of fish are now being compensated for by larger and more efficient trawls, thus maintaining the amount of fish caught per hour's fishing. One of the main types of trawl is the beam trawl. The beam is a rigid bar, which keeps the top of the net above the sea floor and covers the bottom in front of the lower part of the net, which drags along the sea bottom. The flat fish are disturbed by the approaching net, but cannot escape upwards because of the net over them, held up by the beam. This is particularly effective in catching the round demersal fish, such as cod, which tend to escape by swimming upwards. The more modern otter trawl is similar in principle, in that a net bag is drawn across the sea floor. The mouth of the net is kept open by otter boards on the towing lines, while the headline, or upper part of the net, is kept off the bottom by floats. This method requires two towing ropes, one attached to either end of the net, and more accurate towing is required to keep the net in the correct position. The otter trawl is more efficient for catching all the fish except flat fish.

One of the great advances in trawling came with the advent of steam. The efficiency of the trawl was more affected by this development than other fishing methods, which do not depend on power for the actual fishing process. With steam it was possible to fish more distant waters with increasing efficiency, an asset that became more valuable as methods of keeping the fish fresh in ice were developed. At first, when more modern methods were introduced, catches increased, but so did the expenses as more distant grounds were sought, owing to the declining catches on the familiar territory near at hand.

In the nineteenth century the expansion of cod fishing by trawler was taking place largely within the North Sea. In 1833 only the Flemish Bight was trawled by British ships as far north as Yarmouth and central Holland. By 1845 fishing had extended northwards, particularly along the coasts as far as Edinburgh and nearly to southern Denmark. By 1865 the extension had largely taken place in the central part of the sea. By 1875 fishing had extended north along the coast as far as Aberdeen and central Denmark, while by the beginning of the twentieth century the whole North Sea was trawled by British boats; this expansion is shown in figure 7.15. The Grimsby trawlers first went to Iceland in 1891. The Barents Sea was first fished from Hull in 1905. The modern tendency is for the North Sea to be fished largely by seine nets, while the trawls are used in the more distant waters of the Arctic. Use of distant fishing grounds has meant the development of larger trawlers and trawls, which are so expensive that they can only be run by a company. The span of the otter trawl is 24·4 m, compared with 10 m of the older beam trawl. It is, therefore, a more efficient catching device, but requires more power to tow it. Mid-water trawls are also being developed which can catch herrings and other pelagic fish. These are sometimes pulled by two boats.

1.2 Distribution of fishing

The total catch of fish is spread very unevenly over the oceans. The catch in the different zones is shown in figure 7.1 using 1967 data. In 1958 the total world catch of fish amounted to 33,700,000 tons, of which nearly one-third came from the north Atlantic and adjacent seas. The distribution was such that of the total of 21 million tons produced in the northern hemisphere, nearly one-tenth came from the North Sea. Six million tons of fish were caught in tropical waters, and only 1·4 million tons from the southern hemisphere. Of the total 5·3 million tons were caught in fresh water, largely in Asia. Considerable changes in the last few years have taken place in world fisheries. Table 7.2 gives the volume of fish caught by different countries in 1968 and 1969. The total world production fell for the first time in 1969 by 2 per cent compared with the 1968 figure, the decline being entirely in the marine sector. Freshwater fishing was 6,650,000 tons in 1968 and 6,830,000 tons in 1969.

One of the greatest changes in the world fisheries during the last few years has been

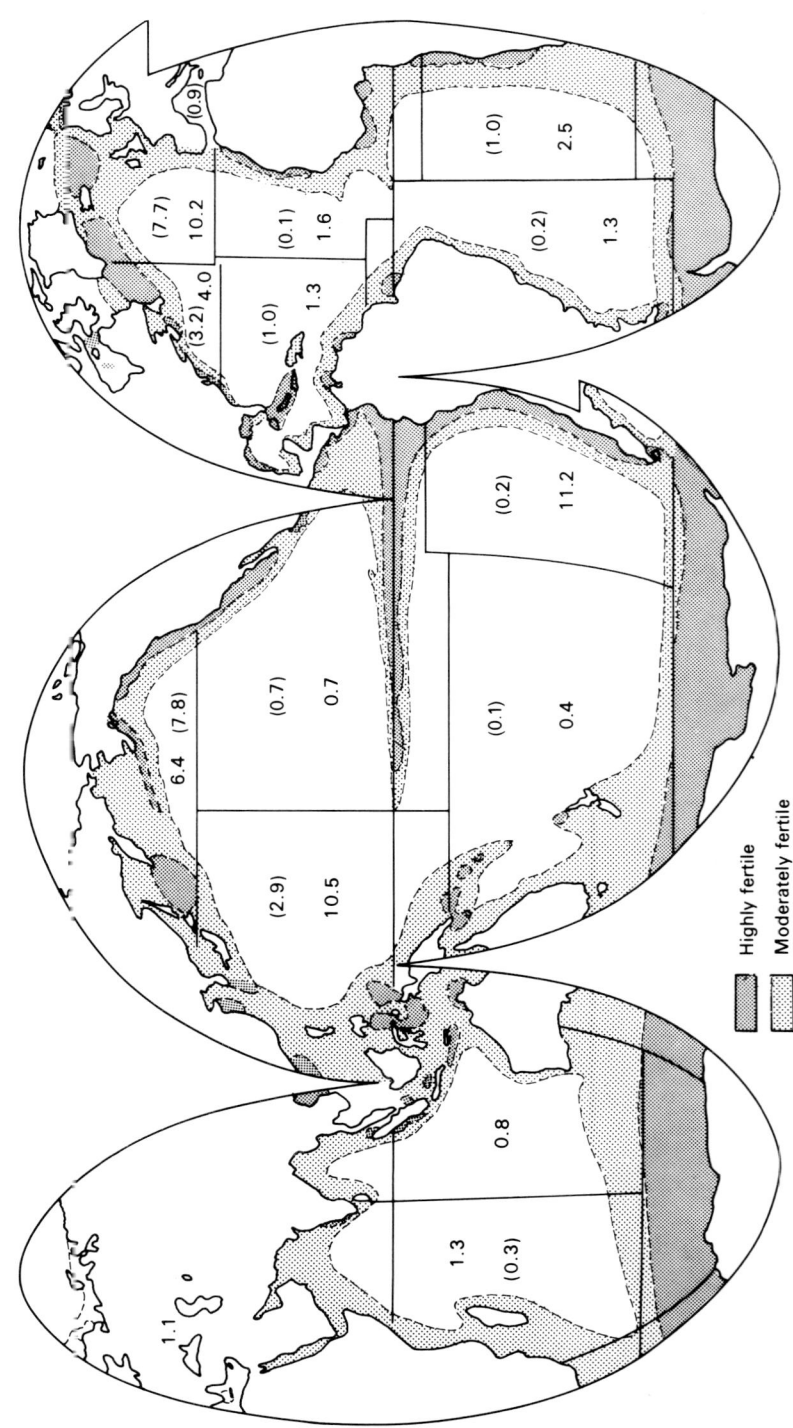

Figure 7.1 Map to show fish catch in 1967 compared with that of 1958 (shown in brackets). The fertility of the oceans is shown, most fertile areas being shaded most heavily. Figures are in million of metric tons.

Highly fertile

Moderately fertile

the spectacular rise in the Peruvian fishery, which has been top of the world's fishing nations for some years now. Peru has exported most fish products during the last five years, mainly in the form of meal derived from the large catches of anchovies. The 1969–70 fish-meal season has been the largest so far, with a total meal production of 1,919,098 tons of fish-meal, an increase of 60,000 tons on the record year of 1968–69. Peru in 1969 produced 45 per cent of the world fish-meal. There are 120 plants to process the fish extending along 1400 miles (2240 km) of coastline. Fishing takes place from September to May, with a one month close season in February–March. An institute to study fish stock

Table 7.2 World fisheries in 1968 and 1969

Country	Volume			Tons		
	1968	1969		Species	1968	1969
Peru	10,520,300	9,223,500		Herring type	20,590,000	18,290,000
Japan	8,670,400	8,623,500		Anchovy		8,960,500
USSR	6,082,100	6,498,400		Cod, hake, haddock	9,460,000	9,820,000
USA	2,441,900	2,495,400		Mackerel, etc.	2,860,000	3,030,000
Norway	2,804,100	2,481,000		Red fishes,		
South Africa	2,200,400	2,130,000		basses, & congers	3,210,000	3,160,000
India	1,526,000	1,605,000		Tunas (mainly		
Spain	1,503,100	1,486,200		yellowfin)	1,620,000	1,670,000
Canada	1,498,700	1,408,400				
Denmark	1,466,800	1,275,400		Pacific		30,200,000
Thailand	1,088,800	1,269,600		Atlantic		22,800,000
Indonesia	1,159,000	1,209,000		Indian		2,700,000
UK	1,040,300	1,083,100				
Chile	1,365,100	1,076,900				
				Total	64,300,000	63,100,000
Asia	24,500,000	24,730,000				
South America	12,930,000	11,310,000				
Europe	11,780,000	11,210,000				
Africa	4,450,000	4,570,000				
Oceania	220,000	200,000				
North and central America	4,630,000	4,560,000				

was set up in 1967 and it has been estimated that the maximum sustainable harvest is about 10 million tons, so that there is not much further expansion possible; the share that the guano birds take makes the calculation doubtful. It is estimated that 40 million birds, including pelicans, cormorants, and gannets, made the 286,733 tons of guano that was collected in 1955. These birds may consume about 1000 or more tons each day of anchovies. Several other types of fish may be exploited in this area, including demersal fish, of which 47,000 tons were caught in 1968, while the estimated potential is 50,000–90,000 tons. Other potential supplies are as follows: merluza 1968 catch 18,800, potential 60,000–100,000; bonito, 1968 catch 54,000, potential 80,000–100,000; Clupeiods 1968 catch 14,000, potential 50,000; mackerel and other pelagic species, 1968 catch 20,800, potential 50,000; and squids, 1968 catch 400, potential 300,000 (all values in tons). In order to protect their fishing interest Peru proclaimed a 200-mile fishing limit in 1947. Peruvian fishery is now in need of management, as are many other fisheries, because over-fishing is becoming a more serious world-wide problem.

Britain's place as a fishing nation has fallen lately. In 1969 it was eleventh in the list of world fishery, and fifth in Europe.

The world production of fish has doubled in the last two decades from 20 million tons in 1948 to more than 50 in 1965. The proportion of animal protein derived from fish is about 10 per cent, but is higher in the Far East. More fish is now used for making meal, as food for land animals. The rise of fish catches means that over-fishing is an increasing danger and problem. Plaice catches in the North Sea, for example, have been declining per fisherman since 1880. The resultant decline in fishing has allowed the stocks to recover somewhat, as shown by the good fishery of 1964; the 1962 year class was very good and this helped. The Pacific halibut fishery shows that control measures can lead to much improved stocks and catches. Diversion of fishing from the North Sea to distant waters

Table 7.3 British fishery in 1965–66

Areas fished	Tons	Percentage		1966	1965
Barents Sea	34,500	4·0	Demersal	700,000	718,500
Norway	30,600	3·6	Pelagic	195,900	157,300
Bear Island	36,000	4·2	Shellfish		
Iceland	187,600	21·9	and crustaceans	32,000	26,200
Greenland	13,000	1·5			
Newfoundland	35,000	4·1			
Faroe	31,700	3·7	Total distant waters	337,000	39·3%
North Sea	327,500	38·2	Total adjacent waters	857,200	
West Scotland	134,600	15·7			
England west	18,300	2·2			
England south	7,700	0·9			

has allowed some recovery of the North Sea stocks. Plaice round Iceland are, however, now declining and cod catches are only maintained with greater fishing effort. The exploitation of pelagic fish, including herrings, has on the whole not been so heavy as that of demersal fish, except in the southern North Sea.

Fishing has recently extended to the west coast of Africa, and southwest Africa was fished by 1965 for hake, Spain taking 118,000 tons, South Africa 87,000, compared with a total of only 39,000 in 1948. Fishing in the north Pacific is being actively extended by Japan and the USSR. Korea has joined the tuna fishing in the Pacific. There is still room for exploitation in the Arabian Sea for Indian mackerel, and oil sardines, while the blue whiting off western Britain could be exploited further. The present rate of extension can only go on for about 10–15 years unless new stocks are exploited.

1.3 Fishery dynamics

The study of fishery dynamics has become complex. Two general approaches have been used, one of which is the Beverton–Holt approach (1957). This is based on the relations considered in the next section, making use of the elements specified in the equation

$$S_2 = S_1 + (A + G) - (C + M)$$

The second method does not attempt to isolate the different elements of the situation, but treats the function, population size, according to the logistic law of population growth. It is called the Schaefer approach (1968). Each can be applied to situations in which certain types of data are available. The Beverton–Holt method assumes constant recruitment and a constant natural mortality rate, but takes into account the number of fish in each year class when they enter the fishery. The method has been applied particularly to the long-observed North Sea haddock and plaice fisheries. Abundance of recruitment of haddock has varied over a 500-fold range, while plaice have only varied 6-fold over the 30-year observation period. Under natural conditions the major fish stocks appear to be more or less stable despite such great fluctuations in the year classes. Rate of recruitment should, therefore, decline as the stocks increase. The Beverton–Holt method was developed to study the effect of mesh size regulations on demersal fish in the North Sea, for which good statistics were available and the determination of age was easy. The Schaefer method was developed primarily to consider the tuna fishery in which statistics on good catch and effort were available. The fishery started from an undisturbed natural state, but age determination was difficult. The Schaefer method with constant parameters suggested that the maximum catch that can be obtained is one-third of the virgin population, whereas, when allowance is made for growth to depend on the population, it occurs at a higher proportion of the virgin population. The value can reach 60 per cent if the relation between stock and recruitment is incorporated into the model at its maximum level. The accuracy of the method depends on the extent to which recruitment is independent of stock.

The theoretical optimum catch for the minimum effort can be calculated, but the major problem is controlling fishing in the open sea where international agreement is essential. This aspect is considered further in the next chapter. Figure 7.2 shows that the curve relating catch to fishing mortality is flat at the top so that a small-to-moderate reduction of effort will lead to very little loss of fish: 98 per cent of the maximum catch

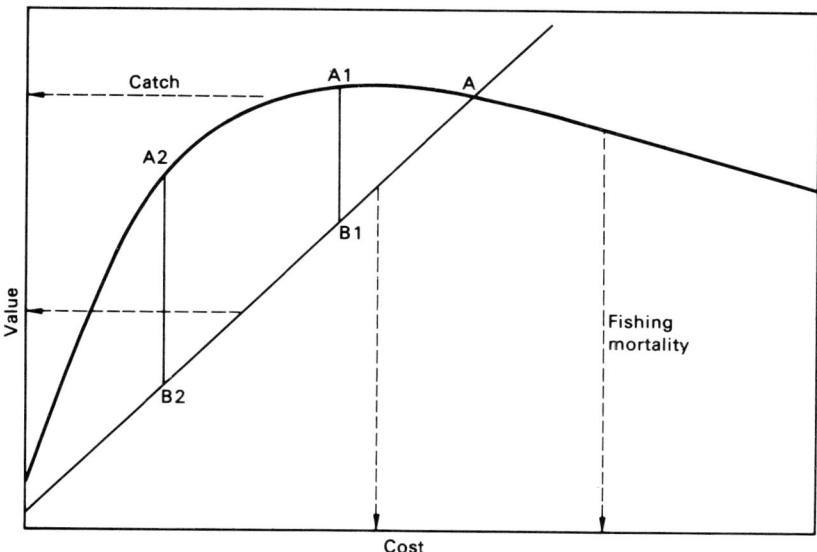

Figure 7.2 Relation between cost and value of fishing, Line A, B₁, B₂, and fishing mortality catch, line A, A₁, A₂. The optimum is between A and A₁. (*After Gullard and Carroz, 1968.*)

can be taken with 80 per cent of the effort. The optimum fishing rate for the maximum return is at the point $A_2 - B_2$. A is the equilibrium point where the line of cost, V (value), cuts the curve. The maximum catch is at point A_1. The curves differ for different countries according to their fishing costs. The combination of different species is a further complication, as different curves may apply to different stocks. For example when trawling for herrings, small haddock may also be taken. Types of regulation used to control fishing include limiting the size of fish retained, closing specific areas, closing some seasons, limiting the gear, limiting the catch, and limiting the total effort.

1.4　Over-fishing

Cod make up 41 per cent of the value of fish catch, haddock 18 per cent, plaice 10 per cent, and herrings only 5 per cent. Herrings in particular are suffering from over-fishing. The decline of herring stock has been noted since 1951, when the forecast for the autumn fishery was incorrect due to lack of recruits to the stock. Decline in the stock is probably due to excess catching going on to the east of the Dogger Bank by Danish trawlers, which are catching many of the immature herrings in this area (Hodgson, 1957). Older fish are also decreasing in numbers so other factors may also be involved. Possibly the herrings of the North Sea are declining in numbers as did those in the Baltic before them. The great capture of immature herring now going on, however, cannot but have an adverse effect on the stocks in the North Sea and over-fishing seems the most likely cause of the decline.

The North Sea, being so close to densely populated countries with a long tradition of fishing, is liable to exploitation as new methods make fishing more efficient. One of the problems concerning over-fishing is that it is an international problem and demands international agreement, which is difficult to achieve. Limited success has been obtained in restricting the size of the mesh, which helps to conserve stock. The forced reduction of fishing during the two world wars provided evidence of over-fishing. The 4·5 years of the First World War produced a remarkable change in the fish caught. At the beginning of the war many small fish were caught, but by the end of it there were plenty of large fish. As fishing continued throughout the 1920s, however, the size of the fish again declined considerably. The Second World War repeated the experiment, with even more striking results, shown in figure 7.3. The amount of fish caught in 1946 was double that of 1938.

The problems of over-fishing are extremely complex; however, essential points can be shown simply in the expression $S_2 = S_1 + (A + G) - (C + M)$. S_1 is the weight of catchable stock at the beginning of the year, S_2 is the weight at the end of the year, A is the weight of young stock coming in during the year, G is the extra growth of the stock S_1 and A, C is the weight of fish caught, and M is the natural mortality. S_2 will be greater or smaller than S_1 according to whether $A + G$ is more than or less than $C + M$. The fluctuation of A will be great from year to year, as it has been shown that some year classes are much more successful than others. G varies mainly with the availability of food and also with temperature. The relative growth of plaice on the Dogger Bank and in their normal nursery ground off Holland illustrates this point. If C, the catching rate, decreases $C + M$ will be smaller, and G will be greater as more fish survive, S_2 will be greater than S_1. If, on the other hand, C increases and G remains the same, S_2 will be less than S_1. The growth rate may not, however, remain the same. If C increases very much, there will be more food left and G may increase as well. If C decreases markedly this might have the opposite effect on G. Similarly a big increase in A would increase S_2 but it might decrease G, so there would be more fish, but they would be smaller.

In an equilibrium area $S_2 = S_1$ and A + G must equal M. If M is small there will be little addition to the stock. This means that there are more likely to be few large, old fish, as these use their food least efficiently to increase weight. Such a ground may well improve in quality as the younger fish have a chance to grow. Thus when the Barents Sea was first

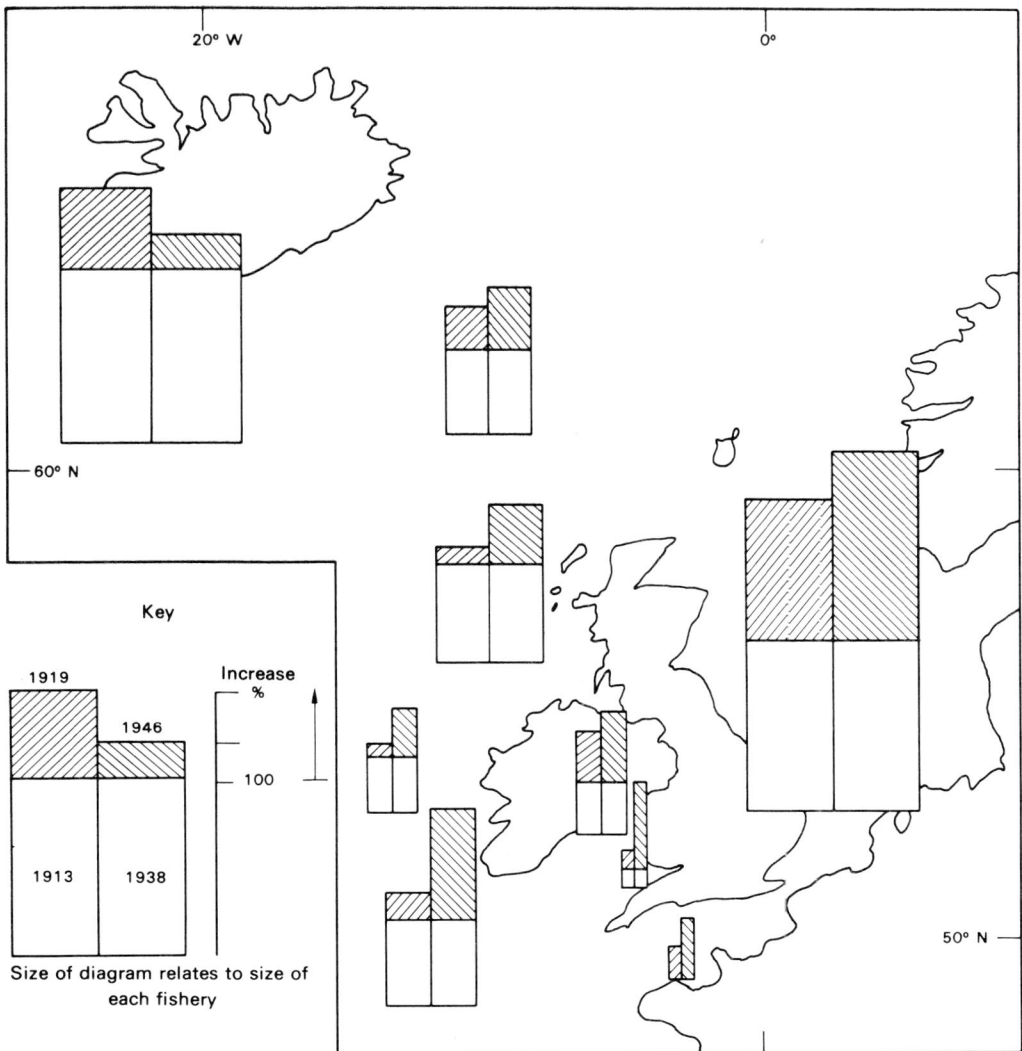

Figure 7.3 The relative size of the fish catches before and after two World Wars around the northeast Atlantic. (*After Graham, 1948.*)

fished in 1907, the size of plaice was much larger than those normally caught in the well-fished North Sea. Nearly all the fish were mature, while many of the North Sea fish were not. To exploit the fishing to the optimum extent S_1 should $= S_2$, and the amount taken depends very much on the conditions of growth and recruiting to the stock.

If too many fish are taken, the effect is increasingly severe. Take, for example, 80 per cent, starting with a stock of 1000 at one year old. This year group will be only 200 strong at 2 years, and 40 at 3 years, dwindling to 2 at 5 years, and the amount caught will also have decreased to 6. With 50 per cent fishing, the original 1000 will be fished out in 11 years, when the stock will be reduced to one, but instead of a total of 1000 caught in 6 years with 80 per cent removal, the same number would keep the fishery going for 11 years at 50 per cent. Although the catch in the first year is greater with the higher fishing rate of 80 per cent, after this the number of fish caught is higher with the 50 per cent rate, because the numbers remaining are higher. After 6 years, taking the weight increase into account, the 80 per cent fishing rate will have yielded 106,102 g as against 161,138 g for the 50 per cent rate. The lower fishing rate thus yields 50 per cent more fish.

As the extent of over-fishing increases it means that more fishing effort is needed for reduced yields. There is an optimum amount of fish that should be caught to keep the stocks in the best condition and yield the most. An increase in the mesh size, which allows the recruits to grow bigger before they are caught, would reduce the yield for a time, but the effect is unlikely to be prolonged as these fish become of catchable size in larger numbers than before. If, however, the rate of catching cannot be controlled then something could be done by increasing the growth rate, or decreasing the mortality due to other predators. Experiments with the plaice shows how this might be achieved for one species of commercial fish.

1.5 Mariculture

Various attempts at farming the seas have been discussed by McKee (1967) (Plate 18). He describes experiments that have been made with artificial reefs off California in depths of about 15 m. In an area where very little benthic life existed, 20 car bodies were put on the bottom in 1958 and two years later 24,000 semi-resident fish occupied the car bodies. Giant kelp became established after 6 months, and cleaner fish, which perform the necessary task of ridding larger fish of parasites, were also established. Experiments were made with trams, car bodies, quarry waste, and concrete fish shelters. The last was the most popular, followed by the quarry waste, car bodies, and trams in that order. The first two are cheaper, but only last 3–5 years off California where the experiments were made, while fish shelters and quarry waste are more expensive but last longer. After one year, 110 different species were recorded, the first to arrive being small barnacles and worms, followed by hydroids, amphipods, bryozoans, and teredos (if wood is available), then slipper snails and abalone, gorganian corals, and fish. The fish species change with time also, starting with sand bass, kelp bass, and perch (who are semi-resident), and then gobies, damsel fish, and rock fish (who are residents). There were 80 species of fish after 3 years. The reefs prevent trawling, but provide excellent sport fishing.

Plate 18 (*opposite*) Fish farming: pearl culture in Japan (*Japanese Information Office*).

Another field of successful fish farming up to a point is oyster cultivation. Around Britain in 1966 there were three categories of ground from this point of view: a) common oyster fishery open to all, although few of these remained, b) a regulated oyster fishery looked after by the local authority, and c) oyster ground privately owned. Artificial reefs can be prepared for oysters and mussels, but their success depends on the environment. Along the south coast of England there is an abrupt change in fertility, with different fauna on either side of the Isle of Wight, owing to the presence of different water types. Shellfish are a useful indication of water type. The Atlantic waters to the west are more fertile than the channel water to the east.

Another method that has been used in an attempt to improve fisheries is the hatchery experiments, which started in the nineteenth century. In Norway hatcheries were set up in 1882 and in 1885 one was started at Woods Hole, Massachusetts, with two further hatcheries in America by 1905. The three American hatcheries were closed down in 1943, 1950, and 1952 as there was no evidence that they improved the fish stock. The three hatcheries produced 3000 million cod fry during their operation. A plaice hatchery was established at Dunbar in Scotland in 1893 and at Port Erin in the Isle of Man in 1902. The principle on which hatcheries are based is an attempt to restore to the sea the fish that are lost in catching females that are about to spawn. The lack of evidence of success is related to the extreme vulnerability of fish larvae at their early stage of development. The larvae of marine fish can be fed on brine shrimps, which hatch out in two days.

In 1951 experiments to rear plaice to the stage of metamorphosis were made. The eggs hatch after 3 weeks; a further period of 7 weeks elapses before the plaice metamorphosis turns the larva to a tiny flat fish. In the first experiments only one fish reached this stage out of 1000 eggs, but improvements in care raised the success rate to 10 per cent by 1960. One problem that was overcome was the lack of proper lighting which prevented the larvae from seeing their prey; the successful solution to this problem raised the success rate to 33 per cent. Another advance was made by the elimination of excess bacteria, which in one test raised the survival rate to 70 per cent in 1962. These experiments show that successful raising can be achieved, but the problem of adaptation to the marine environment from the tanks still remains.

Mortality rate is very high. For example, one mackerel produces 500,000 eggs, but after 62 days only 20 are left, and after 85 days only 2 survived, with the mortality rate at 10–14 per cent/day. Thus one of the fundamental necessities for successful fish farming is to control the predators. Nearly all marine organisms have a larval stage, when they feed on phytoplankton, so all can survive to the metamorphosis stage in the same proportion, according to the supply of phytoplankton. The deposit feeders cause problems because they are not selective in their food supply and can sweep large areas bare, including consuming small shells of mussels and oysters and other bivalves. The brittle stars also eat large numbers of clams and so must be eliminated if the clams are to survive and to provide food in their turn for flat fish, such as flounders. One method of overcoming this problem is to put oyster larvae on a plastic foil sheet, which prevents the predators below the sheet from reaching them. If the young shellfish have even a short time to grow before the predators attack them, then many fewer will be consumed to provide the same amount of food. At present the economics of fish farm experiments are such that the fish produced artificially are much more expensive to rear than their market value.

One possible development is to catch young fish and to place them in ponds heated by the waste heat of radioactive power stations so that their growth rate is enhanced. Experiments are being carried out in Loch Ewe, a Scottish sea water loch, to establish the food chain of different types of fish, but progress is slow and the results are on a very small scale although much valuable basic information is being gathered. Other experiments in Ardtoe Loch are using plaice and other fish bred at the Port Erin hatcheries. One million young plaice, sole, and turbot are being reared each year to place in the loch, which is shallow and has been enclosed. Its environment changes, therefore, and has proved unsuitable to the fish. Open water might provide a better habitat as it is more stable. The problems remain formidable and have not yet been solved.

Some success in farming shrimps has been made in Japan, using the wheel shrimp, *Penaeus japonicus*, a large female up to 23 cm long. The female lays 0·5 million eggs, which are kept in water between 77 and 84°F (25 and 31°C) of the correct salinity. The eggs hatch after 13–14 hours, and the larvae live on their own yoke for 35 hours, after which they are fed a tiny diatom, *Skeletonema costatum*, by causing the water containing the diatoms to drift past them. After a few days they have a meat diet of oyster eggs, larvae and copepods, and brine shrimps; they also tend to be cannabalistic if not enough food is provided. They are then transferred to rearing ponds in which extra oxygen is supplied causing rapid growth. After 10–20 days the shrimps are moved to large ponds to mature, with tidal power being used to circulate the water. They are put in a cooler tank at 13°C to slow down their metabolism and they are then packed in chilled sawdust as insulation and despatched to the markets. Despite high prices such farms only just make a profit and require a very detailed knowledge of all the stages of the shrimp's life. These farming experiments have been more successful than those undertaken with lobsters. This is partly due to the slow growth of lobsters, as they take 6 years to grow from larvae to a weight of 0·5 kg. There is still little information on the early stages of their life cycle, which makes their culture difficult.

Though lobsters cannot be effectively grown from eggs for sea farming, they can be exploited fairly simply by providing suitable habitats in which they can shelter and thus escape from their predators, which off southern England are mainly conger eels. Underwater observations have shown that old tyres attracted lobsters and that when they had been taken they were soon replaced by other lobsters. Thus it appeared that the number of lobsters was partly dependent on the number of shelters available for their protection. This method of sea farming, based on providing a suitable habitat for the lobsters, has the merit of simplicity, of taking place in the open sea, and of exploiting the natural environment and the great natural fecundity of marine creatures, including the lobster. Suitable sites can be chosen where conditions are good for lobsters and favourable from other points of view, such as accessibility at all states of the tide, shallow water, and reasonable distance from land. A suitably hard (but not too rocky) bottom was needed away from an area of trawling, which would damage the lobster structure, which in turn would damage the trawls. An area about 50 by 25 miles (80 by 40 km) is suitable for lobster farming and suitable habitats must be provided, such as open pipes, old tyres, and wrecks. If lobster farming proves profitable then the area exploited in this way can be extended. Experiments have been carried out off Hayling Island in about 3 m of water. Lobsters must be collected by divers who bring them out of the holes provided.

One possible method of increasing the fertility of the sea would be to pump the nutrient-rich deep water to the surface, as discussed by Roels, Gerard, and Bé (1971). The potential fertility of the sea can be measured in upwelling areas; for example, in the Peru current a productivity of $11 \cdot 2$ gC/m^2/day has been measured recently, the equivalent of 4090 gC/m^2/year. This equals 100 tons dry weight plant material/ha/year, if the maximum upwelling and light were available all the year. A crop of phytoplankton of this magnitude would need 5 tons of nitrogen per year. The deep water is rich in phosphates and nitrates and could be pumped up where it occurs close inshore. This situation occurs off Mexico, Panama, Colombia, Brazil, Chile, Peru, Ecuador, in the Caribbean, off parts of Africa, Asia, and Australia. It would be necessary to pump up $12 \cdot 2$ m^3/minute, but the cost of pumping could be offset by the production of fresh water as the deep water is cold (5°C) and electric power could be generated. One hundred tons dry weight could be harvested/ha/year, producing 50 tons fresh weight of secondary producer. A pilot scheme has been established on the north shore of St Croix in the Virgin Islands of the West Indies. The 1000 m deep water is less than a mile offshore at this point. The cold deep water passes through pipes, then through a condenser, intercepting the moisture-saturated air, which, collects as fresh water. The nutrient-rich deep water is then fed into lagoons or ponds where it increases phytoplankton production. A 27-fold increase in carbon fixation was observed in phytoplankton using water raised from a depth of 800 m.

Schmitt and Isaacs (1971) suggest that wind or wave energy could be used for artificial upwelling to increase protein production. Another possibility is to use the waste heat from power stations at coastal locations to enhance the productivity.

In many ways the sea is not so efficient in producing food as the land. Thorson (1971) states that the sea produces about the same as the land although it occupies over 70 per cent of the earth's surface. Total land production is 15,000–20,000 million tons of organic carbon/year. There is a 400-fold difference between primary production in the oceans and the world fish catch, which was 49 million tons in 1966. If productivity is calculated in million tons of carbon, the ratio is 4000-fold, and the best estimate of the ultimate possible fish catch is about 200 million tons/year. It is unlikely that more direct use of primary production can be made without great advances in man's ability to harvest the lower stages of the food pyramid. At present fish and other marine animals are much more efficient than man in catching prey. Fish now provide 89 per cent of the protein obtained from the sea and the other marine creatures only 11 per cent, but there could be much greater use made of shrimps, prawns, and mussels for example. Shrimps could be cultivated to yield 14–15 times increase in two months. Culture ponds in India yield 30–120 tons in a 5-month season. Power plant cooling water could be used to increase the growth rate. Cultivated oyster beds could yield 10 times as much as natural ones by excluding sea stars and boring snails. Some fish are also worth cultivating artificially. Tilapia is an example, as it has a short food chain and can live in warm fresh water. The initial weight increase of 2 pairs was 224 times in 8 months. These fish are very prolific.

In attempting to improve the productivity of the sea, ecological food chain reactions must be taken into account. The Japanese predatory snail, *Rapana thomasiana*, for example was introduced into the Black Sea in 1947. It eats bivalves and has destroyed mussel and oyster beds, and reduced the bottom-feeding fish, that also eat the bivalves. Sand-eels then invaded from the Mediterranean and ate the *Rapana* larvae, followed by the fish

Sargus, which started to eat *Ammodytes cicerellus*. Thus the introduction of one different species can lead to a long chain of subsequent faunal modifications.

Fish can sense a salinity difference of 0·2‰ and temperature changes of 0·003°C. Such knowledge can be useful in locating fish. It has also been established that fish can produce many different sounds and might therefore be attracted to sound sources. Sounds could also be used to frighten dolphins and porpoises from fish traps. Electric shock techniques could revolutionize fishing for shrimps and prawns. Experiments have shown that trawls with electrodes make the fish jump, increasing the catch 5–12 times. Fishing could take place at all seasons and during the day as well as the night. The same ideas could be adapted for crab fishing. Another possible increase in food resources would be the direct catching of the *Euphausiids* in the Antarctic now that the whales that did the job so much more efficiently have been virtually eliminated. Another possibility is the introduction of northern fish to feed on the *Euphausiids*, but this would lead to problems because the environment is different.

The oceans also produce healing substances in marine animals that are used by the medical profession. Weaver fish produce a substance to reduce heart beat, and another species provides a substance that reduces blood pressure. Balloon fish produce tetradotoxin, a pain killer, while antibiotics and chemicals effective against cancer could also be obtained.

Seaweeds were one of the earliest products of the sea to be exploited as food, medicine, and fertilizer. Stone-age man used seaweed for mattresses; fishing by hook and line has also been going on for more than 10,000 years. Fish spears are some of the earliest archaeological relics. Nets and fish traps have also been used for thousands of years BC. Shellfish probably provided one of the first forms of marine exploitation, and nearly all other forms of marine life have been used for many centuries, including mammals, pearl oysters, sponges, and crustacea. The following tables indicate the wide variety of marine organisms that have been harvested. These, however, only form a small proportion of the total number of marine species, amounting to only approximately 0·3 per cent. At present the herring, anchovy, and sardine form the most important contribution, as shown in tables 7.4 and 7.5.

Table 7.4 Proportion of major groups of species harvested from the sea

Type of group	Number of species harvested	Total number of marine species
Fishes	275–300	about 1900
Invertebrates	130	about 160,000
Seaweeds	50	about 4500
Mammals	25	124
Amphibians	3–5	5

The present extent of mariculture is summarized in table 7.6. It shows the importance of oysters in terms of yield, while indicating that other forms of mariculture rarely reach the productivity of land cultivation. Oyster culture is by far the most important form of aquaculture at present. It has the advantage of dealing with sessile forms. Pollution of coastal waters is one drawback to further aquaculture, but the stock of oysters, in particular, nevertheless could be greatly increased. One possibility is to increase phytoplankton by fertilizing the water with sewage. This method, however, would require very careful

Table 7.5 World live catch of marine species by groups of species (metric tons $\times$ 10³)

	1965	1966	1967
Salmon	475	498	436
Smelts	28	30	34
Capelin	281	521	513
Flounders, halibut, sole	960	1090	1200
Cod, haddock, hake	6750	7260	8150
Redfishes, bass, congers	3190	3220	3140
Jack, millets	2110	2090	2030
Herring, sardine, anchovy	16980	18740	19680
Tunas, bonitos, skipjacks	1200	1320	1330
Mackerels, billfishes	1670	2000	2680
Sharks, rays, chimaeras	400	420	440
Other fishes	7620	7960	8290
Crustaceans	1190	1280	1350
Molluscs	2880	2950	3080
Other invertebrates	44	44	50
Blue, fin, sperm, sei whale	6468	57891	51593
Winke and pilot whale	10432	9039	7951
Porpoises	2	6	7
Seals and walruses	—	4	3
Turtles	4	5	12
Pearls, shells, corals, sponges	7	9	11
Aquatic plants	720	750	800

Table 7.6 Aquaculture

		Yield	
Area	Animal	lb/acre	kg/ha
Unfertilized sea water ponds			
Philippines	milkfish	400–980	450–1100
France	grey mullet	300	337
Java	milkfish	40–300	45–337
Indonesia	milkfish	140	157
,,	prawns	46	52
,,	wildfish	23	26
Fertilized sea water ponds, Formosa	milkfish	1000	1120
Fertilized brackish ponds, Palestine	carp (expt.)	755–7970	847–8950
Fertilized commercial ponds	carp	356–4210	400–4730
Oyster culture, US public	oyster	6	6·7
US private grounds	oyster	170–5000	191–5610
France	oyster	320–740	359–830
Australia	oyster	120–4400	135–4940
Philippines (maximum)	oyster	10,000	11,200
Japan (maximum)	oyster	50,000	56,100
Land: cultivated	pigs	450	505
grassland	cattle	5–250	5·6–281

control, since too much sewage causes a deficiency of oxgyen which leads to anaerobic conditions because organic matter uses up the oxygen as it decomposes. Thus over-fertilization of the sea produces noxious results, although small additions can be beneficial.

2   Exploitation: specific

2.1 Demersal fish

The northwest Atlantic has been an important area for demersal fish for a long time, with a long tradition of fishing by the Bretons, Normans, Basques, and Portuguese. The cod of the Newfoundland area were fished in the sixteenth century, both from the shore and the bank fisheries. New England colonists fished the area from the seventeenth century and in 1748 the first cod catch was landed in New England. From 1757 cod were fished from Gloucester, supporting one of the most important industries in England. Cod seemed to be in short supply in the early 1870s on the banks, and this was related (probably wrongly) to adverse effects of dams on herrings, which were the main food of the cod. Attempts were made to restock artificially, but they were not successful in the marine environment, as they can be in fresh water.

Fishery was not regulated until the 1930s. This move was precipitated by the collapse of the New England haddock fishery. Haddock were being increasingly caught instead of cod in the early years of this century, and by 1925, 50,000 metric tons were landed and only 35,000 tons of cod. Haddock catches rose rapidly to 125,000 tons in 1929, but fell sharply to 60,000 tons in 1932, due to stock depletion. In order to overcome the decline a larger mesh was introduced to save the small unwanted fish. The measure could not succeed, however, in international waters, unless it was adopted by all the fishing nations. An international organization to control the fishing in all the northwest Atlantic from Greenland to Rhode Island met in 1949 and initiated ICNAF (International Convention for the Northwest Atlantic Fisheries). The first step was to collect data. The area was divided into 5 subdivisions, for which detailed statistics were obtained. The organization implemented a mesh regulation within two years. The measure, however, failed to improve the seriously over-fished and declining haddock stock. Study of the stock revealed that about 30 per cent increase of yield could be obtained by increasing the mesh size from 5·7 cm to 11·4 cm. Controlled experiments with both mesh sizes shows that the smaller fish were escaping with the larger mesh. The larger mesh also proved more efficient as it caught a higher proportion of the larger fish. The mesh sizes used at present are 13·0 cm in subdivision 1 off west Greenland, and 14·0 cm in the others. Since 1956 other countries have fished the area, including the Soviet Union which in 1967 was second only to Canada, who took over 1 million metric tons, the USSR catch being nearly 600,000 tons. Poland, Germany, and Romania now also fish in the area. These new fisheries have resulted in over-fishing and the total landings of all species exceeded 3 million metric tons after 1965, increasing from just over 1 million in 1953.

Like most fisheries, the year classes vary considerably, and 1963 produced a large class, the largest in the history of the fishery; the 1962 year class was also above average. This population bulge became fishable in 1965. The USSR took advantage of this rise in haddock population, landing 82,000 tons in 6 months in 1965. In this year the catch was 3 times the mean catch for the last 30 years and 3 times the calculated sustainable yield. Total haddock landings were 155,000 tons for 1965 and 127,000 tons for 1966, but after this the inevitable decline set in and continued through 1968. The US landings fell from 57,000 tons in 1965 to 39,500 in 1967 and 28,800 in 1968. Further decreases are expected in the next few years. The years after 1963 have also not been good for spawning. Three poor years in succession have not been recorded before, when good and bad years tended to alternate or occur every third year. There were 5 poor year classes from 1964–68, so no improvement could occur before 1971 at the earliest, and the good 1963 year class will soon no longer benefit the fishery. It is now possible to see that this year class should have been harvested later. Fishing now accounts for two-thirds to three-quarters of the fish mortality each year. The rate of fishing between 1962 and 1965 was such that maximum sustainable yield could not be maintained in both the northeast and northwest Atlantic. A reduction of 30–40 per cent in the fishing mortality rate could increase the long-term yield by 10 per cent. A reduction of effort of 10–20 per cent would result in an increase of 10–25 per cent of average catch per unit of effort. The Georges Bank haddock landings could be increased by 7 per cent and landings/day by 52 per cent with a 30 per cent decrease in effort from the 1963–65 level. This is an example of the complex interplay of biological and economic factors that determine the success of a fishery. The ground fishery in this area has suffered a continuous decline since ICNAF was instituted, due to late measures to control the catch. The decline is partly due to the decline in the numbers of haddock, and now catch limits, closed areas, and seasons, have been established in an attempt to improve the fishery.

The halibut of the northeast Pacific provide an example of a successfully managed stock. They have been managed by Canada and the US since 1923, with the object of obtaining the maximum sustained yield. The fishery started in 1890 and was uncontrolled for the first 35 years, the exploitation being limited by the market. However, 55 years ago the catch could not be maintained although the area fished increased. The first restriction was to impose a winter closed season. The fishery has been extensively investigated during the last 40 years from the biological point of view. A total of $5 million have been spent, but the improvement of the fishery since 1932 has been worth $100 million. The investment has been very worthwhile, and illustrates the value of careful biological investigation, resulting in sensible and effective control. The early years of the fishery saw the usual increase as exploitation increased, followed by levelling off and decline because of over-exploitation.

The management since 1932 has, however, resulted in increased catches up to 1960 (figure 7.4). Four phases can be distinguished. Phase I, from 1930 to 1942, showed an increase of 2 per cent in the annual catch, with increase catch per unit effort. Phase II, from 1941 to 1952, was a period of short fishing seasons, and the permitted yields were held at 1 per cent increases to allow continued build-up. Phase III covered the period to the early 1960s and production was increased as various tests were made to study empirical and model maximum yields. In the fourth phase these high removal rates led to a decline since 1962, which was partly due to a price decline leading to less fishing in 1967 and 1968.

The increase in yields was obtained by allowing the fish to grow to optimum size before they were caught and nursery grounds were closed to protect the young fish. Restrictions on fishing gear were imposed, so that fishing was done by hook and line. This method

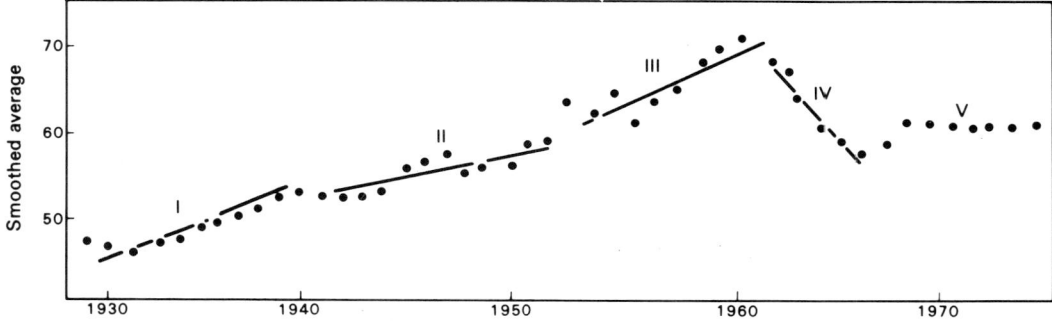

Figure 7.4 Trends in catches of Pacific halibut, shown by three-year moving averages. Five phases in the fishery are indicated. (*After Benson, 1970.*)

allows the halibut to be controlled in isolation from other demersal fish. Setline recoveries yield much larger fish than trawling. Such a separation of fishing methods for different species would not be possible in the north Atlantic. Multi-species trawling would soon destroy the halibut fishery in the Pacific. Figure 7.5 shows how fishing effort, yields, and catch per unit effort have fluctuated and been improved, as has the size of the stock. The fishery has been the major cause of changes in the stock, although short-term fluctuations have been related to variation in year classes. This fishery is one of the few to be closely

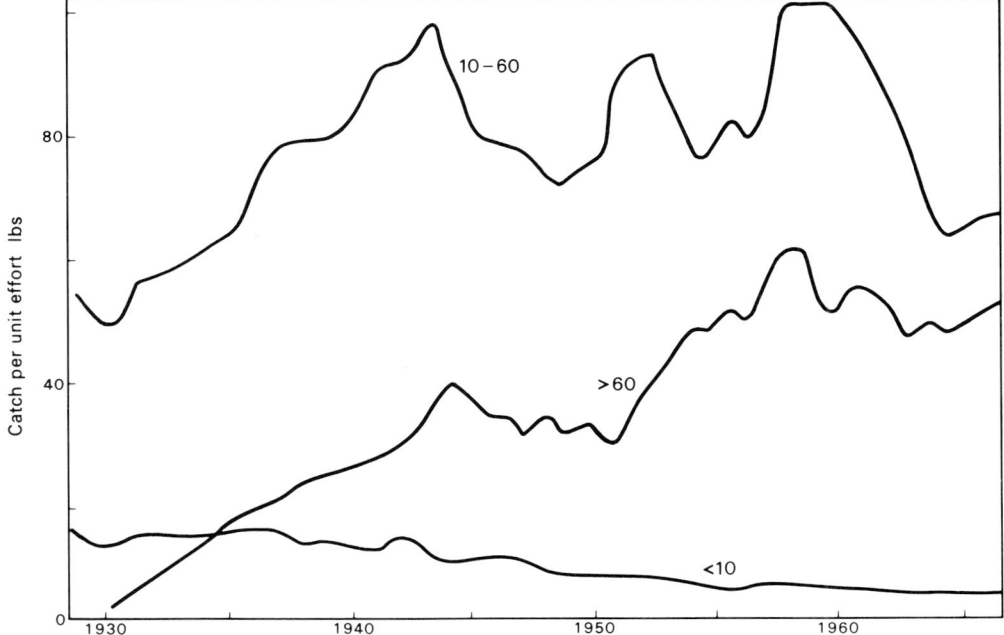

Figure 7.5 Halibut catches in three size groups in terms of catch/unit effort cn the ground west of Cape Spencer, including the Bering Sea. The three lines refer to different sizes of fish in cm. (*After Benson, 1970.*)

controlled. Its continued improvement shows the value of such control. The Pacific halibut stock has been rebuilt to such a size as to be capable of maintaining the maximum sustainable yield, which is 32 million pounds in the southern part of the area and 36 million in the northern. The halibut maintain their geographic homeostasis by a westward drift of the eggs and larvae into the Bering Sea and the emigration eastwards by juveniles and adults. A large fishery in the Bering Sea area could influence the yields given above. The expansion of foreign fishing could also upset the balance set up by the control measures exerted by Canada and the USA. Increased productivity produced by these measures are given in table 7.7. There has been a downward trend in catches in the later 1960s as a result of further trawling developments.

Table 7.7 Northeast Pacific halibut fishery catches

| | Average catch | | | | | |
| | Per vessel | | | Per man | | |
Year	Season	Trip	Day	Season	Trip	Day
Before 1930	309,000	27,000	1,200	33,000	2,900	130
After 1967	367,000	67,000	2,100	57,000	10,500	330
Per cent increase	19	148	75	74	260	153

2.2 Pelagic fish

Pelagic fish are vulnerable to over-fishing as are the demersal fish, and fisheries are liable to sudden decline. In 1967 California stopped all landing of Pacific sardines, thus indicating the end of a viable fishery. This fishery grew rapidly up to 1936–37 and continued at a high level for nearly a decade. It declined first in the northwest Pacific, and progressively south along the coast to southern California. The fish were used mainly for tinning, oil, and meal. As long ago as 1929 the danger of over-fishing was seen and warnings given; a limit of 200,000–300,000 tons was suggested. The causes of decline were over-fishing and poor survival of some year classes, and competition for food from the flourishing population of northern anchovies, *Engraulis mordax*. The maximum sustained yield of the pre-1949 population was estimated at 471,000 tons, given a spawning population of 1 million tons. The actual catch was, however, 570,000 tons, and the spawning stock was variable. The sardines appear to be a recent introduction to the area and are liable to marked fluctuations. There is no trace of them in Pliocene and Pleistocene deposits. During 1952–60 the maximum yield was 57,000 tons from a population of 178,000 tons, but the increase of anchovies has caused a reduction in the sardines, and before the sardines can increase again the anchovies must be controlled.

In the southern Pacific the anchovies off Peru have recently become the basis of the world's largest fishery, as Peru has been top of the fishing nations for the 5 years since 1965 in bulk. The anchoveta that form the basis of this fishery are less than 15 cm long and they live off 99 per cent phytoplankton, mainly diatoms, and 1 per cent zooplankton. They occur off central Chile in 37°04′s to 04°15′s off Peru, extending 48 km offshore in summer and 190 km in winter. They suffer a high natural mortality, being preyed on by larger fish and the many thousands of sea birds that inhabit this coast. It has been estimated that guano birds and bonito consume 3 million metric tons, and other predators bring the total mortality to 6·5 million tons. This natural loss compares with the mortality due to fishing which was 9,760,000 metric tons in 1967. The catch rose in 1968, but fell again in 1969, being 10,520,300 and 9,223,500 respectively in metric tons. There are signs that this prolific fishery may be being over-exploited and controls are being imposed.

The Peruvian anchovy (*Engraulis ringens*) catch has diminished rapidly in the last two years. The peak of nearly 12·3 million metric tons was reached in 1970. In 1971 the catch was only just over 10 million metric tons, and in 1972 it fell to 4·5 million metric tons. The total bulk of fish is probably about 15 to 20 million metric tons at the height of the annual cycle. The anchovies stay in the cool coastal current, which runs close to shore and extends rarely deeper than 200 m. The El Niño of 1957, a severe one, reduced the guano birds from 27 million to 5·5 million, the number recovering to 17 million by the time of the 1964 El Niño. The annual catch of anchovies by the guano birds is estimated at about 2½ million metric tons. Following the 1965 El Niño the guano birds fell to about 4·3 million and they have not recovered since, possibly because of the lack of fish through over-fishing. The sudden failure of the 1972 fish catch will probably be repeated in 1973, when even fewer fish are likely to be caught. There is a suggestion that the stock have been irretrievably damaged. The evidence for this view is, firstly, the size of the standing stock is far smaller than normal, possibly 1 to 2 million metric tons, compared with 15 to 20 in recent years. Secondly, the recruitment of new fish has been the smallest ever observed, barely 13 per cent that of a normal year. The cause may have been the El Niño, in that the 1972 occurrence was one of the most severe on record. The surface temperature in the coastal current rose from its normal 22°c to 30·3°c in February, only falling to the normal value in March, 1973. Salinity fell from 35 to 36‰ to 32·7‰. The guano bird population was reduced to about 1 million. The 12·3 million metric ton catch of 1972 was also a factor as the maximum sustainable yield has been calculated at 10 million metric tons/year. The excess catch is partly the result of too many fishing boats and fishmeal factories. The catch forecast for 1973 is 3 million metric tons (Idyll, 1973).

The fishing of the North Sea herring stock in the inter-war period changed considerably. British fishery declined from 400,000 tons in 1900–13 to 200,000 tons after 1930. Another change was the development of trawling. The total catch rose to about 600,000 tons in the 1950s but the British share declined, a tendency that continued in the 1960s. In 1965 the total North Sea catch exceeded 1 million tons, but the British share was only 40,000 tons in 1962, the difference being caught by Dutch, Swedish, Polish, German, and Danish boats. Trawling in the eastern part of the sea by Denmark and Germany for immature fish for industrial use increased especially, and Norway and the USSR also entered the fishery. The decline of the British herring fishery was first economic and then due to lack of fish. The catch in the southern North Sea and eastern English Channel declined both in total catch and in catch/unit effort from 1947 to 1963 (figure 7.6). The decline in the stock is shown by the decrease in older fish. The total abundance and decline of unit catch/effort has not been found in the central and northern North Sea where the total abundance has not changed so markedly.

The Atlantic–Scandian stock occur in three groups. One is the Norwegian fishery off

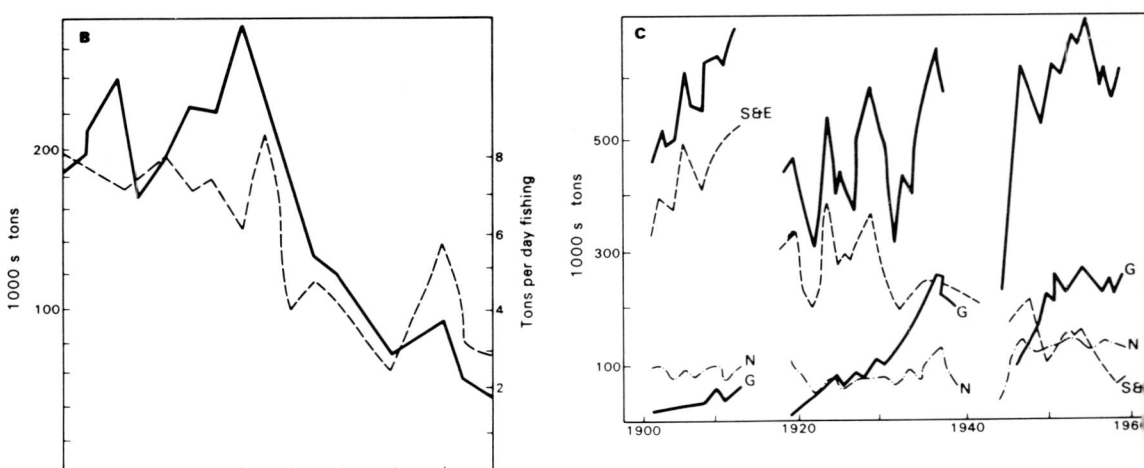

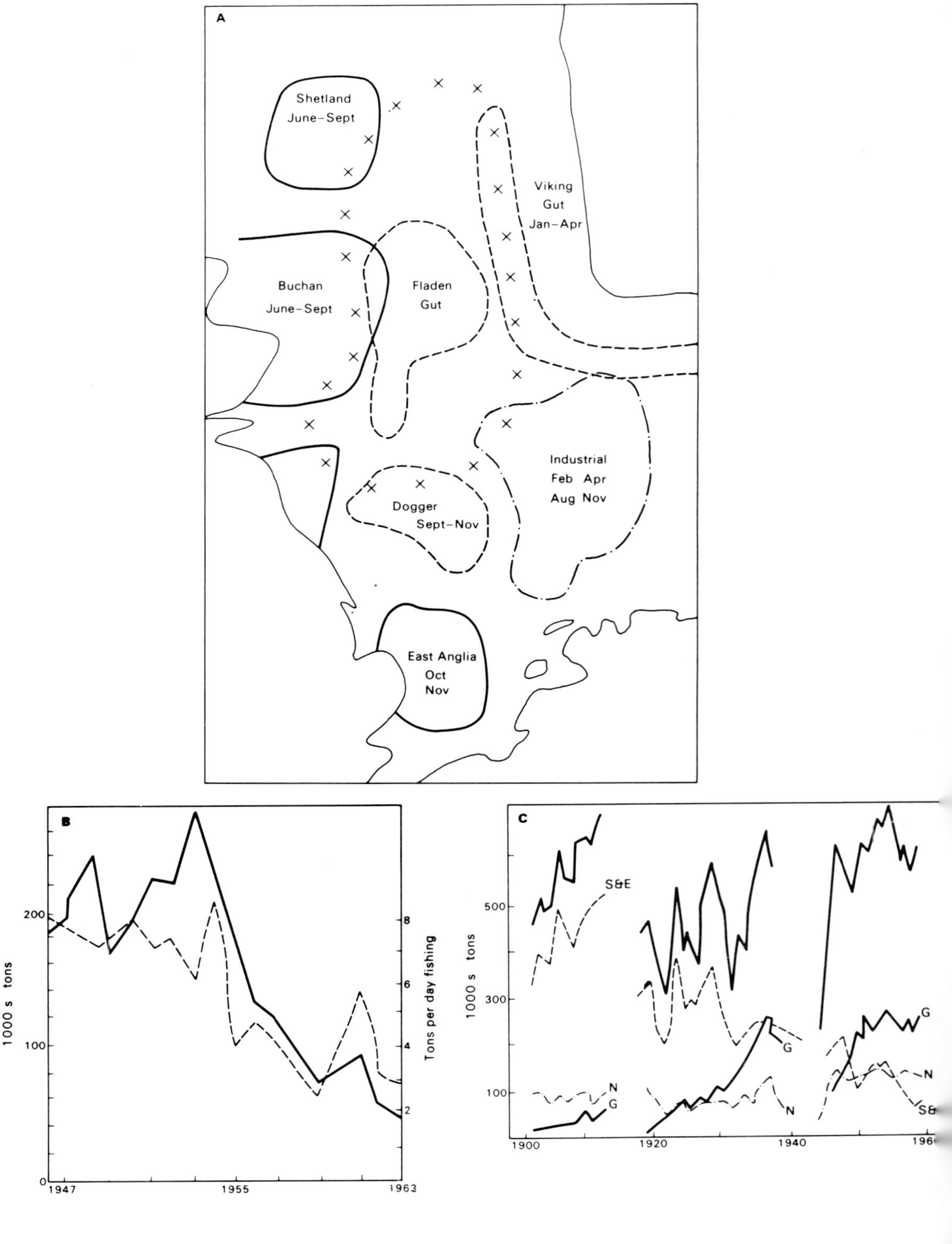

A

Shetland
June–Sept

Buchan
June–Sept

Fladen
Gut

Viking
Gut
Jan–Apr

Industrial
Feb Apr
Aug Nov

Dogger
Sept–Nov

East Anglia
Oct
Nov

B

1000 s tons

Tons per day fishing

200

100

0

1947

1955

1963

8

6

4

2

C

1000 s tons

S&E

G

N

G

G

N

N

S&E

500

300

100

1900

1920

1940

1960

Norway in January to April, the second is the USSR fishery in the Norwegian Sea, and the third is east and north of Iceland. The Icelandic fishery started in 1869 and the Soviet one in 1950. The year classes differ in quantity by about 25 to 30–1. Good year classes occur about every 7 years, producing a cyclic yield. The catch/unit effort of the Soviet fishery has dropped. This was caused by a fall in the spawning stock to one-third of the peak catch in 1961–62. The level of recruitment has been low since 1956.

Herring has been extensively exploited all over the world in their different types, but the Atlantic herring has been less exploited from America than the Pacific herring, *Clupea pallasi*, due partly to supply and demand and partly to the results of fishery management. Since demand for herring has increased in the USA, landings have increased from the Atlantic, but declined from the Pacific. The two species have a similar biology apart from spawning habits. The eggs stick to seaweeds and may be exposed at low tide, in the Pacific herring, which spawn in shallow coastal water, often in the intertidal zone. Spawning takes place in winter and spring, starting in California in December and January. It is progressively later in the more northern population, taking place in April to June in Alaska. The Atlantic herring spawn in deeper water many miles from the coast. They concentrate in the largest number on the Georges Bank, 160 km east of Cape Cod in depths of 25–100 m. Some spawning occurs in very shallow water of 1–2 m in the Gulf of St Lawrence. Most spawning takes place in winter, although in Newfoundland, Quebec, and Nova Scotia waters it takes place in spring. Along the eastern US coast, spring spawning is insignificant, most activity taking place between August and November. Most of the fish are in the 3–6 year classes. Two-year old fish are caught as sardines on the east coast when they are 15–20 cm. long. There is wide fluctuation in catches due to variation in the success of different year classes. At present it is difficult to distinguish between declines brought about by fishing and other factors, including food supply, disease, adverse temperatures, and other environmental factors.

The Pacific herrings are fished from California to Alaska, with Canada being the largest producer. Figure 7.7 shows the fluctuations in the catch between 1937 and 1967. Fishing started in Alaska in 1882, mainly for oil and meal. The fishery reached a peak of 25,000 tons in the early 1920s and fell to less than 10,000 after 1925. Some herring were needed as bait for the halibut fishery. The maximum catch occurred in the 1930s with over 100,000 tons in 1937, 1938, and 1939. Recently the catch has fallen to under 10,000 tons. Many restrictions have been imposed on the fishery including net sizes, exclusion of traps, seines, closed seasons and areas, and quotas. In British Columbia all the coast was being fished by 1950 and the catch exceeded 150,000 tons. By the early 1960s catches exceeded 200,000 tons, but in 1967–68 the catch had declined to 17,000 tons and the fishery was closed in 1968. The catch off Oregon and Washington and California is very small, the fish being used mainly as bait. It has rarely exceeded 1000 tons. Some eggs are taken now on seaweed, but the catch is regulated.

The Atlantic fishery takes place from the Canadian Maritime Provinces and from Maine

Figure 7.6 (*opposite*) Herring fishery in the North Sea. **A**: Areas fished in different ways at different seasons. Full line indicates drift net fishing, dashed line trawling, and dotted line industrial trawl fishing. The crosses indicate the area fished by continental drift lugger. **B**: Total landings in terms of catch/unit effort, dashed line, and total catch, full line. **C**: Proportion of fish caught by the different countries in the southern North Sea and English Channel. (*After Parrish and Saville, 1967.*)

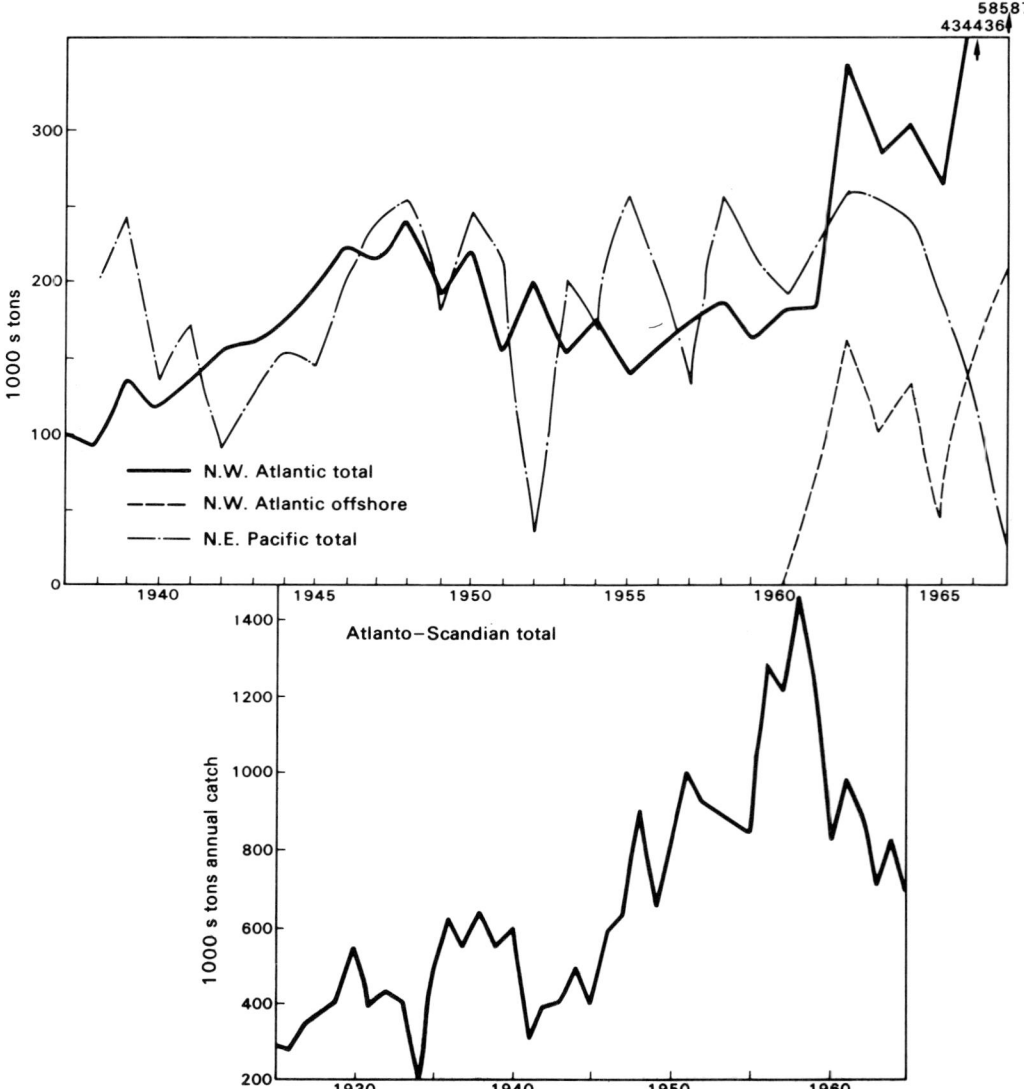

Figure 7.7 The herring fishery on the west and east coasts of North America, and the Atlanto-Scandian fishery on the eastern side of the Atlantic Ocean.

to Virginia; foreign ships operate in the offshore zone. The USSR started fishing in 1961, resulting in a marked increase in the catch, reaching nearly 600,000 tons in 1967 (figure 7.5). The herring industry on this coast was first stimulated to provide bait for long-line fishing and lobster fishing. Sardine canning then became the mainstay of the fishery after 1880, but since 1960 production of meal and oil has surpassed other uses. Fluctuations in catch have been wide. There are more restrictions on the fishery in Maine than elsewhere, and landings here have been less erratic than in some of the other states.

The Pacific fishery has on the whole been more regulated than the Atlantic one. Biological observations have been more careful in the Pacific as the region of fishing is largely inshore, and therefore more amenable to regulation than the open water fishery characteristic of the Atlantic. The reduction of the Alaska fishery seems to be an example of under-utilization, while the high level of present production in the Atlantic Maritime Provinces suggests that the fishery was previously under-used. In British Columbia, on the other hand, the fishery failed when catches exceed 200,000 tons, but could be maintained at 150,000 tons.

The excessive catch in this fishery is probably due to a new technique, the use of light to attract fish. The effect of this measure, however, had not been fully assessed when the fishery failed. It is not certain that over-fishing caused the decline of the fishery. On the

Plate 19 Tuna: a far-ranging, powerful pelagic fish, found mainly in the Pacific Ocean (*Natural History Museum*).

whole management of the American herring fisheries have not been successful owing to lack of basic data on herring mortality, recruitment, and stock abundance.

Tuna fishing in the Pacific provides an example of management of a wide-ranging active pelagic fish of considerable commercial importance (Plate 19). Fishing for tuna from the US west coast started in 1903. Albacore tinning reached over 9000 tons by 1914 from fish caught off California. Tuna are only close offshore in summer and autumn. In 1916 skipjack and yellowfin tuna were also included in the catch, which rose irregularly between the two wars, doubling about every 8 years. The fishing included more skipjack and yellowfin and spread further offshore. Albacore tuna disappeared from off California in the 1925–35 period, and stimulated fishing for the tropical species. During the Second World War fishing was reduced from the US, but fishing from Latin American countries was encouraged. Landings rose rapidly after the war, reaching 154,000 tons in 1950 of yellowfin and skipjack, which are the tropical species. Japanese fleets also expanded eastwards and fished large tuna with long-line gear, and Canada has now entered the fishery. Japanese fishing includes bigeye tuna. The biological basis of the fishery was investigated before the catch became heavy, starting in the 1920s.

Yellowfin tuna is the only species that has yet required regulation to maintain the maximum sustainable yield. Albacore tuna is one population in the north Pacific in which the immature fish live closer inshore and support the American west coast fishery, while the larger mature fish are caught further west in the region of the north equatorial current and the equatorial counter-current. The fish seem to migrate as they grow, and the extent of the fishery has not yet reached the maximum sustainable yield. Some of the bluefin tuna probably migrate across the Pacific. They are taken by purse-seine nets off California and Baja California, where they are young and immature, but there is not any evidence of spawning in the eastern Pacific. They are taken as larger mature fish off Japan. No large increase in the catch in the east Pacific is likely, from the evidence of tagging experiments, although the maximum yield has probably not been reached. Yellowfin tuna and the skipjack are caught over their entire range from California to Chile, the former being the more valuable species, but the smaller and more restricted population, so that it is the only one in need of conservation measures at present. The fish occur continuously in the equatorial Pacific, but two populations exist, with a distinguishable western group. None of the fish tagged in the east were recovered in the west, although 6086 recoveries were made. Catch and fishing effort data are available from 1934. The relationship between catch per days fishing and fishing intensity in terms of total catch indicates that the maximum sustainable yield is about 96,000 tons with 35,000 units of effort, a level that was exceeded in 1961. The fish have a high mortality rate of 0·80, corresponding to 36,000 units of fishing effort. The current age of entry into the fishery is 1·5 years, giving a maximum sustainable yield of 1·4 fishing mortality rate.

The skipjack is a smaller tropical species that occurs with the yellowfin. The catch varies from year to year, but there is no evidence of reaching the maximum sustainable yield, despite an increase in the catch from 10,000 tons in 1934 to 80,000 tons in 1961. The population is a far-ranging one. Some spawning takes place in the eastern Pacific and larger specimens are caught off Hawaii. Spawning is probably more important in the equatorial regions further west. In 1962 a fish tagged off Mexico was recovered off Hawaii. Fewer skipjack recoveries of tagged fish are made, indicating lower fishing than yellowfin. Measures for controlling the catch of yellowfin include minimum weights, while maximum weights were fixed to support the canning industry. In 1962 a catch maximum of 83,000 tons was recommended to control the slightly over-fished tuna, and was accepted in 1966. In 1968 the quota was raised to 106,000 tons as a result of good recruitment. The quota system seems to have worked well. The discovery of mercury in tuna, however, may cause other factors to exert a control of the fishery. One problem is the growing number of vessels which means that each must have a smaller share if the quota is not to exceed the supply. A limitation on the number of vessels may be the most satisfactory answer to this problem.

The mackerel is another commercially important pelagic species off western Britain. Like the herring it swims in shoals and lives on plankton, feeding mainly in spring and summer. The main mackerel shoals are found off southwest England, and are fished from Newlyn in Cornwall. Towards the end of October the fish leave the surface and sink in densely packed groups to hollows on the sea floor. In December they spread out more widely over the bottom, feeding now on benthos. Then towards the end of January or early February they move in shoals again towards the surface and slowly migrate towards their spawning areas, which they reach about April. These areas are mainly situated to the south of Ireland, the fish spawning over the edge of the continental shelf between

April and June. After spawning they disperse towards the coast, feeding on plankton as they move.

2.3 Whaling

The problem of controlling whaling is in some ways more difficult than that of other fisheries. This is partly due to the very slow reproduction of the whale compared with the extremely prolific fish. Another factor is the great investment in the establishment of a whaling fleet with factory ships or shore stations to deal with the catch. The history of the whaling industry and attempts to control it is one of the worst examples of man's greed and short-sightedness in the face of overwhelming scientific evidence of over-exploitation. Only the decline of the stock to uneconomic levels has enforced the acceptance of protection for some of the whale species, but at a date that many scientists consider may be too late to save the largest known animal on earth from becoming extinct. Table 7.8 gives the

Table 7.8 Whale stocks and catches

Species	Original stock	1967 stock	Sustainable yield 1967	Maximum sustainable yield	weight, tons
Blue	>150,000	930–2,790	< 200	6,000	84
Fin	>250,000	39,800	4800	20,000	50
Humpback	c. 20,000	2,000	< 100	1,000	32
Sei	—	70,000–90,000	4400–7000	5,400–6,300	22

estimated numbers of whales and sustainable catches. Recovery time needed to reach the maximum sustainable yield is estimated as at least 50 years for all except the Sei whale. The total maximum sustainable yield would produce 1·85 million metric tons, giving 352,000 tons of whale oil and 800,000 tons of whale meat. The blue whale was the mainstay of the industry until about 1937. After the war when whaling was resumed 8000 were taken in 1948, but the number declined to less than 2000 in 1955, reaching only 1255 in 1962, when biologists estimated the total numbers at between 900 and 3000, a number which had declined to between 650–2000 in 1963. This number may be below that needed for effective reproduction.

At one time the US was one of the main whaling nations, but now the main countries whaling are Japan, the Soviet Union, South Africa, Peru, and Chile. These are the only countries whose catch has increased since the Second World War. Countries whose catch has declined include Norway, UK, Netherlands, Australia, Argentine, Panama, and USA. The southern hemisphere whale stock was the major one to suffer in the inter-war period, and subsequent to the Second World War. International control was evidently required by 1930 and the first whaling convention was signed in 1931.

The present convention was entered into in 1946. The BWU (Blue Whale Unit) system was adopted, whereby one blue whale = 2 fin whales = 2·5 humpback whales = 6 sei whales, and minimum lengths were laid down in proportion to the amount of oil each species was likely to produce. The opening date of the season was fixed to obtain the maximum amount of oil, which increases rapidly as the season advances. The length of the whaling season was limited. Unlike the more rapidly breeding fish, whales did not recover their numbers during the four years without fishing during the Second World War. Whaling nations were influenced by the success of the control measures adopted in the north Pacific halibut fishery and agreed for the 1945–46 and 1946–47 seasons to limit the Antarctic catch to 16,000 BWU, but excluding catches of sperm whales.

The International Whaling Commission was set up in 1946 and holds annual meetings. The commission cannot, however, limit the number of whalers or factory ships, nor allot specific quotas, although it can establish which areas shall be open, the minimum length, the total catch, and the duration of the season. The commission has scientific committees that coordinate research into whale stocks and the best conservation measures. During the post-war years the trends in whale catching indicate the exhaustion of one species after another (figure 7.8). The number of blue whales has continued to decline to vanishing point, while the numbers of fin whales reached a maximum in the late 1950s and early

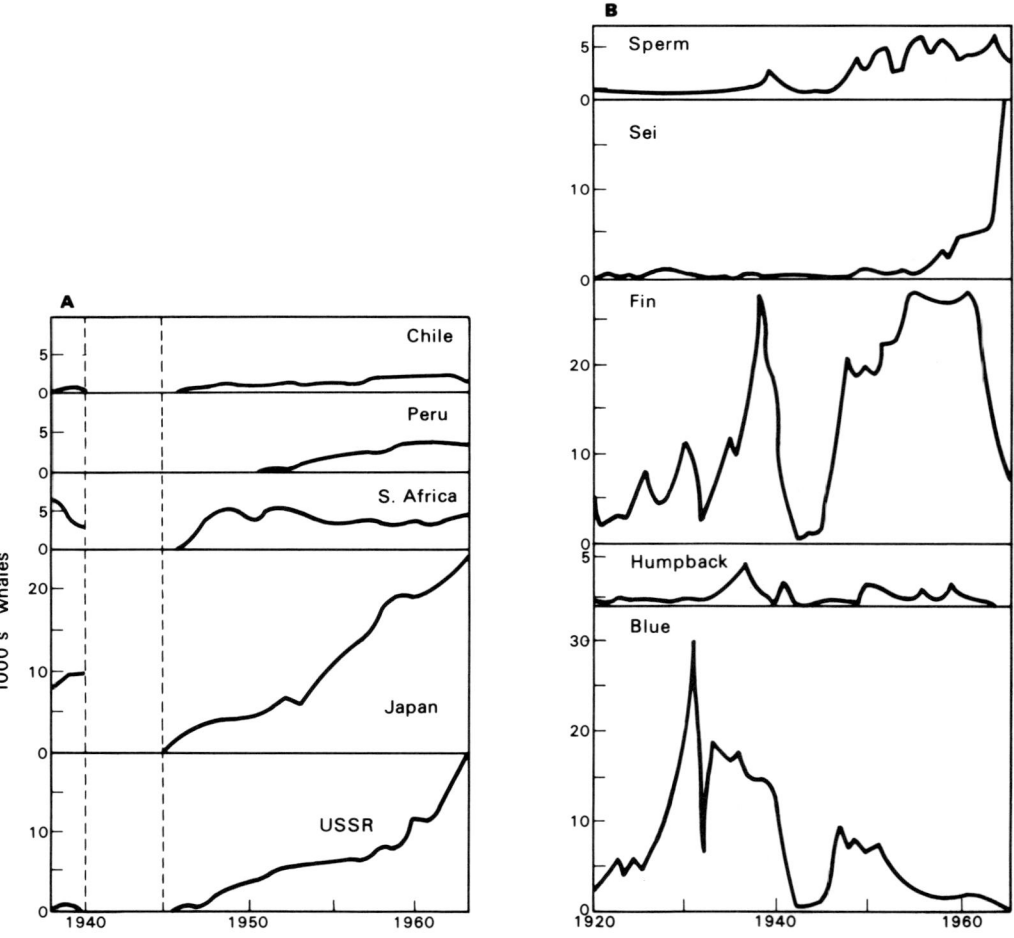

Figure 7.8 A: Countries that have increased their whale catch since the Second World War. **B:** Catches of the different whale species from 1920–60.

1960s. This has been followed by a very rapid decline from about 27,000 to about 7500. At the same time there has been a considerable increase in the percentage of immature whales in the catch. For blue whales the percentage was 35·1 in 1951–52, nearly 40 per cent in 1957–58, and reached 72 per cent in 1960–61, while the figures for fin whales are 17 per cent in 1951–52, 25 per cent in 1956–57, and 32 per cent in 1961–62. This increase is an indication of serious overcatching. As the blue whale and fin whale catches have decreased, so there has been a very rapid increase in the catch of sei whales. Until the early 1960s the number of sei whales caught was under 5000, but the total leapt to over 20,000 in 1964. The catch of sperm whales has been rather steadier during the 1950s at around the 5000 mark. These whales were severely decimated by American whalers during the nineteenth century, but they have slowly recovered, and in 1962 the catch exceeded 23,000.

The adoption of the BWU has had an unfortunate effect on whaling. The effort required to catch any one whale is about the same, which means that the most valuable species will be fished first as the rewards will be greater. This meant that the blue whales were first fished out, followed by the fin whales, and now it is the turn of the sei whales, following the reduction in numbers of the humpback also. The catch of sei whales reached its peak in the 1964–65 season, and these catches are also likely to decline. Since the BWU was fixed there has been a decline in the value of oil and an increase in the value of meat. From this point of view, one BWU is more valuable in the form of 6 sei whales than 2 fin whales, which also accounts for the shift to sei whales.

As the fin, blue, and humpback whales have become scarce in the Antarctic the whaling fleets have moved north in pursuit of the sei whales, who are more abundant in warmer waters. The mid-1960s rate of harvesting of sei whales was at about the maximum sustainable yield, as shown in table 7.8. In 1964 the International Whaling Commission failed to agree on quotas for the Antarctic after several years of quotas higher than could be sustained. In 1965 the whaling fleets failed to reach their privately agreed quote of 8000 BWU, as predicted by the scientists. Further reduction of the quota for the seasons up to 1967–68 were agreed so that the quota in the latter year would be below the sustainable yield, although the quotas required ratification.

There was no decision to manage each species separately, but with virtual protection of humpback and blue whales, and the increase of meat value, the brunt of the catch should fall on the sei whale, virtually amounting to control by species. This agreement shows some advance in conservation. Although economically it would have been advantageous in the short term to catch every whale, some measure of control has been agreed. The commission must now enforce the regulations to conserve north Pacific whaling. If possible regeneration of the Antarctic whales should be allowed, it cannot be done in the life time of anyone now alive. Already three species are being overcaught in the north Pacific— the blue, humpback, and fin whales, although the first two species are now protected. The rise of sperm whale catches is also alarmingly high, and females and calves are not being adequately protected. Present catches could be increased if the stocks were allowed to increase to optimum size, but economic sacrifices now are essential.

Another improvement in the whaling industry would be the general introduction of the electric harpoon to replace the explosive one, which would cause less suffering to the whale and better condition of the meat. The three countries that now dominate the whaling

Table 7.9 Cetacean biomass in tons derived from Antarctic pelagic whaling

Year	Fin	Humpback	Sei	Sperm	Blue	Per cent	Total
1949/50	903,050	56,312	2,328	88,665	526,280	33·4	1,576,635
1950/51	838,752	44,010	7,938	163,599	571,212	35·1	1,625,511
1951/52	1,067,040	40,660	734	184,368	420,168	24·5	1,712,970
1952/53	1,059,850	25,853	2,901	75,382	305,414	20·8	1,469,400
1953/54	1,224,314	15,503	5,274	77,750	220,088	14·3	1,542,929
1954/55	1,242,144	13,558	2,675	182,656	173,040	10·7	1,614,073
1955/56	1,213,872	36,659	6,181	213,311	127,269	8·0	1,597,292
1956/57	1,259,300	19,012	15,844	134,695	117,390	7·6	1,546,341
1957/58	1,210,656	11,128	52,900	182,990	131,352	8·3	1,588,026
1958/59	1,216,013	60,134	31,852	154,954	98,853	6·1	1,611,806
1959/60	1,255,972	36,260	72,442	120,002	92,976	5·9	1,577,652
1960/61	1,315,507	20,104	94,820	131,115	119,683	7·1	1,680,229
1961/62	1,251,723	8,652	104,953	128,535	69,563	4·5	1,556,019
1962/63	868,674	7,641	118,865	130,248	62,155	5·2	1,187,583
1963/64	666,548	51	177,320	168,270	6,270	0·6	1,028,459
1964/65	329,667	0	419,341	108,644	1,069	0·1	858,721

Length of rorquals caught in Antarctic pelagic whaling in feet

	1933	1953	1963
Blue	80·4	77·5	73·1
Fin	68·7	67·3	66·2
Humpback	39·2	40·9	41·6
Sei	—	51·6	49·8

industry are the USSR, Japan, and Norway, but particularly the first two, whose share in whales has risen very rapidly in the post-war period at a steady rate, reaching 25,000 and 20,000 in 1965 for Japan and the USSR, respectively (Figure 7.8).

2.4 Seaweed

Seaweed is being used for an increasing number of purposes and is a commodity that could increase in importance in the future as methods and cultivation improve. Seaweed can be used in animal feeds; it contains iron and iodine, and vitamins A, D, and B_2. Seaweed was first harvested for animal food additive in the United States in 1870, while the industry in Britain did not start until 1948. By 1958 there were five factories in Ireland and four in Scotland, only one of which still survives. The Irish ones, however, expanded between 1960 and 1965 to 12, but some also closed. In Ireland in 1964, 18,000 tons of meal was produced, while in 1959 Scottish production had dropped to 400 tons.

One of the most useful elements in seaweed is alginic acid, which can be made into alginates in combination with sodium, and is used in cooking for thickening and in other processes, such as textile manufacture. World production of alginate is about 15,000 tons/year, half of which is produced in the United States. Table 7.10 gives the world harvest of seaweed, and table 7.11 the world resources. The Pacific kelp beds could yield 3 million tons/year. The standing crop density in tons/ha is 10–16. British Columbia has 750,000 to 1 million tons and the productivity of Nova Scotia exceeds that of Scotland. The standing crop is 1,100,000 tons on 520 km of coastline, and there are 4850 ha of

Table 7.10 World seaweed harvest, fresh weight metric tons $\times$ 10^3

	1963	1964	1965
Argentina	8·3	10·2	19·9
Canada	22·5	19·8	25·0
France	—	12·9	14·8
Japan	425·7	361·0	395·0
Korea	42·9	54·5	n.a.
Mexico	19·5	23·3	17·0
Morocco	2·1	2·2	2·6
Norway	56·9	63·6	74·2
Philippines	0·1	0·1	0·3
Portugal	4·2	4·8	2·7
Ryu Kyu Islands	0·1	0·1	0·1
Scotland	14·5	18·4	21·2
Spain	4·3	5·3	n.a.
Taiwan	0·8	0·8	1·1
Tanzania	0·3	0·4	0·2
United States	3·3	2·2	2·4

Table 7.11 Seaweed resources in 1965, fresh weight

	Resources, million tons	
	Littoral	Sublittoral
Alaska		19·2
British Columbia		1·5
Puget Sound–San Diego		13·8
Nova Scotia	1·2	0·9
San Diego–Mexico		8·5
Sargasso Sea		4–11
Norway		c. 20
Scotland	0·2	c. 10
Russia White Sea		1·5
Russia Black Sea		1–2
New Zealand		0·8
Tasmania		0·35

fucoid and *Laminaria* respectively, while the equivalent figures for Scotland are 1,380,000 tons on 865 km of coast and 15,400 ha. The United Kingdom produces about one-third to one-quarter, and the rest comes from Norway and Japan mainly, with France and Russia providing a small amount. Mechanical harvesters for *Laminaria* have not proved satisfactory, partly owing to the storminess of the seas where it grows. Thus most *Laminaria* is collected from the shore, when it has been washed up. Mechanical harvesting is best developed in California, where 100,000 tons are harvested annually of *Macrocystis*. *A. nodosum* is harvested by cutting at low tide, and 23 cm must be left for regeneration. It takes several years for regeneration if less is left. Harvesting is one of the main problems of the industry, as reserves are large (see table 7.9; Rampton, 1970). Harvesting of *Laminaria* in particular has not been sufficiently developed commercially.

Further reading

ALVERSON, D. L., LANGHURST, A. R. and GULLARD, J. A. 1970: How much food in the sea?
Science **168,** 503–5.

BALLS, R. 1961: *Fish capture*. Buckland lecture for 1959. London: Edward Arnold.

BARDACH, J. 1968: *Harvest of the sea*. London: Allen and Unwin.

BENSON, N. G. (editor). 1970: A century of fisheries in North America. *Sp. Pub.* **7.** Washington: Amer. Fish Soc. 329 pp.

BONEY, A. D. 1965: Aspects of the biology of the seaweeds of economic importance. *Adv. in Mar. Biol.* **3,** 105–253, F. R. Russell (editor). Academic Press.

DILLON, O. W. 1970: The growing importance of commercial fish farming in the United States. *The Amer. Fish Farm.* **1,** 20.

FIRTH, F. E. (editor). 1969: *Encyclopedia of marine resources*. New York: Van Nostrand Reinhold.

GULLARD, J. A. and CARROZ, J. E. 1968: Management of fishery resources. *Adv. in Mar. Biol.* **6,** 1–73.

HULL, S. 1964: *Bountiful sea*. Englewood Cliffs: Prentice-Hall. 340 pp.

MCKEE, A. 1967: *Farming the sea*. London: Souvenir Press. 314 pp.

PARRISH, B. B. and SAVILLE, A. 1967: Changes in the fisheries of the North Sea and Atlanto-Scandian herring stocks and their causes. *Oceanog. and Mar. Biol. Ann. Rev.* **5,** 409–47.

CHAEFER, M. B. 1968: Methods of estimating effects of fishing on fish populations. *Trans. Amer. Fish. Soc.* **97,** 231–41.

8 Uses and problems of the oceans

1 Resources from the oceans

2 Ocean pollution: 2.1 Sewage and fertilizers; 2.2 Chlorinated hydrocarbons and pesticides, including DDT; 2.3 Heavy metals; 2.4 Oil and petroleum products; 2.5 Radioactive substances; 2.6 Plastics

3 Conservation

4 International law at sea

5 International co-operation

Conclusion

One of the major problems associated with the oceans and their use by man is to devise a fair method of distributing the wealth of the ocean, and another major problem is to use the bounty of the sea without detriment to the marine environment. There are many conflicting claims on the oceans, which in the past have been used mainly for fishing,

Table 8.1 Historical uses of the seas

Use or harvest	Approximate beginning	Location
Seaweeds	800–600 BC	China
Pearls	13th century	China
Precious corals	3rd century BC	Greece
Shellfish	1000 BC	China
Sponges	3rd century BC	Greece
Fish	Stone age	Egypt
Turtles	16th century	West Indies
Whales	1200	European coast
Seals	8000 years ago	Bering Sea
Fur seals	1784	South Atlantic
Salts	2200 BC	China
Military ships	4th century BC	Mediterranean
Military divers	3rd century BC	Greece
Diving suit	13th century	England
Diving bell	1707	England
Submarine cable	1848	North Atlantic
Submarine photography	1856	North Atlantic
Modern submarine vessel	1890–1900	North Atlantic
Submarine oil and gas	1959	North Sea
Scientific study	4th century BC	Greece

mineral extraction, transport, military purposes, recreation, and waste disposal. The same uses are likely to be important in the future, but increasing pressure of population in the world will intensify all these demands. It is the developing intensity of oceanic exploitation that is likely to lead to serious and perhaps irreversible damage to the oceanic environment, and this is major cause for concern. In order to assess the potentialities and limits to which oceanic exploitation can be stretched it is essential that more is learnt of the oceanic environment, both physical and biological.

Biological exploitation of the oceans was considered in the last chapter, and the mineral resources are briefly commented upon in this one, which also considers problems of pollution and some of the ecological problems associated with conservation.

1   Resources from the oceans

The oceans promise great returns of many resources, some of which are already exploited in considerable quantities, particularly the living resources. The figures in table 8.2 show

Table 8.2 Oceanic resources

	World		United States	
Resource ($ $ \times 10^9$)	Ocean	Land	Ocean	Land
Chemical	0·3	0·2	0·1	0·1
Biological	6·4	260·0	0·4	45·0
Geological	3·8	73·0	0·9	20·0
Total	10·5	333·2	1·4	65·1
Area, 10^6 km²	361·1	148·9	—	9·4
Continental shelf	26·4	—	3·4	—

Values are in terms of returns to the producers and refer to 1964 data

that for the world as a whole biological resources are of greatest value, but that the ocean supplies a small proportion of the total, apart from chemicals. The continental shelf is the most important part of the ocean from the point of view of resources, particularly geological resources, although it is also important biologically.

Minerals that are economically worth mining from the sea are relatively few. The main ones are listed in table 8.3. The total sea-floor production of the entire world in 1964 was $3.8 billion, one-tenth of the same minerals mined on land. The only field in which sea-floor minerals are of significant proportions are oil and gas, and sand and gravel. Potential resources include phosphorite, which occurs at the edge of the continental shelf of the eastern United States, in areas where land deposits have no access. Deposits also occur off southern California, northwest Mexico, Peru–Chile, and South Africa. They all occur in areas of upwelling, except off southeast USA. There are between 1 and 2 billion tons off the USA, but land reserves amount to at least 10 billion tons, so that marine ones will not be required in the foreseeable future.

Table 8.3 Value of production of geological resources from ocean and land
(s × 10⁶)

	World		United States	
	Ocean	Land	Ocean	Land
Authigenic phosphorite	0	375	0	160
manganese	0	423	0	3
Detrital, sand and gravel	100	2,000?	35	860
Titanium	33	37	9?	11?
Zircon	11	0	1?	0
Tin	5	460	0	0
Diamonds	4	284	0	0
Monazite	1·5	0·3	0?	0?
Iron	0·7	5,300	0	800
Gold	0	1,310	0	50
Organic, oil and gas	3,600	27,500	800	10,500
Sulphur	15	240	15	100

Manganese nodules have been found widely, particularly in the Pacific. They usually occur at depths of about 4000 m, apart from those in the shallow water over the Blake Plateau. Typical contents are given in table 8.4. Tonnages of nodules on the sea floor are of the order of 10^{11}–10^{12}. It is likely that the rate of accretion is greater than the rate of mining of 6 million tons/year. Manganese reserves on land would last for several hundred years, but they are not evenly distributed and the United States produces only 2 per cent of its needs. Commercial mining is not likely for several decades, owing in part to recovery problems. Poor distribution of minerals within the nodules, which would cause over-production of some minerals, is another difficulty that could have economic repercussions.

Sand and gravel are of low value, but abundant in amount. Surveys have been carried out off the eastern USA with a view to using offshore sand for beach stabilization projects. Ample supplies have been located. Recovery and transport costs are the main problems. Offshore sands and gravels frequently have high shell content and are not always suitable for concrete aggregate. Some heavy mineral production from coastal zones has been carried out. Gold has been recovered from drowned beaches offshore, and it has been dredged off Alaska by taking advantage of the stable platform provided by winter ice. It is estimated that 2080 ha leased by one company could yield 300 tons of gold. Some diamonds have also been dredged off South Africa.

Table 8.4 Manganese nodule content

	Pacific Ocean	Blake Plateau
Manganese	24·0	16·0
Iron	14·0	17·0
Silicon	9·0	11·0
Nickel	1·0	0·4
Copper	0·5	0·2
Cobalt	0·4	0·3
Ignition loss	26·0	24·0

By far the most important mineral production from the sea is that of oil and gas, representing more than 90 per cent of the value of minerals obtained from the sea. Potentially also they are the most important. In non-communist countries 17 per cent of the oil and 6 per cent of the natural gas comes from offshore. By 1980 probably one-third of the oil produced will be from offshore, amounting to 4 times the 1969 output of 6·5 million barrels/ day. Twenty-eight countries now obtain oil and gas from offshore, while 50 are undertaking exploratory surveys. Since 1946, 10,000 wells have been drilled off the USA. More than 6·5 million acres of the outer continental shelf had been leased by 1969, although only 10 per cent of the shelf has yet been surveyed in detail. Important recent finds have been made off North Alaska, Mexico, Trinidad, Brazil, Dahomey, Australia, Taiwan, and in the North Sea.

Some of the oil and gas of Japan's reserves are associated with salt domes, which are also common in the Gulf of Mexico and the North Sea. Salt forms dome structures, owing to its low density. The dome provide suitable traps for gas and oil, where the overlying strata are impermeable. Most of the present production is from the shallow water of the continental shelf. Future supplies may be deeper, coming from the continental rise in depths of 1525–5500 m. Drilling has already been carried out in 3570 m in the Gulf of Mexico. To develop deeper sources a floating platform will be essential. At present platforms anchored to the bottom by huge legs 12·2 m across can operate in water up to 100 m deep. The total height of one installed off Africa in 1967 was 180 m high; it weighed 8500 tons and occupied 1 acre (0·405 ha) of space. Eventually it may be possible to lower structures to the bottom and to work from there. A semi-submersible platform is a first stage; it gives greater stability than a ship for deep drilling. The average annual rate of oil and gas accumulation is estimated at 1×10^{10} kg cal or 1×10^{-11} of the influx of energy from the sun. The rate of extraction is 3×10^{16} kg cal or 3 million times the accumulation rate. Thus further resources will continually be needed to maintain the supply, which will eventually run out.

Chemical resources that have been extracted from the ocean water include salt, of which 35 million tons/year are obtained by evaporation, being one-third of total world production. Magnesium and bromine have also been extracted. The greatest potential use of the ocean, however, is probably for fresh water by desalination. Cost at present is about 0·85 dollars/100 gallons, a rate that is not yet competitive with land-derived water in most areas. With increasing water demand, desalination could well be necessary. Technology may then make the distillation process more efficient. Fresh water is already produced in some arid areas, such as Israel and Kuwait.

Metal-rich sediments are known from the ocean rises, but only since the *Glomar Challenger* drillings have they been located on older surfaces (Cronan *et al.*, 1972). Leg 16, which was in the eastern equatorial Pacific, drilled through iron-rich sediments that are chemically similar to those currently forming on the East Pacific Rise. They were found just above the basement at several sites in the area. The sediments probably collected when the site was near the crest of the rise and have moved to their present location by sea-floor spreading. The sediments extend to 140°W, 12°N near the Clarion Fault zone and date from early Miocene in the east to mid-Eocene further west.

2 Ocean pollution

The oceans are the ultimate sink into which all wastes in rivers and much of that in the air are likely to find their way. In addition there is some direct oceanic pollution. There are many forms in which pollution can occur; for convenience the discussion will focus on six broad types.

2.1 Sewage and fertilizers

The effect of sewage in the sea depends on the extent to which it is treated before being released. Complete treatment includes removing the sludge and suspended solid particles, oxidizing polluted organic material, sterilizing effluent, and removing plant fertilizing compounds. Most treatment plants do not achieve the last process effectively. Some authorities get over the problem of sludge by dumping it at sea by means of tankers or through pipe lines. The environmental conditions then determine whether the sludge will sink or disperse according to the relative densities and turbulence. One effect of sludge

Table 8.5 Nutrient budget of the world's oceans

Millions of metric tons	Nitrogen	Phosphorus	Silicon
Reserve in ocean	920,000	120,000	4,000,000
Annual use by phytoplankton	9,600	1,300	—
Annual contribution by rivers	19	14	4,300
Dissolved	19	12	150
Suspended	0	12	4,150
Annual contribution by rain	59	0	0
Annual loss to sediments	9	13	3,800

disposal is the decrease of light penetration, and a resulting decrease of photosynthetic activity and plant production. Fish may be polluted and benthic fauna destroyed as the sludge settles. Such events are, however, usually local and occur close to the disposal sites. Chemical and biological degradation of the organic material in sewage poses much more serious and widespread problems of which the most critical consequence is the reduction in oxygen. When oxygen has been used up entirely, hydrogen sulphide may appear and this substance is poisonous. Phosphorus and nitrogen remain, and these are fertilizers which lead to excessive growth of phytoplankton on the surface. Several effects may follow. Bacteria that break down the surplus growth themselves use up too much oxygen, and the blooms of some dinoflagellates release toxic material that kills other organisms, including larval and adult fish, and shellfish. These build up toxic substances within their bodies and can in their turn become poisonous to humans. The result of these processes includes the departure of adult fish; if adult fish remain and spawn there may be insufficient oxygen for eggs to hatch, prey animals can become depleted, and profuse growth of bacteria can prevent egg hatching and larval survival. The water may be rendered turbid to the extent of influencing development of eggs and larvae. The excessive phytoplankton soon die, and more phosphorus and nitrogen are released, thus adding much more of these elements to the water than that in the original sewage, even if this was untreated. The effect also becomes much more widespread in area. Over-fertilization can cause red tides. Filter feeders may not be killed themselves, but they can concentrate the poisons that kill other organisms and man, who eat the filter feeders. Some fertilizer can be beneficial if the amount is added in the correct quantity. The fertility of coastal waters is often due to this cause, and rivers supply nutrients by the same means. The ocean is nearly in a steady state in respect to nutrients, including nitrogen, phosphorus and silicon, the latter being necessary for the diatoms in the pelagic oceans. Only 1 per cent of the nutrient reserve is used by phytoplankton (table 8.5). One method that is being developed to study the effect of sewage on phytoplankton is the development of computer models that can be of any desired complexity. A possible development is the processing of sewage in the euphotic marine environment. This cannot be undertaken, however, until more is known concerning the growth of harmful blooms. Field experiments may be cautious, and artificial sea water ponds could be used to study blooms. The use of sewage as fertilizer could be potentially valuable, if blooms can be controlled.

Domestic sewage is biodegradable if there is enough oxygen. The North Sea has 54,000 km^3 of water. The volume is not sufficient to prevent harmful results of waste deposits, as shown by waste deposited off Holland that moved into the Wadden Sea and killed mussels (Korringa, 1971). The problem of sewage is that it can lead to excessive plankton, but it is more likely to lead to eutrophication, which creates an excess of nutrients that can result in red tides. Filter feeders concentrate certain minerals from industrial waste. Oysters, for example, can contain 100,000 times the amount of copper in a similar weight of sea water at times. The same applies to zinc and mercury.

2.2 Chlorinated hydrocarbons and pesticides, including DDT

The term chlorinated hydrocarbon is better than pesticide, although these substances are often referred to as biocides, insecticides, or simply as pollutants. Most of these substances (of which over a billion pounds/year are made in the USA alone) are readily degraded in water, and only a few hydrocarbons have been traced in sea water. DDE is probably the most abundant synthetic pollutant of the sea, while PCB is also abundant and widespread in marine ecosystems. PCB is not a pesticide, but it is a hydrocarbon and is derived from industrial waste. In order to control these pollutants the rates and routes of entry must be established, and their distribution and effects on marine communities must be determined. At present there are relatively few data available on these matters. Damage only occurs when the material is present in high concentration. For comparison the amount of inorganic carbon incorporated annually into organic matter by phytoplankton is a useful measure; it is about $2 \cdot 0 \times 10^{16}$ g/year. A pollutant introduced at 10^{10} g/year has a potential capacity to inflict damage. Pollutants may be expected to have adverse effects when concentrations exceed about 1 ppm. The problem is intensified by the concentration of pollutants in organisms, which can raise concentrations to damaging levels as shown in tables 8.6 and 8.7. Concentration of DDT of various types is very high in marine birds, and many are heavily contaminated with DDE and PCB, some containing 1000 ppm or more DDE and several hundred ppm PCB. DDE residues as high as 2500 ppm have been recorded in lipid. The residue is mainly in the form of DDE, being up to 80 per cent, as DDE is more persistent than DDT, from which it is derived. Marine bacteria do not seem to degrade DDE and PCB.

Table 8.6 Concentration of total DDT compounds and PCB in marine fish

Species and locality	Number	DDT net weight ppm		PCB net weight ppm
Northern anchovy				
San Francisco Bay	17	0·59 ±	0·11	NM (Not measured)
San Francisco Bay	29	0·33	0·04	NM
Monterey	30	0·90	0·22	NM
Morro Bay	29	0·74	0·22	NM
Port Hueneme	15	3·04	1·00	NM
Terminal Island	44	14·00	1·90	1·0
English sole				
San Francisco Bay	18	0·55	0·07	0·11
San Francisco Bay	33	0·55	0·12	0·11
San Francisco light-ship	15	0·19	0·04	0·05
Monterey	15	0·76	0·16	0·04
Shiner perch				
San Francisco Bay	14	1·00	0·10	1·20
San Francisco Bay	10	1·40	0·30	0·40
San Francisco Bay	15	1·10	0·10	0·90
Jack Mackerel				
Channel Islands	31	0·56	1·00	0·02
Hake				
Puget Sound	22	0·18	0·05	0·16
Channel Islands	6	1·80	1·10	0·12
Bluefin tuna				
Isla Geronimo, Mexico	7	0·56	0·24	0·04
Yellowfin tuna				
Galapagos Island	13	0·07	0·02	NM
Central America	13	0·62	0·19	0·04
Herring				
Baltic	18	0·68		0·27
Plaice				
Baltic	6	0·02		0·02
Cod				
Baltic	5	0·06		0·03
Salmon				
Baltic	11	3·40		0·30

Table 8.7 Concentration of DDT in the food chain

	ppm
River water	0·000003
Estuary water	0·00005
Zooplankton	0·04
Shrimps	0·16
Insects—diptera	0·30
Minnows	0·50
Fundulus	1·24
Needlefish	2·00
Tern	2·80–5·17
Cormorant	26·40
Immature gull	75·50

The world production of DDT is about 10^{11} g/year, or about five orders of magnitude lower than the primary productivity of the ocean. Much of the material probably reaches the sea by air transport, in view of its widespread dissemination, which includes the snow of the Antarctic and Greenland. The relative importance of rivers and air transport is not yet accurately known. Rivers with low silt loads seem to be low in chlorinated hydrocarbons. Streams in the USA contained 0·005 ppb of DDT compounds compared with 0·08 ppb in British rainfall. The highest recorded river transport was with a high silt content and was only 0·12 ppb. San Francisco Bay receives 1900 kg/year from rivers draining into it, while the Mississippi carries 10,000 kg/year to the Gulf of Mexico. The estimated total for rivers in the USA is about 20,000 kg/year, or about 0·1 per cent of the total consumption. The maximum possible river contribution, based on maximum USA river concentration, would be about 4×10^9 g/year, or 4 per cent of the annual production. The true value is probably nearer 10^8 g/year. Relatively low concentration is due to low solubility and the affinity of DDT with soil and silt particles.

Estimates of total aerial fallout of DDT compounds yield a value of $2·4 \times 10^{10}$ g/year, based on concentration in rainfall. This is about 25 per cent of world production. Dry fallout also occurs, as indicated by collections of dust. If the total production of DDT of 10^{12} g were uniformly distributed throughout the sea the concentration would be 0·001 ppb, which is not trivial in view of organic concentration of the pollutant. The estimated total weight of marine fish is about $2·4 \times 10^{14}$ g, which would give an amount of DDT compounds in them of 2×10^8 g. The effect of DDT is to cause reproductive failure in both fish and sea birds. Sea trout in Laguna Madre in Texas have declined from 30 fish/acre in 1964 to 0·2 fish/acre in 1969, and no juvenile fish have been observed as a result of DDT pollution. Crustaceans are also being killed by DDT concentrations of 0·2 ppb, and molluscs are affected by 0·1 ppb. In the open sea, gray and sperm whales contain up to 0·4 and 6 ppm respectively in their blubber. Some marine mammals have a concentration up to 800 ppm. Oysters contain up to 5·4 ppm. The total amount of the standing crop of plankton has been estimated at 3×10^9 metric tons, and of fish 6×10^8 tons. The concentration of DDT in plankton averages 0·01 ppm and in fish 1·0 ppm. The total of DDT in plankton would be 3 metric tons and in fish 600 metric tons. This is an insignificant fraction of the total annual input of these residues into the marine environment, which is estimated at 10^5 tons. DDT, however, is likely to be concentrated on the surface, where its effect is more marked. Its impact on the marine environment is, nevertheless, considerable.

One of the most noticeable effects is that on sea birds. Their egg shells become thinned to the extent that reproduction fails in the fish-eating birds. Peregrine falcons and ospreys have suffered severely. Significant correlations have been demonstrated between egg shell thinning and DDT residues and experimental evidence has confirmed the connection. The addition of DDT of the amount liable to be found in the environment produced shell thinning in mallard and kestrel eggs. Pelicans off Baja California have suffered serious loss through egg breakage, and those in several other areas have also been affected. In 1970, 500 nesting attempts on Anacapa produced no young birds. Double-crested cormorants off southern California also failed to breed in 1969 and 1970, due to collapse of the eggs. PCB may also have other effects on birds, by unbalancing their metabolism.

There is a connection between DDT and oil pollution: DDT and Dieldrin are highly soluble in oil films and so become concentrated in the upper layers in which the oil floats. For example the amount of Dieldrin in the top 1 mm of water in an oil slick was more than 10,000 times higher than that in the underlying water. Phytoplankton and zooplankton are often in this layer at night and they concentrate the DDT still further.

2.3 Heavy metals

One of the heavy metal pollutants of both atmosphere and ocean is lead, which is mainly derived from vehicles burning fuel that contains lead. Thus at present most of the lead in the ocean enters it via the atmosphere. In water below 500 m the concentration of lead lies between 0·02 and 0·04 μg Pb/kg sea water. In the upper layers the amount of lead is much more variable, being much higher near intense sources of industrial lead aerosols. Near Bermuda the value is 0·07, in the western Mediterranean it is about 0·20, but rises

Table 8.8 Enrichment factors for the trace element compositions of shellfish compared with the marine environment (Merlini, 1971)

Element	Enrichment factor		
	Scallop	Oyster	Mussel
Ag	2,300	18,700	330
Cd	2,260,000	318,000	100,000
Cr	200,000	60,000	320,000
Cu	3,000	13,700	3,000
Fe	291,500	68,200	196,000
Mn	55,500	4,000	13,500
Mo	90	30	60
Ni	12,000	4,000	14,000
Pb	5,300	3,300	4,000
V	4,500	1,500	2,500
Zn	28,000	110,300	9,100

to 0·35 μg Pb/kg in the Pacific off the California coast. Before lead was used in petroleum fuels it is likely that the concentration near the surface was about half of that at greater depths, a situation that is now found in the concentration of barium and radium. Thus the concentration of lead in the upper layers has increased from about 0·01–0·02 to about its mean present value of 0·07 μg Pb/kg. The increase of 0·05 could have been added to the upper layers by polluted rain in only a few places. The residence time of lead in the upper ocean is very brief, thus keeping the values at a fairly low level, although there are problems of concentration in organisms of this highly toxic substance (Patterson, 1971).

Heavy metal pollution includes copper, zinc, chromium, cadmium, nickel, and lead (Merlini, 1971). Mercury is another very toxic pollutant. Pollution has until recently been most severe in land waters, but is likely to extend more and more to the oceans. Examples of pollution include the concentration of copper on the beaches of Holland in 1965 that caused the death of about 100,000 fish; the concentration of copper sulphate reached 500 μg/l, instead of the normal North Sea value of 3 μg/l. Mussels were also killed. One problem concerning heavy metals is their indestructibility in ocean water, and filter feeders have a strong tendency to concentrate the metals from very dilute solutions (see Table 8.8).

The vastness of the oceans is not a safeguard against pollution because heavy metal pollutants are concentrated mainly in those areas where they can do most damage to shallow water benthic and other organisms, while the longevity of heavy metals adds to the pollution problem, especially through concentration in living organisms through the food web. Animals can also increase the dispersal of the pollutants and increase the ecological disturbance. A great deal needs to be learnt concerning cycling of metals in the marine environment.

Mercury is one of the most poisonous heavy metals to enter the ocean. It has been suggested that eventually half the world production may reach the sea. The Baltic is already so heavily contaminated with mercury that its fish are inedible. The concentration of mercury in shellfish led to many deaths in Japan in the coastal area worst affected by this form of pollution. A plant using mercury allowed it to enter the bay in which the shellfish lived. The problem of chemical and other pollution is made worse in the ocean by active circulation and the unitary character of the ocean–atmosphere system. No area is exempt from pollution. Water deposited in the Atlantic by the Atomic Energy Company, for example, was trawled up by fishermen on the coast of Oregon in the Pacific Ocean.

2.4 Oil and petroleum products

Petroleum consists of hydrocarbons, which are stable in the absence of oxygen. Under natural conditions a stable state exists between the input from oil from natural seeps and removal mechanisms. Oil provides one of the most conspicuous forms of pollution when a large tanker runs aground and spills its cargo unintentionally, or when an offshore oil source discharges in an uncontrolled way. An example of the first of these disasters occurred off Cornwall when the *Torrey Canyon* went aground on the Sevenstones Reef in 1967, and of the second off Santa Barbara in 1969 when an offshore oil blow-out occurred, covering 30 miles of California beaches in oil.

It has been calculated that about one million tons of oil are spilled into the ocean each year. Some of this is due to the washing out of tanks at sea, although there are rulings as to the areas in which this is allowed. Altogether a total of about 13 million tons of evaporated petrol, waste oil, and other solvents are probably added to the oceans annually. One of the main problems of this form of pollution is that the oil floats on the sea surface. This means that evaporation is reduced, light is cut out, and oxygen penetration is prevented. The aromatic hydrocarbons are also toxic. The effect of the *Torrey Canyon* disaster was made worse by the detergents that were used to disperse the oil. The detergents had been dissolved in aromatics and were themselves highly poisonous to many forms of marine life. Detergents can also destroy the waterproofing of birds' feathers and the skin of other marine creatures so that they can no longer remain afloat (Plate 20). There are also long-term effects of oil pollution caused by disasters such as the *Torrey Canyon* wreckage. Many fish rely on a sense of smell to catch prey or escape predators and this sense is destroyed by oil pollution. One method of dealing with oil spillages is to sink the oil by use of chalk, but this way of dealing with the problem has adverse effects on benthic fauna (Plate 21).

Oil exploitation at sea is bound to cause further accidental pollution as it becomes more widespread, and accidental pollution such as the *Torrey Canyon* disaster gets potentially more serious as the size of the tankers grows larger. Four tankers of 327,000 tons are now in use, ones of 500,000 tons are under construction, and tankers up to 800,000 tons are

Plate 20 Marine pollution: a shag smothered in oil (*G. W. Potts*).

projected. At present oil pollution is very widespread at sea as shown by the survey made by *Chain* in the Sargasso Sea when recording plankton. Thor Heyerdal also recorded oil across much of the Atlantic during the voyage of *Ra* from Africa to America. The plankton tow nets of the vessel *Chain* were encrusted with oil. Coastal waters are badly affected by oil, and many beaches near oil refineries have very reduced faunas as the oil chokes shallow-water benthic fauna and other shore life. The Arctic is another difficult environment in which oil pollution would be particularly difficult to counteract. This problem has given rise to serious concern about the extraction of oil from the north Alaskan oil fields, as indicated by the experimental cruise of the *Manhattan* through the northwest passage and Canadian Arctic waters.

Attempts have been made to control oil pollution at sea. The International Convention for the Prevention of the Pollution of the Sea by Oil was set up in 1954 and amended in 1962. The convention is administered by the Governmental Maritime Consulting Organization. It urged more effective control in 1969, with larger fines being imposed for deliberate pollution. Accidental pollution will, however, still remain a problem. By law oily waters must be discharged more than 80 km offshore and 160 km off the northeast USA seaboard in order to protect the fisheries of this area. One source of oil pollution that cannot be controlled is that from sunken ships. This source adds appreciably to the oil pollution problem and it has been estimated by Ikard (1967) that 200 million gallons of oil in sunk ships is at the bottom under the US flag alone. Another source of oil is natural

Plate 21 Oil pollution on the rocky foreshore of southwest England (*G. W. Potts*).

seepage, of which seven sources in the Caribbean and two off California at least are known. The Santa Barbara oil seep accident released 10^9 g, forming a slick covering 800 square miles (2700 km²) and lasted about 2 weeks, although effects were apparent for 2–3 months. This blow-out of oil while drilling was in progress illustrates the difficulty of arriving at agreed solutions to such pollution problems. After the blow-out the public urged all drilling to cease, while the President's Special Committee advised further drilling to relieve the pressure in the rock structure that caused the blow-out, thus preventing further similar events.

Oil as a serious form of oceanic pollution started during the First World War. Since 1914 most ships have burnt oil, and oil is transported by sea to an increasing extent, the increase being exponential as more and more machines and engines use oil. The sources of oil pollution are partly natural, and oil seepages have always occurred, although their contribution to the oil in the oceans is not thought to be significant. The main sources of pollution are vessels; accidents to them have contributed substantially to the pollution problem, while oil wells offshore are a potential source of trouble, but at present ships are the major source of pollution. Ballasting of tankers by sea water and other ships creates oil loss to the sea, as the tanks are used alternately for oil or other substances and water. In 1967, 55 per cent of all ocean transport of goods consisted of petroleum, 60 per cent of this being crude oil and 40 per cent petroleum products. Oil was first transported by sea from Philadelphia to England in 1859 in leaky wooden barrels. A ship exclusively designed to carry oil was launched in 1886, the SS *Gluckauf,* which was 100 m long and had a dead weight tonnage of 3000. Between 1900 and 1966 the number of tankers increased from 109 to 3524, and the dead weight tonnage from 530,725 to 102,908,000. The increase in size was even greater: in 1967 there were more than 50 tankers with dwt of over 100,000 tons, a year later there were 6 tankers of 276,000 tons, with an overall length of 346 m, 6 tankers of 312,000 tons were near completion, the first becoming operational in 1968. Such larger tankers, and the even larger ones projected, can enter very few ports, and none in the USA. They must discharge their cargo at sea-loading terminals, where they are moored to buoys in deep water; the oil is pumped ashore through hoses and pipe lines. The estimated amount of oil entering the ocean because of cleaning and ballasting operations is about 400,000–500,000 tons/year in 1963 at a maximum. Welding instead of riveting of ships has greatly reduced oil leakage, but there are still problems in the disposal of waste oil from cleaning operations.

One of the difficulties of ending oil pollution is the lack of understanding of the problem. It is first necessary to determine the conditions under which oil pollution is injurious, and these conditions are not yet known. Sources of pollution must be attacked in order of importance, and discharge facilities for pumping waste ashore must be provided. A list of oil spills made by the Battelle Memorial Institute, Oil Spillage Study, Department of Commerce, Clearing House, Washington, DC in 1968 gives useful information. Between 1926 and 1968, 1.8×10^{11} barrels of crude oil were produced over the whole world, or about 2.4×10^{10} long tons. Of this tonnage 34 per cent was moved by sea, or about 8.3×10^9 long tons. The amount lost by cleaning and deballasting (about 8.5×10^6 tons) was about 0·1 per cent of the cargo and was valued at approximately $213 million, a sum worth making the cleaning process more efficient as well as reducing the amount of pollution.

There has been a close correlation in the development of ocean technology and international law relating to the oceans. Offshore oil is one of the most important oceanic resources, and hence laws relating to its extraction are essential. At present about 30 nations produce oil and gas from offshore; with 20 per cent of the earth's reserves or about 85 million barrels, present production is 16 per cent of the world total or about 6·5 million barrels/day. The offshore reserves will become increasingly important, and by 1980 about 30 per cent of oil and 40 per cent of gas for the USA will come from offshore. Laws at present relate mainly to exploration and exploitation and only recently have spillage and pollution aspects been developed. The 1958 convention, as well as defining the zone for exploration and exploitation on the continental shelf, also gave rulings on safety, regulations concerning oil discharge by ships, and measures to prevent pollution and aid conservation of marine life. There was an international convention for the prevention of pollution of the sea by oil in 1964, which governed the discharge of oil by ships only, while the 1969 international legal conference was concerned with damage due to marine pollution. It was convened by the international maritime consultative organizations, but it did not extend the scope of the injunctions.

The problem of Arctic oil spills is related to the difficulty of oxidation where pack ice imposes lack of mixing and tides are often nil. Low temperatures also reduce degradation. In normal climates a well-dispersed oil slick probably has little effect on the productivity of an area. The transport of oil from the north Alaskan oil fields poses difficult problems. The voyage of the *Manhattan* to test the feasibility of using the northwest passage for transporting oil by tanker illustrates the problems. The delivery of oil to the east coast would be facilitated by use of the northwest passage. The *Manhattan* is 286 m long and has a weight of 142,500 tons. The hull was strengthened for ice-breaking by the addition of 9000 tons of steel, and elaborate navigation equipment was installed. The ship was altered in four separate parts which were then reassembled in 1969. The voyage through the northwest passage and back to New York covered 14,000 km; the ship was accompanied by two ice-breakers. She reached Prudhoe Bay in September 1969, and returned to New York on 18 November 1969. It is thought that pollution by this route is not a serious problem, owing to the lack of other shipping, but there is always the danger of ice and running aground in this delicate environment. The voyage from New York to Prudhoe Bay and back would take 2·5 months with ships of 250,000 tons. Successful commercial development of the northwest passage would be a considerable advantage to the development of northern Canada and Alaska, but there would be serious dangers to the environment in this Arctic climate.

The *Torrey Canyon* is the most publicized oil pollution disaster. About 50,000 tons of her cargo of 119,000 tons was lost in 10 days, and later a further 50,000 tons were lost when the ship broke up. One of the main advantages of the disaster was the gain in knowledge from the experiments carried out in the recovery programme. Many methods were tried, such as bombing and the use of detergents, the latter producing more damage than the oil it was meant to disperse. Spills are most troublesome in calm conditions and near shores, where birds are badly affected. Frequent spills can cause oxygen depletion, while cumulative effects include the build-up of heavy toxic metals, such as nickel. Crude oil may contain many components. Oil spills emulsify fairly quickly and gentle stirring helps to disperse the oil, which in the laboratory breaks up in about 10 days at room temperature, but poisoned or unstirred slicks are stable for long periods. Once emulsified oil will sink.

The microbiology of the oceans is closely connected with oil pollution, as the oceans are nature's large septic tank. The biochemical degradation of wastes in the sea is a vital matter in connection with oil and other forms of pollution. Until the processes of biochemical degradation are better understood a scientific control of pollution will not be possible. The amount of oil being moved by sea is so large that some is bound to enter the sea through accidents and normal processes. Bacteria can degrade petroleum and this fact must be considered in methods of dealing with oil spillages. Laboratory experiments suggest that oil might be oxidized at rates as high as 100–960 mg/m³/day, or 350 g/m³/year, but these rates may be too high under natural conditions. They may also be too slow to be effective in controlling pollution. The oil must be well mixed with water for maximum biochemical action. Rapid degradation cannot take place where the oil becomes concentrated into large lumps; the longer the weathering, which increases the size of the lumps, the slower the degradation. Hydrocarbons can be converted into protein by bacteria, a process that reduces one-third of the petroleum to food suitable for fish. Oil companies are experimenting with protein production from petroleum. Fifty million tons of petroleum could be used to double the world production of protein of 25 million tons/year. This would be only 2·5 per cent of the world crude oil production. The oil waste lost into the sea between 1926 and 1968 would have produced $4\cdot3 \times 10^6$ tons of protein, enough to have doubled the protein in the diets of $1\cdot3 \times 10^8$ people in underdeveloped countries for one year.

Because oil is biologically degraded in the sea all states through which it passes are transient, time of degradation being controlled by bacterial count, oxygen, temperature, hydrocarbon dispersal, and distribution. Bacteria are essential to the degradation process, but their operation is not understood. Research on this and related problems should be financed by the oil companies. In the case of a stranding, petroleum can be ignited, but crude oils are virtually impossible to ignite, and after 30–45 minutes weathering prevents ignition. At this stage the slick should be emulsified by boats running back and forth through it, to blend the oil and water, thus rendering it less harmful to birds. Slow release of oil into the sea from a stranded tanker allows time for biological degradation. A tanker sailing at full speed is higher at the bows than the stern, so that when stranding takes place the bows settle further onto the rocks or bottom as the ship becomes level. Oil reaching the beaches should be pushed back into the sea to allow degradation to take place. Larger tankers should produce less pollution as 75 per cent of the accidents takes place in pilot waters, which are shallow and which large tankers will not therefore be able to enter. Fewer journeys will be required to transport the same amount of oil as the size of tankers increases.

Oil has a considerable effect on the intertidal ecology, but oil pollution should not prove insuperable as natural remedial processes exist, and some of these are quite rapid. The volatile fractions of crude oil evaporate rapidly, except in low-temperature calm conditions, such as occur in the Arctic. Some of the non-volatile residue is converted into an-oil-in water emulsion, which mixes with sea water and soon disperses. Much, however, is converted into a water-in-oil emulsion, which does not mix with sea water and floats, forming what is called sometimes 'chocolate mousse'. In this condition the oil persists and gradually hardens, although over months these patches are degraded by bacterial organisms. Some oil sinks as it mixes with silt or barnacles settle on it. The floating globules are widely dispersed, indicating their slow decomposition, and much is eventually driven ashore. Oxidation and decomposition depend on many variables, including the nature of the crude oil, the type of slick, sunlight intensity, temperature, silt concentration, and the nutrients available for the bacteria. Oxidation is more rapid in sunlight. Crude oil with nickel and vanadium oxidizes more rapidly with ultra-violet light. Other processes associated with bacterial action render the oil into a form in which it can be used as food for marine fauna. Many species co-exist in a complex ecological system, and each prefers one type of oil, although some oils with more than 30 carbon atoms are less digestible. A new oil slick becomes innoculated with bacteria in 10 days, and with good conditions the bacterial population doubles every 3–4 days. Protozoa prey on the bacteria and in turn provide food for plankton and other fauna. The speed with which these processes operate affects the rate of bacterial decomposition. Samples of oily sand from Bovisand, Devon, from a mid-tide position, contained more than 400 million decomposing organisms in each cubic centimetre, a quantity greater than the normal on a beach or in the open sea. A common quantity in clear Atlantic waters is $2000/cm^3$. Non-volatiles in oil may remain for two years, but three-quarters of the original material has usually disappeared in this period. The sunny tropics provide the most rapid decomposition conditions, especially when the oil is spread in a thin film. A layer $1/10,000$ inch ($1/25,000$ cm) thick can be decomposed in a week, in sunshine by the process of oxidation. This is the equivalent of 2000 kg/km². Bacteria could account for one-third of the non-volatiles in a crude oil spillage in 6 months, but the processes slow down when the digestible matter has been used up. If one-quarter of the crude oil spilled turned to tar, then one-half of the tar would be destroyed by oxidation and through bacterial activity in about one year. At present there are about 80,000 tons of tar floating in the Atlantic and Mediterranean, there being about 1 kg/km² in the Atlantic and 20 kg/km² in the Mediterranean. The major problem concerning oil pollution is whether the natural processes can be relied on to continue dealing with the problem, or whether the seas will become dead in the same way as some lakes are already dead, in that bacterial action is no longer possible because the bacteria have been killed (Pilpel, 1972).

2.5   Radioactive substances

The large nuclear tests made in the equatorial north Pacific in 1954 initiated the pollution of the oceans by radioactive waste. Contamination of the oceans in this way can be brought about from three sources: 1) from nuclear plants on land, 2) from nuclear-powered ships, and 3) from nuclear weapon tests. The wastes from the first source are disposed of either as liquid effluents on the coast or packages that are disposed of in the deep ocean, usually below 2000 m. The packages seal the radioactive waste in concrete, which last from 10 to 20 years before leakage takes place although developments may prolong the period before leakage occurs. Ship waste is disposed of at least 19 km from the shore, while waste from weapon tests enters the sea from three sources, immediate fallout near the test site, tropospheric fallout, and stratospheric fallout. The residence time in the troposphere is 0·1 year and in the stratosphere it is 1–2 years. Contamination from the latter two sources is mainly through precipitation, amounting to 90 per cent.

Most of radioactive materials enter the ocean at its margins, either the air–sea interface or the land–water interface or from the sea floor. Careful monitoring has shown an increase in some radioactive elements in clam shells after 1955, while migratory fish carry the contaminants further afield. Ocean currents also disperse the pollutants that are dissolved in the water. An example is the spread of pollutants from the Bikini–Eniwetok tests in 1954. One month after the tests ended, maximum nuclear activity occurred 450 km west of Bikini, and moderate activity had spread over 2000 km from the site in the north equatorial current. By March 1955, the zone of maximum activity was situated around 20°N, 122–132°E, west-northwest of the original site, while later in the summer by August, the pollutant had spread in the Kuroshio current along the coast of Japan, showing that the Kuroshio current is a direct continuation of the north equatorial current. The rate of dilution in the western north Pacific was rapid until 1961, when renewed testing caused a further slight rise, which has since declined and was small in 1967. By 1964 the concentration in the east and west north Pacific was equal, showing that mixing had been thorough in the 10-year period. In the vertical sense, the radioactivity remained essentially above the shallow thermocline at about 100 m depth for at least 1300 km horizontally from the site, although after 8 months weak activity was recorded at a depth of 500 m due to vertical turbulent diffusion. Another method of vertical distribution is by means of organisms, in which the activity is concentrated by a factor of 10^3, but this effect is probably small in the tropical Pacific as organisms only take up 0·3 per cent of the total activity. Some activity reached a depth of 6000 m only 3–5 years after the tests, which provides useful data on vertical diffusion rates. The vertical diffusion coefficient between 100 and 1000 m in the equatorial north Pacific was about 40 cm²/sec. Observations in the Atlantic give a rate of 200 cm²/sec at 500 m depth, decreasing to 50 cm²/sec at 1000 m, and increasing again to 150 cm²/sec at 2000 m, while a further decrease to 100 cm²/sec took place to the bottom. The sinking of water down to 500–1000 m occurred at a maximum rate of 0·0007 cm/sec, while a slow upwelling of 0·0001 cm/sec occurred at greater depths. These rates of spreading are higher than those obtained through estimates of the 'age concept'. It would take about 100 years to obtain equal distribution of radioactivity at all depths at this rate of diffusion. It is expected that with a doubling of demand for nuclear power by the twenty-first century, $5\cdot6 \times 10^3$ MCi of ^{90}Sr (Strontium 90) will be released, which is 560 times the amount released up to the end of 1962. The maximum permissible concentration of ^{90}Sr is 0·8 pCi/l, and to keep below this limit not more than 1/2000 of the total fission products should be put in the ocean, even assuming an uniform spread throughout the ocean; allowing for local concentrations the amount should be only $1/10^4$–$1/10^5$. Even this amount could cause dangerous levels of activity of some nuclear products in areas of concentration. Problems of concentrations in organisms of waste from nuclear-powered ships, of which there may be over 500 by the end of the century, are also severe. Fish, particularly sardines and mackerel, swallow charged waste resins and this could cause widespread threat of contamination, as the fish migrate over large distances. Rates of concentration of radioisotopes vary greatly in different organisms and with different substances (see table 8.9).

One of the major danger zones from the point of view of nuclear contamination are

L

Table 8.9 Approximate concentration factors for radioisotopes of probable significance in the marine biosphere

	Algae	Crustacea	Molluscs	Fish
^{14}C	4000	3600	4700	5400
^{90}Sr	50	2	1	0·2
^{131}I	5000	30	50	10
^{55}Fe	2×10^4	2500	10^4	1500
^{239}Pu	1300	3	200	5
^{45}Ca	2	120	0·4	1·2

the estuaries where there are high concentrations of phytoplankton and zooplankton and where much of the contamination from nuclear power plants will be likely to arise. The relative concentrations of phytoplankton and zooplankton in estuaries and the open ocean are evident in the following figures (Rice and Wolfe, 1971):

	Mean standing crop, mg C/m³	
	Phytoplankton	Zooplankton
Estuaries, North Carolina	187	9·66
Open ocean, tropical eastern Pacific	5·9	0·29

The high proportion of fishing that takes place in the shallow-water zone also adds to the problem, while many of the shellfish that are exploited from this zone also have a high capacity to concentrate nuclear material. Bivalve molluscs are especially sensitive to radioactivity. Marine organisms are affected differently at various stages of their development and primitive organisms are usually more resistant than more complex animals, but different species of the same genera can differ markedly. The sensitivity of animals often decreases with age, the eggs being particularly sensitive. Thus the fertility of species is directly affected and this in turn affects the population and ecological balance. Table 8.10 gives the time required to halve a population. There is little definite evidence of genetic effects in marine organisms as a result of radioactive pollution, but marine creatures are probably considerably less sensitive than humans so that a safe human level would also

Table 8.10 Time (years) required to halve populations on basis of per cent eggs destroyed year

Per cent eggs destroyed	Species Mullet	Horse mackerel	Anchovy
5	55	68	100
10	29	34	50
20	15	18	25
30	11	13	17
40	8	10	13
50	7	8	10

protect the natural marine environment. Maximum permissible concentrations have been laid down by the International Commission on Radiological Protection in 1959, for 240 radionuclides, for different circumstances including the marine environment. The MCP for sea water is based on that for sea food, divided by a concentration factor, so that the value varies for each organism. The concentrations factor is also affected by environmental variables (Rice and Wolfe, 1971).

Radionuclides fall out from the atmosphere and enter the ocean, where there is a danger of concentration. Seaweeds play an important part in the ecosystem from this point of view. Seaweed contains sodium alginate, and this substance can purge Strontium 90 from organic bones, leaving calcium.

2.6 Plastics

The longevity of plastic is well demonstrated by its concentration in the Sargasso Sea (Carpenter and Smith, 1972). In the western part of the sea, concentration of 3500 pieces and 290 g/km^2 is widespread. The range was between 50 and 12,000 pieces in the form of pellets, which were brittle and had diameters between 0·25 and 0·5 cm. Diatoms and hydroids were attached to them. They could be the source of polychlorinated byphenyls that are observed in oceanic organisms. The mean concentration was 1 particle/280 m^2, with a maximum of 1 particle/80 m^2. The nearest station of land was 240 km northeast of Bermuda, and to the mainland 900 km southeast of New York. The source could have been from passing ships.

Instructions for the disposal of noxious materials at sea do not necessarily produce the desired result. Greve (1971), for example, cites the instance of the appearance of drums found in the North Sea in numbers of tens of thousands. The drums had been deposited in the Atlantic in a minimum depth of 2000 m and 160 km from the coast.

Other forms of environmental disturbance that are not strictly polluting in the narrow sense include artificial coastal changes brought about, for example, by dredging, the construction of coastal facilities for shipping, such as harbours, breakwaters and piers, and the building of other amenities for recreation. The increasing amount of offshore sand and gravel extraction also causes some coastal pollution. Many of these and other forms of pollution, as well as making the ocean water dirty and fouling beaches, have a serious effect on the complex ecosystems of the oceans. Some examples of the way in which the natural balance of the ecosystems has been disturbed in the ocean are mentioned in connection with conservation.

3 Conservation

Man is having an increasing effect on the marine ecosystems. He has caused widespread dissemination of some species, while others have been eliminated. One example of dissemination of an undesirable species from the human point of view is the spread of wood-boring animals with the introduction of wood into the marine environment on a large scale in the form of ships, docks, break-waters, and groynes. Before any major operations are undertaken, it is essential to consider the ecological and physical repercussions of these man-made structures. Some major operations have been suggested that might well have very great impact on the marine life of the areas affected. An example is the proposal to blast a sea-level canal through the Panama Isthmus. This could result in the spread of warm water over cold on the Pacific side of the isthmus, and cause more severe and frequent occurrences of the El Niño in Ecuador and Peruvian coastal waters. At present it is not possible to assess the ecological results of such an operation, so that research is essential before such projects are undertaken.

There is a general similarity of fauna on either side of the Panama Canal. Only one species of fish has penetrated the present canal due to the inhibiting effect of the fresh-water lakes along it. There are, however, differences of habitat on either side of the canal, with sandy beaches and mangrove swamps characteristic of the Atlantic side and volcanic shores on the Pacific side. The tidal range is much greater on the Pacific side, where it reaches 6 m, while it is only 30 cm on the Atlantic side, where the water character is also more constant. There are no corals on the Pacific side, because of the upwelling of cold water along this shore. The degree of speciation on either side would determine the possible interbreeding of species on either side if a sea-level connection were to be made. Experiments suggest that interbreeding between the fish population on either side would still be possible. The physical changes resulting from a sea-level canal would probably be local. Warm water from the Atlantic would spread into the Pacific, for example, but the biological effects could upset the balance of populations and the population dynamics. A chain effect could influence a large area and extinction of species could occur. Results are difficult to foresee in detail.

There are fields in which knowledge about the oceans is still lacking and this makes measures of conservation difficult to formulate and implement. The concentration of carbon dioxide in the ocean, for example, is not known. Monitoring of the oceanic environment is, therefore, urgently needed. The data collected would provide a baseline for future comparison. A minimum of 1000 sampling sites is suggested for the whole ocean area, and these should be distributed both horizontally and vertically (Massachusetts Institute of Technology, 1970). The suggested pattern includes samples from surface film, a shallow array above the main level of density increase, and a deep array at 1000 m, 2000 m, 3000 m, and near the bottom. Deep array samples should be taken near the centre of the major ocean gyres and in enclosed seas such as the Mediterranean, Bering, and Japan Seas. The equatorial area of all oceans should also be sampled. The shallow array should be taken in the western and eastern boundary currents. Surface film samples should be widely distributed and include areas of maximum evaporation, precipitation, and major shipping routes. The organisms and the amount of man-made pollutants in them should be assessed, taking into account the following: 1) the geographical distribution of primary production, 2) the geographical distribution of fisheries, 3) sites of major river outflow, 4) the general pattern of ocean circulation, 5) sites of major upwelling and downwelling, 6) centripetal centres of the major gyres, 7) the desired coverage of organisms, 8) ease of sampling, and 9) the cost. Sampling stations should include the high productivity areas of the continental shelves. There should be 42 composite samples of demersal fish, molluscs, pelagic fish, and 12 of benthic crustacea. Samples of plankton and flying fish in the open ocean should number 142 and 40 respectively, giving a total of 124 fish samples, 42 molluscs, 12 crustaceans, and 142 plankton, a total of 320. Plankton samples should be evenly spread throughout the oceans in all latitudes.

Problems of ocean conservation stem from a number of factors; the complexity of the ecological system is one. The sampling pattern suggested above would give at least some indication of the present position. Another difficulty is the complexity of the multiple uses made of the marine environment by so many people and organizations with conflicting purposes. Conservation is made difficult because no overall plan is available and no single body is responsible. For example, in California the Fish and Game Commission tries to preserve the kelp forests from the outfalls approved by the Water Quality Control Board. In Florida one organization, such as the Bureau of Commercial Fisheries, tries to preserve the estuaries, while the Corps of Engineers excavates deep channels through the estuaries for deep-draught vessels. There are a great many unco-ordinated bodies involved in the exploitation and conservation of the marine environment.

In 1966, the US Congress, in an attempt to overcome this problem, set up a Commission on Marine Science, Engineering, and Resources to try and find ways of co-ordinating oceanic activity. A National Council on Marine Resources and Engineering Development was established with the Vice-President in charge. This council has a Committee on Multiple Uses of the Coastal Zone, including conservation measures for erosion and ecology, and pollution abatement. Marine parks have been recommended in California and Florida. The need for consideration of local amenity, as well as commercial gain is imperative. Conservation groups are gaining momentum, such as the American Cetacean Society, which is concerned with protection of the gray whale, and on the east coast, the American Littoral Society is looking after the estuaries. In Britain, the Nature Conservancy and the County Naturalist Trusts perform a somewhat similar function in their attempts to preserve coastal amenities and coastal ecology.

Conservation measures would be facilitated by the setting up of regional fish commissions, which could deal with problems of mining, pollution, and conservation. Such bodies could prevent the apportionment of the ocean bed for military purposes.

The complex reactions of marine ecosystems, brought about by the interdependence of so many species, provide many difficult problems concerned with marine life. One of these is the interaction between the Crown of Thorns starfish, *Acanthaster planci*, and its prey, the corals of many Pacific ocean coral reefs. The Crown of Thorns starfish underwent a population explosion in 1969. The cause is not known, but it could have been due to mortality among the planktonic predators of the Crown of Thorns larvae because of organo-chlorine pollutants. At present the Crown of Thorns starfish are destroying large areas of Pacific ocean corals. They have destroyed 260 km^2 of the Great Barrier Reef off Australia, nearly 40 km of coral reef on Guam, and have affected Borneo, Fiji, Palau, Saipan, Wake, and Midway Islands. The loss of the coral affects all the reef ecology, and destroyed reefs have not regenerated. The loss of animals that prey on the Crown of Thorns could have been caused by DDT or radioactive residues from atomic tests in the Pacific ocean.

Chesher (1969) has recorded the depredation of *Acanthaster planci* in Guam. The Crown of Thorns starfish, which has 16 arms, appeared simultaneously over a wide area of the Indo-Pacific. The several population explosions were not short-term fluctuations. The species was rare until 1963, when the Great Barrier Reef near Cairns was affected by large numbers. Since 1967 over 90 per cent of living coral along the coast of Guam has been killed on a 38 km stretch of coast between low spring tide level and a depth of 65 m.

Algae grow over the dead coral and most of the fish leave. The coral was killed at a mean rate of 378 cm²/animal/day, or 1 m²/month. Thus if the density of *Acanthaster planci* is 1/m² the whole area is killed in one month. The predators moved from the dead areas to the living coral areas where suitable ground could be crossed. They were stopped by stretches of sand that was continually being stirred by surges, but depth was no barrier.

The depletion of the triton shell, *Charonia tritonis*, is a possible cause of the damage, as this species preys on *Acanthaster planci*, although most of the *Acanthaster* escape and can regenerate. A more likely cause of the population explosion is the destruction of filter feeders because of coral blasting. This removes the predators of *Acanthaster planci* in its larval stage. The entire loss of coral would be an ecological disaster, and would be followed by the loss of fish, wave damage to the unprotected coast, and general disturbance of the ecology of the reefs.

One method of control that has been attempted is to kill *Acanthaster planci* over short stretches of coast, and to starve areas between these before new larval infection can occur. Starvation of the starfish in the intervening areas takes place as the coral is killed. Control, however, will only be possible if the original cause was brought about by blasting, which by destroying predators on *Acanthaster planci*, allows their effective larval growth. If the population explosion of *Acanthaster planci* is a basic change in their life history, then control would probably not be possible. In the past rugose corals have become extinct, and now Madreporean corals may also become extinct. Thus if the ecological damage is man-induced it may be contained, provided the cause can be found and removed. It may, however, be a natural biological process, whereby species periodically suffer a decline due to an imbalance in the prey–predator relationships.

Seaweeds are also susceptible to ecological disturbance. The giant kelp forests of California showed marked degeneration as a result of a population explosion of sea-urchins that feed upon the kelp. Study showed that neither over-harvesting nor sewage pollution were responsible for the decline in the kelp, but that the sea-urchins were responsible. One way of controlling the damage was to apply quicklime, which killed the sea-urchins, but a complete control of the problem requires a knowledge of the cause of the sea-urchin population explosion, which has yet to be explained. It must involve other ecological considerations.

The red tide is another form of ecological imbalance. This phenomenon is the result of a bloom of one particular phytoplankton species. It has occurred in different places for a long time, and has given its name to the Red Sea. The *Iliad*, for instance, speaks of the Mediterranean changing colour. The sea also at times turns white off Ceylon, yellow off Brazil, green off Spain, black off West Africa, dull red off California, and red off Peru when the El Niño or El Pintor occurs (Marx, 1967).

In recent years the coast of Florida has suffered frequently from a red tide. Previously red tides occurred occasionally, as in 1916, 1932, and 1946, the latter continuing over a period of 10 months. The over-production of plankton that causes the red tide leads to depletion of oxygen and the death of fish in large numbers. The species responsible for the red tide off Florida is *Gymnodinium breve*, which is a dinoflagellate. It increased from 1000/litre to 60 million/litre during the red tide. The toxic effect of *Gymnodinium breve* is due to a waste product that is poisonous to fish nerves. The dead fish provide more nutrient for *Gymnodinium breve*, but at the same time deplete the oxygen, causing further deaths.

Since 1947 the red tides have occurred more frequently—in 1952, 1953, 1954, 1957, 1958, 1959, 1960, 1961, 1962, 1963, and 1964. The outbreak in 1954 covered an area of 24 km by 80 km. Red tides are worse in years of heavy rainfall. *Gymnodinium breve* appears to be stimulated by the presence of vitamin B_{12} and tannic acid in the river water, which is increased in time of heavy precipitation. An attempt to control the red tide was made by the application of copper sulphate, but its effect was short-lived because it sank. The toxic property of *Gymnodinium breve* is concentrated by shellfish in amounts that could be harmful to man. Southern California also experiences red tides, which are due to a bloom of *Gonyaulax polyhedra*, another dinoflagellate, one that also occurs off Peru.

Ecological problems of coastal areas include insects in salt marshes. Their control has resulted in much marsh reclamation and draining, but more recently spraying by DDT has been used until its side effects led to its banning. Flooding of mosquito areas has become a more acceptable control mechanism, and could be compatible with aquaculture. The preservation of coastal wetlands for wildlife is growing in importance. In the last five years the first national park was instituted on the coastal islands and reefs of the Florida Keys, and many other state and local parks have been established.

In dealing with decision-making processes it is necessary to distinguish between the coastal area, the continental shelf, and the deep sea uses of the ocean. The coastal area is the most direct concern of the largest number of people with different interests and needs. Thus zoning in this area is the most difficult because conflict between different uses is most intense. The continental shelf is most important from the industrial and military point of view, but less intimate than the coastal area, which directly affects many individuals. International competition is most likely to be severe on the continental shelf, and legal problems associated with it are the most difficult. The deep sea and sea bottom are, however, continually growing in importance. Drilling such as that carried out by the *Glomar Challenger* indicates that increasing use will be made of this zone as technology advances, and land-based resources are depleted.

In 1970 the United Nations Sea-bed Committee met in New York to consider the extent of limits of national jurisdiction, the development of sea-bed resources, and the preservation of environmental quality. Proposals included the following objectives: a) to assure that exploration and exploitation of sea-bed mineral resources will be carried out in a manner that will protect human life, prevent conflicts between users of the sea-bed, safeguard other uses of the ocean environment against undue interference, avoid irreparable damage to the environment and its resources, and promote the use of sound conservation practices; and b) to provide terms and procedures governing liability for damage resulting from exploration and exploitation of sea-bed minerals so that damage will be adequately repaired or compensated.

Canada proposed legislation in April 1970 to establish an Arctic Pollution Control Zone, extending seaward from the nearest Canadian land above the 60th parallel for a distance of 100 nautical miles. Within this zone it would be prohibited to discharge any substance that would be detrimental to any form of life. On 23 May 1970, President Nixon proposed a new US oceans policy, which suggested that beyond the 200 m depth line ocean resources be regarded as the common heritage of mankind. They would be administered by the neighbouring state as trustees on behalf of developing countries. The proposals in their present form require further development, but they could provide a basis for a reasonable settlement of some difficult problems. The provisions of the Outer Continental Shelf Lands Act of 1964 included the leasing of all offshore lands beyond the 3 mile territorial sea under US jurisdiction. The act thus covers an area far larger than the geological shelf, and could extend to the base of the continental slope or a depth of 2500 m. This rather aggressive approach has now been modified by the Nixon proposal. The oil industry, as the main exploiter in these areas, must show an appreciation of the influence of its activities on other people and other uses to which the areas in which it is operating could be put. The ecological balance must be protected. Conservation rather than exploitation must become the main aim of offshore operations.

4   International law at sea

The basic law concerning the use of the ocean has been that of freedom of the seas, for trade, travel, fishing, and war from time immemorial. Later such operations as submarine cable installation and research have been added to the traditional freedoms (Henkin, 1968). There are, however, laws appertaining to the high seas, such as those against piracy, slave-running, and wartime limitations on trade between neutral and belligerent states. Problems now arise as to the ownership and exploitation of the sea bed and its subsoil. Coastal waters and the land beneath them belong to the coastal states. They have full rights over their territorial waters, apart from some 'historic' rights to fish in some instances. Claims for territorial waters vary from 320 km (200 miles) by some Latin American countries to 4·8 km (3 miles) by the United States, for example. Conferences were held in 1958 and 1960 to discuss this matter, but no settlement was reached. A compromise of 9·6 km (6 miles) was suggested, but failed to be adopted by a narrow margin. The great majority of states agreed to a 19·2 km (12 miles) maximum. The main purpose of the wider zone was to exclude foreign fishing fleets. The US claimed 19·2 km for fishing, but only 4·8 km as territorial waters. Some states have also claimed continuous zones for customs, immigration, and sanitary matters, or to prevent broadcasting by pirate radio stations.

The law of the continental shelf is an important new addition to international law at sea. The 1958 convention gave coastal nations rights to the continental shelf for exploitation of resources of sedentary fish and other marine life. For mining, the shelf is considered a continuation of territorial waters, but for other purposes it is 'high sea' and thus free. The concept of freedom has led to increasing conflict of interests. Rules of navigation and sea lanes have been established for safer travel. Interference with navigation has been resisted and thus has had priority. Cable-laying has also been allowed. As far as military uses go, the policy has been laissez-faire. Research and fishing suffer from legal uncertainties and lack of international co-operation. It is not clear when research ends and exploration begins in the search for minerals, for example. Fishing could be improved by more legal controls, co-operation, and conservation.

Laws relating to subsoil exploitation need most clarification. The major problem concerning the shelf is its definition. Its legal definition is not the same as its geological one. The shelf is defined as '. . . . the seabed and subsoil of the submarine areas adjacent to the coast but outside the area of territorial sea, to a depth of 200 m, or beyond that limit, to where the depth of the superjacent water admits of the exploitation of the natural resources of the said area. . . .' The geological shelf in depths greater than 200 m is excluded. The depth of 200 m was considered to be that to which mining could technically extend, but now exploitation could take place in virtually any depth and so the definition leads to uncertainty as to where the rights end. They could extend to the centres of the oceans, where adjoining state's 'shelves' would meet. There is a view that exploitation should stop at the edge of the geological shelf. The clause concerning depth beyond 200 m was added for the benefit of nations who have little or no geological shelf. Eventually the shelf must be redefined, taking into account developments in technology, otherwise conflict will increase.

Deep sea exploitation is not governed by law. It is not clear whether a nation can acquire and claim sovereignty of the sea bed, nor is it clear whether it is lawful for states to dig for minerals and keep what they extract. Uncertainty is the only certainty regarding the law of the deep sea, and works both ways. Exploration is inhibited by the possibility of interference by others, and also by the great expense of deep sea projects. On the other hand adventurous entrepreneurs would be able to attempt such operations. President Johnson in 1966 warned against colonizing on the sea bed, and stressed the need to maintain freedom for the benefit of all.

The military use of the sea can be affected by general laws and also by agreements on the control of arms. One suggestion has been a sea-wide network for tracking submarines, operated by some international authority. Demilitarization has also been suggested, such as the exclusion of certain weapons from the sea bed, following similar agreements for outer space. Antarctica has already been declared an area where no military purposes including nuclear tests will be allowed. The problem of the oceans is complicated by the fact that it has been used for warfare for a very long time and strategic policy is related largely to submarine-based missiles. It is unlikely that powers would agree to give up these strategic forces until complete disarmament is achieved. Nuclear testing has already been banned by the treaty of 1963. It would be possible to demilitarize the sea bed, while leaving the sea undisturbed from this point of view. Bans on sea bed military installations pose enforcement and inspection problems. It seems unlikely that the US would agree to demilitarizing the sea bed, and the granting of UN sovereignty would also lead to problems, as control would then be determined by majority votes in the General Assembly.

Problems also occur in the field of scientific research, particularly on the continental shelf. The adoption of the 19·2 km limit for fishing rights could improve control, and therefore, efficiency and total catch. Some views that have been suggested include: 1) Comprehensive internationalization of world fisheries. 2) Establishment of an international agency to devise and enforce conservation regulations. 3) Negotiation of a treaty among different nations defining their fishing rights, with an international agency to enforce the treaty. 4) The right of innocent fishing, allowing all nations to fish anywhere in coastal waters. None of these is likely to be attempted in the near future. The first and fourth seem out of the question, while the second and third seem the most hopeful approach. An international body to develop and enforce conservation measures seems a sensible and essential measure. The principle of licensing could be applied in this field and also for mineral exploitation. This is a new field to which much legal thought has been devoted. One of the main essentials is a better legal definition of the continental shelf. Definition both in terms of width and depth have been suggested. The proportion of total sea bed to a depth of 200 m is only 7·5 per cent, to 2000 m it is 16·3 per cent, and to 3000 m it is 24·8 per cent, thus leaving plenty of unclaimed deep water. The problem of the shelf around small islands also causes difficulties.

Suggestions concerning the deep sea can be divided into the two broad classes—those that advocate individual countries obtaining rights to specified areas of sea bed, and those that consider that the sea bed should remain international and be used for the benefit of all. The obvious international body to deal with the matter would be the United Nations, but their authority to do so has been questioned, even if they could agree to a course of action, which is doubtful. It is, however, possible that an agreement to give the UN sovereign rights to explore and exploit the deep sea bed could be made, and then could have legal effect. Such an agreement would not include military uses of the deep sea. There has been a suggestion that payments could be made to the UN as royalties for deep sea exploitation and that the proceeds be used to close the gap between the richer and poorer nations. Some agreement appears to offer the best solution to these problems.

5   International co-operation

The freedom of the sea implies that there is possibility of conflict in the uses of the sea, while at the same time there is also much opportunity for, and need of, international co-operation (Skolnikoff, 1968). International organization is essential for the sensible conservation and regulation of the use of marine resources, especially the biological resources that are self-maintaining, given suitable conservation measures. International co-operation is required not only for basic research but for safe oceanic transport, communications, and exploitation.

The International Council for the Exploration of the Sea was established in 1902, mainly for investigation in the north Atlantic Ocean. The 1958 conference included a convention on Fishing and Conservation of the Living Resources of the High Seas, which came into force in 1966, Fishing on the high sea is free for all, but it was soon realized that control was necessary. The 1966 attempt was the first effort at world conservation measures. Most of the present international fishery bodies were established at conventions after the Second World War, under the FAO. There are now 22 such international bodies, mostly dealing with sea fisheries, and much of the ocean surface is under the jurisdiction of at least one organization. The detailed workings of some of these was mentioned in the previous chapter. Some of these bodies deal mainly with research, such as the International Council for the Exploration of the Sea. Others attempt to control exploitation; the International Whaling Commission is now trying to regulate catch by adoption of national quotas. Control measures have benefited individual fishermen, but a view has been expressed that conservation would be improved if a system of tax or licensing were introduced to reduce profits and hence discourage new participants. The income could be used in part for better management and research, while some could go to those actually engaged in fishing, assuming that the managing body paying the licence did not actually do the fishing. Some new approach to international co-operation will be needed if fisheries are to maintain and improve the present catch. It is in the long-term interests of all to manage the stock as efficiently as possible, rather than to obtain the maximum short-term gain, thereby depleting the stock for future use. It seems all too likely that, in the case of the whales, permanent and irremediable damage will be caused by the extinction of the species.

The development of highly sophisticated technologies means that the natural environment can be changed to an increasing degree, affecting a larger number of people. International control thus becomes essential. Multi-national ocean research first took place on a large scale in the International Geophysical Year of 1957, organized by the IGGU, from which developed the Special Committee on Oceanic Research. This organization instituted the International Indian Ocean Research Expedition in which 40 ships took part in 180 cruises from 1959 to 1965; 23 countries were involved. UNESCO established the Intergovernmental Oceanographical Commission in 1960, which took over the work of SCOR in the Indian Ocean and also works to the control of oceanic pollution.

Most UN agencies deal with special aspects of the oceans, such as meteorology, while the fishing commissions deal with specific regions or species. The existing organizations have no enforcement powers. Fisheries need conservation, which requires enforceable regulations. The extraction of minerals also needs control. One suggestion that has been made is that all the ocean bed beyond a definitely defined continental shelf should be under international control and developed and used for the benefit of less developed and poorer countries. There is also need for research into all aspects of oceanography. The US Congress passed Public Law 89–454, the Marine Resources and Engineering Act, in 1966. Scientific research was to be stimulated and marine resources development accelerated under four committees which are: 1) marine research, education, and facilities, 2) ocean exploration and environmental services, 3) food from the sea, 4) multiple use of the coastal zone. A budget of $450 million was spent on these projects in 1968.

One aspect of the oceans that must be considered throughout the research on the development of marine resources is the complex interaction of all elements of the oceanic environment. These elements include the air above the oceans, the water within them, the surface beneath them, as well as the living organisms associated with them. The whole is one complex ecological system, with many feedback loops involved in the interaction of the different elements. Thus before the oceans are further exploited the interaction of these various aspects must be studied. The effects of pollution on the waters and the use of the oceans for waste disposal must be considered in terms of their repercussions on marine ecology, while the exploitation of the living resources of the sea also requires a detailed knowledge of the interrelationships among marine organisms and their environment.

The oceans provide one of the best opportunities for true international co-operation, and it is to be hoped that mankind will make the best use of these bountiful resources, with maximum benefit both to all people and to the environment. The need is for long-term planning for a profitable future, and not for a short-term gain at the expense of the future, a state that has been demonstrated so clearly in the sad story of whale exploitation. That careful control can increase production has been demonstrated by the Pacific halibut fishery, and it is to be hoped that more results of this type will be achieved. The more that is known about the oceans, the more likely that they will eventually be able to provide maximum gain. In the research involved in this acquisition of knowledge there is also an opportunity for international co-operation of the type exemplified by the International Indian Ocean Expedition. Such work is likely to be beneficial from many points of view, including co-operation among marine scientists. Their results benefit from such interaction, while the resources discovered can be exploited for the benefit of the people of the area.

Conclusion

The subject of oceanography is so wide that it has not been possible to do more than indicate some of the interesting and important fields that are now being actively explored. A few of the ways in which the oceans influence and play their part in human life on earth have been briefly mentioned, but the interaction of the oceans and the land are so intricately interwoven that it is impossible to unravel all the threads. It is hoped, however, that enough has been said to show how vital are the oceans to life on earth and how wide are the topics and interest of the oceans themselves. They play a much greater part on the earth than the attention normally given to them would suggest, and exert an influence either direct or indirect over the whole surface of the globe, of which they cover over two-thirds. Within them are still hidden many secrets, but elaborate and detailed research work is gradually revealing the unknown and intriguing processes operating in the oceans, on their surface, at their edge, where they meet the land, and on and under their floor. Because they are so important, the study of the oceans is of increasing value and necessity.

Further reading

CARTER, L. J. 1968: Continental shelf: scramble for federal oil-lease revenues. *Science* **160,** 1431–2.

CHESHER, R. H. 1969: Destruction of Pacific corals by the Sea Star, *Acanthaster planci*. *Science* **165,** 280–2.

GEORGE, J. D. 1971: Can the seas survive? *Ecologist* **1** (9), 4–9.

GULLION, E. A. (editor). 1968: *Uses of the seas*. The American Assembly, Columbia University. Englewood Cliffs: Prentice-Hall.

HOOD, D. W. (editor). 1971: *Impingement of man of the oceans*. New York: Wiley–Interscience.

IKARD, F. N. 1967: Oil wastes at sea. *Science* **157,** 625.

LOFTAS, T. 1969: *The last resources*. London: Hamilton. 256 pp.

MARX, W. 1967: *The frail ocean*. New York: Ballantine Books. 274 pp.

MASSACHUSETTS INSTITUTE OF TECHNOLOGY. 1970: *Man's impact on the global environment*. Cambridge: MIT.

MERO, J. L. 1965: *Mineral resources of the sea*. Amsterdam: Elsevier. 312 pp.

PILPEL, N. 1972: Oil pollution of the sea. *Ecologist* **2** (3), 4–7.

RUBIOFF, I. 1968: Sea level canal across central America. *Science* **161,** 857–61.

WENK, E. 1969: The physical resources of the oceans. *Sci. Amer.* **221,** 166–176.

WOOSTER, W. S. 1969: The ocean and man. *Sci. Amer.* **221,** 218–314.

Appendix
Law of the sea and ocean resources

A short summary of the international legal problems relating to the uses of the seas

Edgar Gold[1]

1 Introduction

The subject of this essay is as vast as the oceans of the world themselves. The purpose in writing about this huge topic is not to give a detailed analysis of the public international law of the sea in all its multi-faceted areas; rather it is to give natural scientists a short overview of the background, difficulties and significant developments in the law of the sea. The inclusion of this essay in a scientific treatise is significant—for too long the oceans have been discussed and investigated in the distinctly separate and purposely separated areas of social and natural science with the result that both disciplines have been losers. Legal, precedent-oriented thinking has been slow in meeting the challenge of rapidly changing demands and is required to become much more 'science oriented'. On the other hand, scientific research has often chosen to neglect the most basic norms of law and other social sciences.

It should be pointed out that this is not only a rapid overview but also that it is written on the 'eve' of the Third United Nations Law of the Sea Conference which is to convene in 1974. This Conference may change international law of the sea quite radically to the extent of making much of what is to be said here redundant. On the other hand it may not, and even if it does the changes will take some considerable time to implement and international law of the sea as it stands today will still form the basis for much which is to come, in whatever way it may come.

2 The law of the sea: a problem of jurisdiction

Since the end of World World Two the law of the sea has undergone transformations which have been as revolutionary and as significant as any changes which have occurred for the previous three millennia. At present, these developments show no sign of abatement; indeed, acceleration of problems, claims, counterclaims and proposed solutions seem to be the trend.

The law of the sea, like virtually all international law, has two main sources—custom

[1] Edgar Gold (Master Mariner, MRIN, MNI) is a barrister and solicitor (Canada). He is currently holder of a research associate grant from the International Development Research Center, Ottawa, Canada, for work in the Department of Maritime Studies, University of Wales Institute of Science and Technology, Cardiff.

and treaty (Colombos, 1967, 7). Customary international law has been defined as follows:

> Whenever and as soon as a line of international conduct frequently adopted by states is considered legally right, the rule which may be extracted from such conduct is a rule of customary international law (Oppenheim, 1955, 27).

Customary law has from the beginnings of history proved to be by far the greatest source of law of the sea. However, whenever a sufficient body of customary international law has developed, it is often formalized by treaties or conventions resulting in law of a more binding nature. There have, of course, been various other sources of the law of the sea, such as decisions of national or international courts and arbitrators, the opinions of publicists, and even the consequences of war. However, even these sources are generally rooted in custom or prior agreement.

While it is thus clear and accepted that the law of the sea has developed from a number of sources, it must be remembered in considering everything which follows that whatever the source of the rules, it is the interests of the maritime and the coastal nations that ultimately control. The most important example illustrating this is the principle of the 'freedom of the seas'. Indeed it has been most ably argued that the whole history of modern international law of the sea can be best understood by considering it as a continual conflict between two diametrically opposed, yet co-existing, even complementary concepts— territorial sovereignty and the freedom of the seas.[2] Freedom of the seas, is a natural result of the interest in the free flow of commerce between nations with a minimum of friction, danger and constraint. On the other hand the concept of territorial sovereignty as expressed, for example, in the principle of the territorial sea, in opposition to freedom of the sea, is an outgrowth of natural interests of coastal states in military security and other commercial interests. There has rarely been a stable boundary between the two concepts and the actual position of any one at a particular time has been largely determined in accordance with the self-centred interests of the most powerful maritime nations of the day.

The origins of the first rules of the law of the sea are lost in history, except as those rules have been adopted by later civilizations as customary law. It is reasonably clear that many of the ancient Mediterranean people and nations considered the sea free and open to any legal and legitimate use; however, even at that early stage, there is some evidence that the maritime nations of antiquity had at least a limited concept of territorial jurisdiction over coastal waters.[3]

The earliest comprehensive code of the law of the sea that has survived is known as the *Rhodian Maritime Code*. Rhodes conducted some of the most widespread and comprehensive maritime commerce in the ancient world, and thus developed a successful code of sea law. The general consensus is that the great part of our modern private international law of the sea (commercial maritime/admiralty law) has descended from that code—being

---

[2] E. D. Brown, Maritime Zones: A Survey of Claims, in R. Churchill, K. R. Simmonds and J. Welch, *New Directions in the Law of the Sea* (London, 1973), Vol. III, p. 157.

[3] Carthage, for example, is known to have restricted severely the use of her port by foreign vessels, while the City State of Rhodes encouraged the use of what it considered its own waters.

passed on through Roman law, the basis of modern international public and private law of the sea. The important factor of the *Rhodian Code* is that it recognizes the right of all nations to use the seas for legitimate commerce, and this right was made part of Roman law. Rome, of course, exercised a broad measure of control over the Mediterranean and adjacent seas, but did not claim exclusive ownership thereof (Fenn, 1926, 5).

With the collapse of the Roman Empire, law, both national and international, virtually ceased to exist. From this time until about the mid-fifteenth century, there was no law of the sea beyond survival of the fittest. Thus, a highly developed and established body of law relating to the seas was lost for almost a thousand years, until necessity required its resurrection.

The necessity was brought about by the huge claims which were made to parts of the oceans in the Middle Ages. Medieval Venice claimed the entire Adriatic Sea. English rulers claimed enormous belts of water around the British Isles. The culmination of this 'territorial expansionism' was probably reached in 1493 when Pope Alexander VI Borgia divided the Atlantic Ocean from pole to pole between Spain and Portugal (Colombos, 1967, 49). The Elizabethan period of discovery really resulted in the first clear articulation of the concept of 'freedom of the seas'. The Elizabethans rejected the Iberian oceanic expansionism and received support from the United Provinces of the Netherlands. In 1609, the first legal treatise concerning the law of the sea of any consequence, *Mare Liberum*, was written and published by Hugo Grotius (Grotius, 1916). Clearly advancing the interests of his native Holland, Grotius fully supported the Elizabethan theory that the seas belonged to all nations and are not subject to unilateral appropriation by any nation. *Mare Liberum* was, of course, a work of enormous importance based heavily on Rhodian/Roman Law and current Mediterranean practice and was to be used as a standard source of law until modern times. In 1635, John Selden of England replied to Grotius in his work *Mare Clausum* in which he advocated coastal maritime states had the right to appropriate large sections of the sea (Brierly, 1963, 305). By this time, of course, the Elizabethan period of discovery and its use of the concept of the freedom of the seas had given way to the preoccupation of James I with the British fisheries in opposition to the superior sea power of the Netherlands (Colombos, 1967, 55).

Despite the passage of some 350 years since the doctrinal clash between Hugo Grotius and John Selden and, despite the fact that technological and scientific advances since then have expanded our functional notion of the sea, the fundamental juridical–political problem—the extent of the limits of national jurisdiction—remains the same today. The nations of the world have been, and remain, divided over the question where to draw the line of national jurisdiction. One side of the argument, the advocacy for the concept of 'freedom of the seas', has prevailed, but up to the eighteenth century there was hardly a part of the European seas which was free from proprietary claims by individual powers and such claims were made in varying degrees over most other seas (Colombos, 1967, 48). From the beginning of the nineteenth century, Great Britain, the leading maritime state, pursued and consolidated the concept of freedom of the seas on a world-wide scale. Most, if not all, other maritime nations followed suit. The concept thus became established and is considered by many to be the single most important accomplishment of the law of nations developed over the last three-and-a-half centuries (Friedmann, 1971, 3). Others see the concept in considerably less favourable terms by stating that it is only a thinly disguised freedom allowing the most powerful states to pursue their own selfish interests at the expense of the weaker—by being free to deplete fish stocks, free to exploit minerals wherever they may be, free to carry out nuclear tests and military activities, and free to pollute (Legault, 1971). However, originally the concept included only four 'freedoms': (i) that of unimpeded navigation on the high seas; (ii) that of fishing on the high seas; (iii) that of laying submarine cables and (iv) that of aerial circulation over the high seas.[4]

In the last fifty years three international conferences on the law of the sea have tried to reach agreement on the opposed fundamental principles of freedom of the seas and territorial sovereignty. The meetings resulted in a failure to agree but in some compromise solutions to the early conflict. This was reached by a relatively small number of nations then comprising the world maritime community and only after an evolution of customary rules of international law over a period of approximately 100 years. The Hague Conference of 1930 consisted of delegations from 38 nations and the 1958 and 1960 Geneva Conferences of delegations from 86 states (Colombos, 1967, 103–4, 110). Today with a United Nations community of over 130 states and a world community of over 150 nations the situation is vastly changed. Almost every nation, whether coastal or landlocked,[5] has an interest in the sea of varying complexity. Once again the jurisdictional question is the most pressing problem due to the present confusion of state practice and the horizontal expansion of the international community since 1960. The technological revolution of the past decade has enhanced the awareness of individual states of the potential open to them in and under the sea.

The Geneva Conventions on the Law of the Sea, as adopted in 1958[6] had been in force for only a few years when a new 'decade of the sea' was ushered in. This was in 1967, when Arvid Pardo, the distinguished ambassador of Malta, first raised the question of the seabed in the United Nations. Ambassador Pardo in a speech and draft resolution advocated: (i) the exclusion of the seabed and the ocean floor 'beyond the limit of present national jurisdiction' from national appropriation, and (ii) the establishment of an international agency to regulate, supervise and control all ocean-bed activities beyond the limits of national jurisdiction.[7] At the time there was no indication that discussion of this question would reopen considerations of the law of the sea in general, but it very quickly became evident that the ultimate resolution of the seabed issues required nothing less than a complete reworking of the law of the sea—one that, unlike the Geneva Conventions, treats the sea conceptually as part of the whole marine environment. Accordingly in December 1967, the United Nations General Assembly established an *ad hoc* Committee 'to study the scope and various aspects of the Pardo idea'.[8] One year later the same forum established a permanent forty-two state committee for the peaceful uses of the

[4] See, Convention on the High Seas, Geneva 1958, Art. 2. The 1958 Geneva Conventions are reproduced in Lay, Churchill and Nordquist, *supra*, note 2.

[5] *On the particular needs of land-locked countries, see for example: Report of the Committee on the Peaceful Uses of the Sea-Bed and the Ocean Floor beyond the Limits of National Jurisdiction.* U.N. General Assembly Official Records: XXVIIth Session Supplement No. 21 (A/8721), pp. 30–31.

[6] Final Act of the United Nations Conference on the Law of the Sea. U.N. Doc. A/CONF. 13/L.58, 30 April 1958. See, *supra* note 4.

[7] See, U.N. Doc. A/6695 U.N. General Assembly XXIInd Session, 18 August 1967.

[8] See, U.N. Doc. A/RES/2340 (XXII) U.N. General Assembly, 28 December 1967.

seabed and the ocean floor beyond the limits of national jurisdiction.[9] In 1970, the 25th Assembly of the United Nations moved from the relatively narrow focus of the seabed issue to the broader field of preparations for a new conference on the law of the sea, in a decisive and historic resolution. The resolution called for the convening of a Conference on the Law of the Sea in 1973 to deal with a wide spectrum of marine issues:

> . . . the establishment of an equitable international regime—including an international machinery—for the area in the resources of the seabed and the ocean floor and the subsoil thereof beyond the limits of national jurisdiction, a precise definition of the area, and a broad range of related issues including those concerning the regimes of the high seas, the Continental Shelf, the territorial sea (including the question of its breadth and the question of international straits) and contiguous zone, fishing and conservation of living resources of the high seas (including the question of the preferential rights of coastal states), the preservation of the marine environment (including, *inter alia*, the prevention of pollution) and scientific research.[10]

The reasons for this far-reaching proposal by the general assembly were given as the necessity of considering problems of ocean space as a whole; the political and economic realities which, together with modern scientific and technological advances, have established the need for progressive development of the law of the sea—and the fact that many of the present UN members did not take part in previous conferences on the law of the sea.

The same UN resolution also expanded the International Seabed Committee set up as a result of the Pardo proposals from its original membership to eighty-six members. This was virtually a new committee and became the preparatory committee for the new law of the sea conference. The work of the committee which, since its formation, has met twice annually in 1971, 1972 and 1973[11] in Geneva and New York has been much more politically oriented and is strikingly different from the preparatory work for the 1958 Geneva Conference which was carried out primarily by the International Law Commission (ILC) (McDougal and Burke, 1962, 6ff.) in an atmosphere of scholarly deliberation with much narrower terms of reference. The new 'Seabed Committee', now really a misnomer, established three subcommittees: Subcommittee I—dealing with the proposals for an international seabed regime; Subcommittee II—dealing with the List of Issues for the conference including the Territorial Sea, Innocent Passage, Functional Zones, High Seas, Fisheries, Straits etc.; and Subcommittee III—dealing with Marine Pollution and Scientific Research.

During the summer meeting of the Seabed Committee in Geneva in 1972, it was decided that the Third UN Conference on the Law of the Sea should commence with a short preparatory session, mainly to discuss procedural and logistic matters in New York in November/December 1973. The substantive sessions of the Conference were then to be held in Santiago, Chile, in April/May 1974. The choice of Santiago was due to the fact that many states, particularly in the developing world, felt that this new, most important Conference should be held in a developing country rather than in Geneva. Chile, as

[9] See, U.N. Doc. A/RES/2467 (XXIII) U.N. General Assembly, 14 January 1969.
[10] See, U.N. Doc. A/RES/2750 (XXV) U.N. General Assembly, 17 December 1971.
[11] See, *supra*, note 5 for 1972. For 1971—Doc. (A/8421); for 1973 (A/9041 Vols. I–IV).

a Latin-American developing country, with a democratically elected Marxist government, with its long history of contributions to the law of the sea, seemed ideally suited for most nations to be the venue of the Conference. However, the subsequent political events in Chile precluded holding the Conference there. After the New York preparatory meeting of the Seabed Committee in December 1973 the General Assembly of the UN decided to hold the Conference in Caracas, Venezuela, from June to August of 1974.[12]

This then is a very rapid overview of the law of the sea on the eve of the most important meeting on the subject in history.

3   The jurisdictional zones and related uses of the sea

We will now attempt to examine the problem areas in the law of the sea a little more specifically. The overall problem facing the Conference will, of course, be the age-old jurisdictional question which is directly related to the acquisitive interests of all nations and whether expressed in functional or spatial terms means simply—how much sea area is within the jurisdiction of a coastal state? The answer to this question will also answer the amount of control a coastal state can exercise in 'its' sea area over living and non-living resources, navigation, scientific research and all other uses of the sea. The jurisdictional problem can best be examined through the division of the sea into five jurisdictional zones: (i) Internal Waters; (ii) Territorial Sea; (iii) Functional Zones; (iv) Continental Shelf; (v) High Seas, and their interrelationships to particular problems in the law of the sea. In addition the jurisdictional question also affects two important areas in the law of the sea which overlap into two or more of these five zones. These areas are the law relating to living resources of the sea and the law relating to marine pollution.

3.1   Internal waters

The internal waters can be defined as comprising all water areas, salt and fresh, which lie within the baseline of territorial waters (Smith, 1950, 7). The main feature of all internal waters is that, with certain exceptions, the nation owning them has the same sovereignty over them that it has over land areas within its borders. The only limits on the rights of the 'owning' nation are those that it voluntarily surrenders by treaty or other agreement, and these treaty limits are in fact manifestations of sovereign rights.

The rules governing the outer limits of internal waters—that is the baseline from which the territorial sea is measured—are now quite well established. The rules are codified in the 'Geneva Convention on the Territorial Sea and Contiguous Zones, 1958'[13] and are probably generally accepted even by non-parties to that Convention. It is thus unnecessary to discuss these rules here but it should be borne in mind that the problem of archipelagoes is a formidable one which will have to be dealt with by the forthcoming Conference on the Law of the Sea.[14]

[12] See, U.N. Doc. A/RES/3067 (XXVII) U.N. General Assembly, 16 November 1973.
[13] See, *supra*, note 6.
[14] See, for example, Brown, *supra*, note 2.

3.2 Territorial sea

The territorial sea of a state extends outwards from the limits of the inland water/shoreline of the coast. At present, state practice exhibits such a variety of claims and such a lack of stability, that it is not even possible to give proper details except to say that no nation claims less than 3 nautical miles.[15] It is in this area of the law of the sea that the most considerable difficulties have always existed. Generally, a coastal state possesses the same sovereign rights in territorial waters that it has in its internal waters, the main and important exception being that territorial waters are subject to a right of free and innocent passage by merchant vessels of all other nations. Ships passing through this zone must conform to the laws and regulations of the state claiming the waters and are within that nation's jurisdiction, but nevertheless they have the right to pass.

The concept of the territorial sea was already well established in international law at the beginning of the twentieth century with the adoption of the 3-mile zone by Great Britain, the USA, Germany, France, Netherlands, China and many other coastal states (Colombos, 1967, 99). The original 3-mile zone grew out of the ancient 'cannon shot' rule which effectively gave to the coastal state those areas of the sea that could actually be covered by gun fire from the shoreline (Colombos, 1967, 92). Although the cannons of the seventeenth and eighteenth century were not capable of firing their balls for 3 miles, the rule nevertheless helped to establish in international law the legal proposition that in order to preserve certain national rights and interests, nations could exercise a measure of control over certain areas of coastal waters. At the same time another school of thought developed which, basing its assertions on the idea that the territorial sea remained essentially a part of the high seas, claimed that the coastal belt was subject only to certain limited and well-defined rights in the coastal state. It can be seen that the two schools, one of actual ownership, and the other of limited guardianship, were quite opposed. The former, however, appeared to prevail and became the accepted norm by the end of the first quarter of the present century. The big question which remained unsolved was, however, the breadth of the territorial sea. There has been a consistent failure to reach any sort of lasting agreement on this issue and the differing views seem to grow farther apart with each attempt at settlement. There have been three major international conferences on the law of the sea and many smaller ones, but each one of these, despite progress in other areas, resulted in a deadlock over the question of the breadth of the territorial sea.

As already stated, at the beginning of the twentieth century, most maritime nations accepted a limit of 3 miles as the maximum extent of territorial waters, though at times several asserted wider claims without effect. Within a very short time, however, and for a variety of reasons, the situation had changed and many states extended their claims to distances up to 12 miles, some claiming absolute, others limited jurisdiction in these new areas. This led to considerable confusion and conflict and eventually the League of Nations called an international conference in 1930 in order to codify a general rule for the breadth of the territorial sea. The delegations from the thirty-eight nations, despite considerable effort, were unable to reach any agreement. The main point of conflict was the wish of many coastal states to include in the new code national rights to exercise limited jurisdiction in a zone contiguous to the territorial sea. However, powerful maritime states,

[15] *Ibid.*

particularly Great Britain, took the type of unyielding stand which made even compromise impossible.[16] The Conference ended up deadlocked without any result on the breadth of the territorial sea, although there was general agreement on the legal status and rights related to territorial waters (Colombos, 1967, 105).

There was little serious effort to come to grips with the problem between 1930 and 1945 because of the unsettled international situation which resulted in World War Two. In 1945, however, the USA, long a supporter of Great Britain and the 3-mile zone, appeared to be opting for a wider acceptance and recognition of the existence of contiguous zones. In that year the USA declared that a fishing conservation zone existed outside of the 3-mile territorial zone off the USA and that similar claims by other nations to a limit of 12 miles from the coast would be recognized by the USA.[17] Almost at once a large number of states followed suit, although some went far beyond the 12 miles established by the USA (Colombos, 1967, 153).

The problems related to the territorial sea were, of course, exacerbated by the general political situation prevailing in the post-World War Two period. Firstly the world had been divided into the various political camps of the cold war, making it increasingly difficult to reach any reasonable settlement on anything of such world-wide importance as the breadth of the territorial sea. Secondly, this period also saw the end of the colonial era and the emergence of a large number of new nations intent on asserting their new-found political rights as sovereign nations.

In 1945, the task begun by the League of Nations in 1930 in attempting to codify the laws of territorial waters was placed in the hands of the International Law Commission (ILC) of the United Nations. By 1951, the ILC, declaring that the legal regime of the territorial sea was one of the most important matters needing codifications in international law, urged immediate action. Finally, in 1958 an international conference on the law of the sea under the aegis of the UN was commenced in Geneva. Eighty-six nations participated in the Conference which was comparatively successful and resulted in four international conventions being signed. The four conventions, one each on fisheries,[18] the high seas,[19] the Continental Shelf,[20] and the territorial sea and contiguous zone[21] have all become effective (Burke, 1969, 114). As in 1930, however, no satisfactory settlement could be reached on the breadth of the territorial sea, even though the major underlying problems had been solved to a certain extent by the four conventions that were adopted.

Two years later at the Second UN Conference on the Law of the Sea, Geneva, 1960, the territorial sea was again the main subject for debate. The Conference resulted in deadlock and failure. At both Conferences however, notable changes in position were apparent. Great Britain, for example, had proposed a 6-mile limit which was not considered favourably by most of the nations present. A compromise by the USA and

[16] See, *Conference for the Codification of International Law, The Hague* (1930), Minutes of the Second Committee, p. 123 ff.

[17] See, U.S. Federal Regulations, Vol. 10, No. 12,304 (1945).

[18] In force since 20 March 1966. *Supra*, note 6.

[19] In force since 30 September 1962. *Supra*, note 6.

[20] In force since 10 June 1964. *Supra*, note 6.

[21] In force since 10 September 1964. *Supra*, note 6.

Canada—the now famous 'six plus six' proposal—failed by only one vote. This proposal involved a 6-mile territorial sea similar to that proposed by Great Britain, but added an additional 6-mile contiguous zone for fishery control. This proposal basically failed due to the block voting practices of the East European socialist states, the Arab states and some of the newly independent states.

The international situation at the end of these two Conferences was one of even greater confusion than before. One definite result was the virtual end of serious adherence to the traditional concept of the 3-mile limit. However, even at the present time and on the eve of a new law of the sea conference the situation is as confused as ever. The most commonly asserted basis for extension of the territorial sea and the wide contiguous zones have been enforcement of customs and fiscal legislation, protection of fisheries resources, and national security. This side of the argument has in recent years, however, been considerably reinforced by the discovery of offshore minerals which make the 'ownership' of a wide coastal belt of sea attractive to many coastal states. Also, greater controls over coastal sea areas have been advocated by many and implemented by some coastal states to combat marine pollution.

On the other hand, the supporters of the severely limited territorial sea and contiguous zone will rely on the need for free and efficient flow of seaborne commerce, efficient exploitation of the resources of the sea for all nations, and the problems which coastal states would experience in properly exercising functions of neutrality, customs, policing and others. However, particularly for the major powers, questions of security appear to dictate an adherence to the principles of a narrow territorial sea.

An important area is, of course, the problem of navigable straits. A broader territorial sea would place many important straits within the control of the coastal state or states nearest to the strait. This is opposed by many of the larger maritime nations which fear that the coastal states might place excessive and unacceptable controls on their shipping.

Another area of considerable debate is the question of scientific ocean research. A broader territorial sea might give coastal states much greater control over scientific research carried out by other nations in 'their' sea area. This is a point which is of particular interest to many developing coastal states who are at present precluded from this type of research which is basically limited only to the advanced developed states for reasons of cost and experience. The question of whether ocean science should be allowed to proceed unimpeded by any national controls, other than those of the states which are sponsoring or financing the work, is a delicate one. Many states, particularly in the developing world, feel that it is difficult to draw the line between pure research of benefit to science in general and commercially and militarily-oriented research.

As the Geneva Convention on the Territorial Sea and Contiguous Zone was left open-ended on the breadth of the territorial sea, many states decided to expand their territorial sea unilaterally. At the end of 1972, for example, only 25 states still adhered to the 3-mile limit, 6 states had opted for a 6-mile limit, but 59 states had broadened their territorial sea to 12 miles or over. The latter group included 7 states who had declared that their territorial sea was to be 200 miles wide.[22] Even the USA, so long a supporter of the 3-mile limit, had now indicated that it would be prepared to accept 12 miles provided the rights of innocent passage were maintained at least in their present form and provided

[22] Brown, *supra*, note 2.

that passage through international straits (approximately seventy of which would be affected by such a limit) be kept free and open for American merchant and war ships. There is strong support for this stand from many other maritime nations, particularly the USSR and Japan. However, one of the important limitations which will be put on the traditional rights of innocent passage can be expected to be in the area of marine pollution which will be discussed below.

3.3 The functional zones

The whole history of the law of the sea in relation to the jurisdictional problem has been one of spatial concern. A functional approach to the problem is of very recent vintage. The contiguous zone as proposed in the Geneva Conferences considered by many to be the first rational approach at the international level is, however, based on many earlier proposals along these lines. For example, soon after the 1945 Truman Declaration on the Continental Shelf by the USA,[23] several Latin-American states began to enunciate claims to extensive maritime jurisdiction (Garcia Amador, 1963, 73). El Salvador, for example, constitutionally enshrined a territorial sea with a breadth of 200 miles. Argentina and Mexico claimed the 'epicontinental sea', which encompassed not only the Continental Shelf but also the superjacent waters. More famous than these unilateral initiatives was the trilateral decision in 1952 of Chile, Ecuador and Peru to establish a 'maritime zone' which was declared to extend not less than 200 miles from their shores. These and other Latin-American claims to extend maritime jurisdiction varied significantly from one another, both in form and content, but most, unlike the Salvadorean legislation, purported to claim something less than complete jurisdiction. In contemporary language they could be said to be early proposals for a multi-purpose functional zone within which the coastal state would exercise exclusive jurisdiction for designated purposes, but allegedly without prejudice to existing rights of navigation and associated rights under the regime of the high seas. Most of the reasoning behind the Latin-American claims of the 1950s was not only cultural in its reference but economic in its purpose. The 1952 Santiago Declaration on the Maritime Zone (Lay, Churchill and Nordquist, 1973, 23), rectified by Chile, Ecuador and Peru and acceded to by Costa Rica stated quite clearly and, with hindsight, quite progressively—that governments of coastal states had a duty to preserve and conserve natural ocean resources and the right to protect and control these resources which are a means of developing the economies of these states. This thesis was further developed by Chile, Ecuador and Peru during the next year and found support at the Lima Conference of 1954 and the Mexico City meeting of the Inter-American Council of Jurists in 1956 (Garcia Amador, 1963, 74). At the latter Conference, for example, a resolution was adopted, with one dissenting vote (USA), which recognized 'that each state is competent to establish its territorial waters within reasonable limits, taking into account geographical, geological, and biological factors, as well as the economic needs of its population, and its security and defence' (McDougal and Burke, 1962, 443).

However, at the Geneva Conferences these types of proposals by Chile, Ecuador and Peru merited little debate and were treated with almost disdainful disregard, being considered as too extreme by most of the represented states. Peru, for example, in withdrawing its proposals charged the Conference in 1958, as well as in 1960, with having

[23] Presidential Proclamation No. 2667. U.S. Fed. Regs. Vol. 10, No. 12,303 (1945).

'failed to study adequately the technical, biological and economic aspects of the law of the sea'.[24] In retrospect, this charge may very well have been quite true.

The 1960s undoubtedly were the years when economic concern was to supersede all other considerations. The many newly independent nations were all, without exception, developing nations faced with considerable economic problems in a competitive, unequally divided world. Despite foreign-aid efforts the economic gulf separating rich and poor was widening steadily. The United Nations in convening the United Nations Conference on Trade and Development (UNCTAD) which had met in Geneva in 1964, in New Delhi in 1968 and in Santiago in 1972, attempted to come to grips with many of these economic difficulties. The results were, as is well known to all, less than satisfactory. In the intervening years the sea and particularly the seabed had once again emerged into the international arena. The catalyst had been Malta's action in the United Nations asking that the resources of the seabed be reserved for the 'common heritage of all mankind'.[25] As has been pointed out above, resource orientation and environmental concern quickly brought the whole question of sea and seabed back into the forefront of international concern generally culminating in the call for a new law of the sea conference.

Many states, particularly in the developing world, dissatisfied with the results of previous law of the sea conferences, had taken their own unilateral action in spatially enlarging their coastal ocean areas. Amongst these were, of course, many of the Latin-American nations who had put into unilateral practice what they had preached in the 1950s (Garcia Amador, 1972). Most of this new legislation since 1960 deals with fishing regulations and conservation, reflecting a more sophisticated understanding of national fishery needs and problems. A similar wave of national legislation for the prevention and control of marine pollution was also under way. The growing importance of living resources of the sea and the potential wealth of non-living resources of the seabed, which had become scientifically discernible and technically exploitable, broke down the barriers to man's greatest untapped resource. Not surprisingly, the next move was to come from the developing world.

In December 1971, the Organization for African Unity (OAU) endorsed a recommendation for member states to extend their territorial sea to 200 miles with the result that '212 nautical miles would thus constitute the national economic limit in the oceans surrounding Africa'.[26] Emphasis was placed on the word 'economic' giving a completely functional connotation to the proposed zone—an 'economic zone' (EZ) which was to serve the economic development of African coastal states, rather than jurisdictional expansionism. In June 1972, these proposals were incorporated into unanimous recommendations made by the African States Regional Seminar on the Law of the Sea in Yaoundé, Cameroon, which was attended by representatives of sixteen African states.[27] Also in June 1972, a Specialized Conference of the Caribbean countries on Problems of the Sea, met in Santo Domingo and was attended by fifteen states of the region and

[24] U.N. Conference on the Law of the Sea, Geneva 1958, Official Records, Vol. 3, p. 176.

[25] *Supra*, note 7.

[26] See, *Asian–African Legal Consultative Committee, Report of the 13th Session*, Lagos, January 1972, Appendix VI, p. 452.

[27] *Report of the African States Regional Seminar on the Law of the Sea*, Yaoundé, 1971. U.N. Doc. A/AC 138/79.

observers from other Latin-American countries. This Conference adopted the Santo Domingo Declaration by a vote of 10 : 0 with five abstentions.[28] The main principles of the Declaration referred to a 12-mile territorial sea; a 'patrimonial' sea not exceeding 200 miles, with sovereign rights over all resources, the right to regulate scientific research and adopt pollution prevention measures. The 'patrimonial sea' is thus another name for the EZ idea.

During the July 1972 session of the UN Seabed Committee, the delegation of Kenya submitted the *Draft Articles on Exclusive Economic Zone Concept*[29] which basically formalized the whole EZ concept. There is little doubt that the EZ idea quickly became the 'hottest' new law of the sea issue. The purpose of the EZ has been most adequately stated by Mr F. X. Njenga, the distinguished delegate from Kenya who introduced the above draft article:

> . . . the economic zone concept offers a good basis for resolving the impasse between those who believe in a narrow and those who believe in a broad belt of territorial sea. Basically, the purpose of the exclusive economic zone concept is to safeguard the economic interests of the coastal states in the waters and sea-beds adjacent to their coasts without unduly interfering with other legitimate uses of the sea by other states.

Several other states have been highly active in the area of functional zones for a variety of reasons. States concerned with fishing, conservation and protection have been negotiating zones to protect fisheries, although unilateral action in this area, such as taken by Iceland, has not always been successful. Canada unilaterally established a 100-mile anti-pollution zone in its Arctic areas.[30]

There is little doubt that the whole concept of functional zones will be the area in the law of the sea where agreement and compromise must be found at the Third UN Conference on the Law of the Sea. On the other hand there is ample opposition to the concept, particularly from many of the highly developed nations and long-distance fishing nations who are liable to be excluded from large areas of the oceans if the protagonists for the zone concept prevail.

3.4 The Continental Shelf

The limitation of the Continental Shelf was another one of the unfinished items on the agenda of the 1958 Geneva Conference. The definition as set out in Article I of the 1958 Convention is as follows:

> . . . for the purpose of these Articles, the term 'Continental Shelf' is used as referring (a) to the sea-bed and subsoil of the submarine areas adjacent to the coast but outside the area of the territorial sea, to a depth of 200 metres or, beyond that limit to where the superjacent waters admits of the exploitation of the natural resources of the said areas; (b) to the sea-bed and subsoil of similar submarine areas adjacent to the coast of islands.[31]

[28] *Declaration of Santo Domingo*, 1972. U.N. Doc. A/AC 138/80.

[29] U.N. Doc. A/AC 138, S.C. II, L.10.

[30] E. Gold, Pollution of the Sea and International Law: A Canadian Perspective, *J. of Maritime Law and Commerce*, 1971, III, 13, at 35.

[31] *Supra*, note 6.

The Convention on the Continental Shelf was basically a codification of the 1945 Truman Declaration with which the USA unilaterally claimed the Continental Shelf contiguous to its coast subject to its jurisdiction and control.[32] However, the open-endedness of the definition cited above combined with the rapid development of technology undoubtedly were principle motivating factors in Ambassador Pardo's initiative in 1967. There was genuine concern that unchecked, the most powerful nations could easily have staked claims with some legitimacy to all the ocean floors of the world. At stake was, of course, the promise of an immense reserve of natural resources waiting to be tapped by an energy-hungry and resource-starved world.

The new proposals which have been put forward at the UN Seabed Committee meetings have been many and varied, each influenced by the particular situations of the states putting them forward. For example, the 1970 Nixon Proposals on the Continental Shelf call for a limitation on claims to the 200-metre isobath set out in the above definition or at a point 12 miles from the shore whichever was farthest. Beyond that point, the coastal state would hold the remainder of the Continental Shelf to the edge of the slope as an international trustee. An examination of the US coast, however, will disclose that most of the US Continental Shelf fits handily within these limits as prescribed by the US President. There are many other claims, counter-claims and proposals. Canada, for example, claims control of the Continental Shelf to the Continental Margin in order to obtain the 300–400 miles needed to gain jurisdiction of the Grand Banks. There is a growing group of states which want a 200-mile limit on the Shelf—other claims and proposals range from 12 to 100 miles.

The EZ proposals as described above, if accepted in some form or other, would greatly facilitate agreement in this area. Many states with a very narrow Continental Shelf feel that they should not be penalized due to their geological shortcomings and consider Continental Shelf limitations highly arbitrary. On the other hand, those states with broad Continental Shelves and exploitable living and non-living resources have, of course, considerable self-interest at stake. A good example is the Continental Shelf in the North Sea which has already been the cause of argument before the International Court of Justice.[33] The energy crisis and the search for oil and gas on all Continental Shelves will not provide this problem with easy solutions at the forthcoming Law of the Sea Conference.

3.5 The high seas

This is the sea-area where the concept of 'freedom of the seas' has been historically applicable as it is defined as that part of the sea which is not included in the territorial sea or in the internal waters of a state (Colombos, 1967, 47). The Geneva Convention on the High Seas codified many of the customary rights, obligations and principles related to this part of the sea which had been handed down through 3000 years of history. There is basically little dispute with the principles of the Convention—what is in dispute, however, is the area of the high seas. The age-old argument of freedom versus sovereignty comes to its full conclusion in this area. The area of the high seas is directly dependent upon the breadth of the territorial sea, the width of the functional zones and, to a certain extent, the amount of control a coastal state can exercise over 'its' Continental Shelf.

[32] *Supra*, note 23.
[33] *The North Sea Continental Shelf Cases*, International Court of Justice Reports, 1967, 6–7.

The high seas will, of course, include the seabed beyond the limit of national jurisdiction. Once the Continental Shelf limits are set the world community will have to deal with the management and exploitation of the remainder of the seabed. Spurred on by the need for new sources of ocean-bed minerals many of the technologically advanced states have already commenced to explore and exploit this vast area, taking advantage of the lack of international legislation before the Third UN Conference on the Law of the Sea. Other nations have proposed a 'Moratorium' on all exploration and exploitation until such time as international law has been agreed on at the forthcoming Conference.[34] For obvious reasons, this proposal has met with only very limited success.

Subcommittee I of the UN Seabed Committee has had the task of dealing with proposals for the management of the international seabed area. The various proposals which have been put forward for management as well as orderly exploitation range, for example, from an autonomous international licensing authority with power to grant leases or rights to exploit[35] to a UN agency system.[36] Such arrangements would probably call for royalties to be paid into an international fund which would be used for the benefit of all states of the world including those which are landlocked, and could be used to finance under-developed nations in particular. The type of agency best suited to handle this enormous task will undoubtedly be of considerable issue at Caracas, involving painstaking and arduous debate.

3.6   Living resources of the sea

The unrestricted right to fish on the high seas has been settled for many centuries by the states of the world as a basic principle of international law which dates back before the dawn of history.[37] Once again, however, technology and the increasing population pressures of a protein-hungry world have turned this right into a virtual licence to destroy. Modern fishing methods and traditional fishing habits have caused many species of fish to become endangered to the point where in the case of some species very little commercial exploitation is presently possible. There is no need to dwell on this as it must be a fact of which the scientific community is well aware.

It also follows, of course, from the doctrine of freedom of the seas, that fishing everywhere on the high seas is open to the subjects of all states. In the territorial waters fishing is theoretically reserved for the coastal state 'controlling' the territorial waters. In practice, however, traditional fishing rights are often recognized by the coastal state which allows other states to take advantage of these rights in its coastal areas.

In the interests of the preservation of fish, many states have, for a long time, made regulations for the control of fishing on the seas, although basically, such rules are only binding on their own nationals. It has also been recognized for a long time that unlimited fishing may seriously deplete fish stocks and that it is necessary to regulate the exercise of fishing rights on the high seas between the states whose nationals are engaged in the

[34] Adopted by the 24th Session of the U.N. General Assembly by a vote of 62 : 28 with 28 abstentions. U.N. Doc. A/RES/2574 (XXIV), 15 December 1969.

[35] U.N. Doc. A/AC 138/25.

[36] U.N. Doc. A/AC 138/33.

[37] The most wide-ranging exposition of the international law related to fisheries including its history can be found in: D. M. Johnston, *The International Law of Fisheries* (New Haven, 1965).

fishing (Koers, 1973). Much of this type of agreement has taken place on a regional level resulting in bilateral and multi-lateral treaties between states in a specific area. This was dictated at the time by the great variation in fisheries throughout the world which precluded the necessity of overall fisheries agreements for the whole world. Most of these agreements were, of course, limited to the developed western world, particularly the northeast and northwest Atlantic arena and date back to the early nineteenth century. There were, of course, some agreements and regulations in other sea areas, but these were basically to protect a particular type of living resource—such as the regulations protecting the Australian Pearl Fisheries; protection of the Behring Sea Seal Fisheries, the Halibut Fisheries in the Pacific (Colombos, 1967, 401 ff.), etc. This regional approach to fisheries has worked basically quite well and is used to this day (Lay, Churchill and Nordquist, 1973, 352–502). This is not to say that there are no areas of friction. Far from it. The fisheries issue is undoubtedly going to be one of the hardest fought problems at the forth-coming law of the sea conference. This is due to the fact that the interests of the various states of the world are widely divergent in the area of fisheries. At stake is, of course, a resource which is of vital necessity in feeding a large segment of the population of the world, as well as being an enormous source of revenue for commercial exploitation to many states. Since 1955 the commercial catch has increased at a compounded rate of more than 6 per cent per annum.[38] This 'harvest from the sea' has induced many nations, but in particular the USSR, Germany DR, Poland, Korea and Japan to invest heavily in distant-water fishing technology. The advances in fishing technology. the new factory-ships, large trawlers and their ability to make very long voyages and process vast quantities of fish in foreign waters have caused many coastal states to be very concerned about these foreign vessels reaping this marine harvest off their coast, even outside the 12-mile limit. Argentina, Brazil, Chile, Ecuador and Peru, for instance, have extended their territorial sea claims to 200 miles, basically as a reaction to the practices of distant-water fishing off their shores (Garcia Amador, 1972). The Iceland–Great Britain dispute is another example of these difficulties, although in this case Iceland being almost completely dependent on its coastal fisheries took action against the vastly increasing fleets of countries which had traditionally fished off the coasts of Iceland.

In another area, the International Whaling Commission which is charged with setting quotas for whale catches, has been unable to prevent Japan and the USSR, who between them take seven-eighths of all whales caught, from pursuing ruthless policies that are likely to lead to the extinction of certain whale species.

Many of the principles related to fisheries which have become customary rules of international law or which have become accepted due to the various regional treaties which have been in force for many years were also discussed at the 1958 Geneva Confer-ence. The Conference codified these and various new principles in a brand new 'Conven-tion on Fishing and the Conservation of the Living Resources of the High Seas', which is considered by many to be the most successful piece of international legislation to emerge from the 1958 Conference. As its title implies the Convention laid down basic ground rules for effective conservation of the world's fishery resources which would at the same time give maximum benefits to all nations. The Convention in being concerned with

[38] See, *Yearbook of Fishery Statistics*, Vol. 29, New York, 1970.

conservation of resources as well as national rights on the high seas does not directly consider the question of territorial seas, but does settle many of the problems which caused the fisheries question to be related to the territorial sea problem. Although the Convention was accepted by many states, its practical implementation was another matter. Conservation requirements of the Convention have, as a general rule, not been followed although the UN specialized agency in this area, the Food and Agriculture Organization (FAO), has been highly active in this field for some time (Bowett, 1970, 102–3).

The combination of ever-greater demand for the resources of the seas, the basic lack of conservation methods except on a limited regional basis, the new fishing technologies and methods and, to a certain extent, the environmental concern, dictate new approaches in this area for the new law of the sea conference. Once again, there have been many proposals which can only be summarized here. Because most of the fish are taken near land masses, it has been suggested by many countries that coastal states which are adjacent to the fishing zones should be given authority to conserve the fisheries and manage them and that this authority ought to go beyond the very limited special interests granted by the 1958 'Convention on Fishing and the Conservation of Living Resources of the Sea.' This will, of course, bring in the jurisdictional question once again which will be countered by the functional-zone argument such as is expressed in the EZ proposals. The 'species' approach taken by a group of developed states, which divides the resource into sedentary, coastal, anadromous and wide-ranging species and urges a separate specific allocation of authority for each, is another much-discussed proposal (Christy, 1973, 65).

Whatever the outcome of the Conference in this area, it can be said with some certainty that the Conference will have a hard time denying preferential rights to coastal states over their fisheries. Regardless of whether such rights are incorporated in an 'economic zone', a 'patrimonial sea' or a custodial area, they will be tied to a duty to exploit and manage stocks in order that the world demand for protein, particularly from the developing world, can continue to be met. In addition a new global fisheries convention which will have certain consequences on the present regional arrangements, many of which have worked successfully in the past, will probably be tabled and debated. There is little doubt that such a new convention is now required in order that the present 'vacuum-cleaner' methods of exploitation of one of the world's most vital and natural living resources can be controlled to the advantage of the community of nations as a whole.

3.7 Marine pollution

So far, in the previous six sea areas and associated uses of the sea, the law of the sea has attempted to regulate rights over resources and privileges over uses of the seas. In other words, in those areas, there was always some state, or group of states, which could gain advantages from the sea. In the area of marine pollution, the unpleasant by-product of man's advances in modern technology, everybody loses. The basic question here is to spread the loss and prevent its occurrence as equitably as possible.

As much as 80 per cent of all pollutants in the ocean originate on land and is a part of the environmental deterioration of the earth in general (Johnston, 1972–3, 65). It is not to be considered here. In any case, the main thrust of the international law to combat marine pollution has been directed at the remaining 20 per cent, that is, the forms of pollution that originate from the dumping, discharging and other activities connected with seaborne trade. It is this area which will be taken up here.

The degradation of the marine environment caused by a variety of ocean pollutants is considered to be a serious danger to the oceans today. It will not be necessary to elaborate on this danger in a book on oceanography. The danger has been accepted and it will only be necessary to show how the law of the sea is attempting to cope with it.

The territorial sea and the internal waters of coastal states are the areas most affected by marine pollution. Marine pollution has been of concern to the maritime nations since the 1920s (Colombos, 1967, 430–31) but little, if any, positive action was taken until the 1950s. The renewed concern at that time resulted in the 'International Convention for the Prevention of Pollution by Oil' which was concluded in 1954, and came into force in 1958, amended in 1962 and 1969 (Lay, Churchill and Nordquist, 1973, 557–80), and which is the only existing International Convention dealing solely with the prevention of a major potential source of marine pollution prior to its commission. The Convention prohibits discharges of oil and oily mixtures between certain geographically defined areas but leaves enforcement to the state whose flag is being flown by the offending vessel. This has, of course, caused great difficulties and has virtually rendered the Convention unenforceable.

The question of pollution of the high seas was also considered, though not in depth, by the Geneva Conference, 1958.[39] 1958 was also the first year of operation of the Inter-Governmental Maritime Consultative Organization (IMCO) which is a specialized agency of the United Nations charged with the control of marine pollution. IMCO was basically formed as a technical, non-legal organization with the objective to facilitate co-operation among governments in technical matters of all kinds affecting shipping (Bowett, 1970, 105). Legal requirements related to shipping had heretofore always been decided upon by the Comité Maritime International (CMI), which is a non-governmental international shipping organization. Since 1867, the CMI has been the main influence in shaping international maritime law and which in turn had reflected the interests of the major shipping nations fully, as it was the international spokesman of the various national law associations representing the shipping, cargo, insurance, chartering and other related shipping interests.[40] The *Torrey Canyon* disaster which resulted in widespread oil pollution of the British and French coastal and sea areas (Nanda, 1967, 400) and which caught a totally unprepared world completely off-guard, was a 'blessing in disguise' for those concerned with the protection of the environment. This disaster not only electrified public opinion but also resulted in speedy action by a number of international organizations and governments. For example, IMCO which had been rather strictly limited to technical matters affecting international shipping was suddenly entrusted with combating the marine pollution problem by the convening of an IMCO *ad hoc* legal committee. This meant that for the first time an international organization supposedly without special interests had become involved in the pollution problem. This resulted in the 1969 amendments to the 1954 Convention and the convening of a diplomatic conference in Brussels in 1969 by IMCO. The results were two new conventions: the 'International Convention for Civil Liability for Oil Pollution Damage' (Lay, Churchill and Nordquist,

[39] Geneva Convention on the High Seas, Art. 24.
[40] Gold, *supra*, note 30.

1974, 557–80), which was to deal with the considerable difficulty surrounding the question of liability for pollution. This is basically a private international law problem related directly to the enormous clean-up costs incurred by coastal states, particularly after shipping accidents. The shipping industry maintains that it is unable to obtain sufficient insurance to cover these types of claims and thus wishes to limit its liability. This Convention sets out limitations which have not been accepted by many coastal states.

The other Convention was in the area of public international law and was entitled 'International Convention Relating to the Intervention on the High Seas in Cases of Oil Pollution Casualties' (Lay, Churchill and Nordquist, 1974, 602). This Convention indicates a complete departure from the hallowed 'freedom of the seas' principle by stating that coastal states may interfere against foreign ships upon the high seas to prevent or minimize major pollution damage where a marine accident threatening or causing oil pollution has already occurred. This latter reservation showing only the remedial nature of the legislation has been considered to be unacceptable to several coastal states who feel that a convention of this type should also have preventive principles and powers. At a further Conference in 1971, the 'International Convention for the Establishment of an International Fund for Compensation for Oil Pollution Damage' (Lay, Churchill and Nordquist, 1974, 602) was concluded. This Convention was to augment the amount of compensation set out in the 1969 Civil Liabilities Convention.

Until 1971, despite much activity by coastal states there had been no international legal measures adopted which purported to combat marine pollution by substances other than oil. The first such initiative was the adoption of the 'Convention on the Control of Marine Pollution by Dumping from Ships and Aircraft', Oslo 1971 (Lay, Churchill and Nordquist, 1964, 670), which was a regional initiative taken by the major shipping states of the northeast Atlantic area. The contracting states pledged themselves 'to take all possible steps to prevent the pollution of the sea by substances that are liable to create hazards to human health, to harm living resources and marine life, to damage amenities or to interfere with other legitimate uses of the sea'.

In the interim, of course, global environmental concern had resulted in the preparations for the United Nations Conference on the Human Environment, which was eventually held in Stockholm in 1972.[41] This Conference and its preparation were what has been called the culmination of international environmental policy seen in a global perspective. Certain recommendations of this Conference deal specifically with marine pollution calling upon governments to co-operate in various ways with one another and with international organizations such as the Joint Group of Experts on the Scientific Aspects of Marine Pollution (GESAMP); the Food and Agriculture Organization (FAO); the World Health Organization (WHO); the International Oceanographic Commission (IOC); the International Atomic Energy Agency (IAEA); the World Meteorological Organization (WMO); and other organizations. An ultimate result of the Stockholm Conference was the establishment of a new United Nations specialized agency UNEP (United Nations Environment Programme) with headquarters in Nairobi, Kenya, which is to co-ordinate all environmental efforts on an international scale.

In November 1973, the 'International Convention for the Protection of Pollution from

[41] See, *Report of the United Nations Conference on the Human Environment*, Stockholm, 5–16 June 1972. U.N. Doc. A/CONF.48/14/REV.1.

Ships, 1973',[42] was concluded under the auspices of IMCO in London. Although all forms of pollution are included in the new Convention, the power and interests of the major shipping nations once again prevented many progressive proposals to combat marine pollution from being implemented.

It should be pointed out that only the 1954 Convention and its 1962 amendments are presently in effect. This delay in the acceptance of vital international measures plus the ever-increasing volume of seaborne trade particularly in the carriage of pollutants such as petroleum and petroleum products has prompted states, once again, to take unilateral action and propose radically different approaches to this area of the law of the sea. Canada's 100-mile Arctic Pollution Zone[43] and the marine pollution suggestions in the various EZ proposals[44] are examples of this concern which is shared particularly by developed and developing coastal states. In the preparatory work which has taken place 'n Subcommittee III of the United Nations Seabed Committee, the old jurisdictional problem has frequently appeared, setting the interests of the coastal states against the interests of the large maritime powers—'freedom of the seas' versus 'territorial sovereignty'! However, the protection of the marine environment is indeed an exercise in functionality and will be another area of heated debate in the forthcoming Conference on the law of the sea.

4 Conclusion

The selection of the foregoing list of issues has obviously been far from comprehensive and has been entirely personal to the writer. Many important areas have been superficially treated or left out entirely. We have tried to pick out what we personally considered to be the most pressing problems in the law of the sea *vis-à-vis* the historically evolved concept of the 'freedom of the seas' which we consider to be the one most vital area in which all law of the sea agreement or failure will occur. This area will also be of the greatest concern to the ocean scientist.

The brief look at history should point out that the notion of 'freedom' is basically an ideological tool wielded by national interests and requirements. 'Freedoms' are often claimed by the strong and powerful just as 'protection' is often demanded by the weaker to allow them to maintain a threatened privileged position. Freedom was helpful to the Elizabethans in challenging the Iberian monopolies on the sea but it was not useful when it later precluded the English from excluding the Dutch fishery from their shores. The analogy with the modern situation should be obvious. Unfortunately, international lawyers are inclined to look for historical legal precedents and former treaties which have solved problems similar to those which face them today. This makes the solution to our current problems more difficult. The magic words 'freedom of the seas' are uttered by those concerned with problems in the law of the sea today in sterile reverence and with hasty misquotation and show a complete misunderstanding of what these words meant in the early part of the seventeenth century. The adjectives 'closed' and 'open' are today

[42] See, IMCO Press Release—IMCO/11/1973, 2 November 1973.
[43] Gold, *supra*, note 30.
[44] D. M. Johnson and E. Gold, *The Economic Zone in the Law of the Sea: Survey, Analysis and Appraisal of Current Trends*. Kingston, R.I., 1972.

mere slogans and quite unrelated to the very real problems at hand. It should not be beyond the ability of the 'family' of 150 sovereign nations to find new answers to old problems and approach the sea and its manifold resources in a much more functional rather than sterile and legalistic way. Undoubtedly the Third United Nations Conference on the Law of the Sea at Caracas, will not be able to solve all the problems in a period of 10 weeks. The law of the sea will occupy many people for many years and the Conference will probably become an on-going process, meeting periodically, with intensive work by committees in between. However, the oceans covering 70 per cent of the world surface deserve nothing less.

References

These references include an additional selection of works (marked *) which might assist the ocean scientist who wishes to pursue some further research into the area of international law of the sea. It is not meant to be a comprehensive list.

*ALEXANDER, L. M. (editor) 1967: *The law of the sea: offshore boundaries and zones.* Proceedings of the 1st Annual Conference of the Law of the Sea Institute, University of Rhode Island, 1966. Columbus, Ohio: Ohio State University Press.

*   1968: *The law of the sea: the future of the sea's resources.* Proceedings of the 2nd Annual Conference of the Law of the Sea Institute, University of Rhode Island, 1967. Kingston, RI.

*   1969: *The law of the sea: international rules and organization for the sea.* Proceedings of the 3rd Annual Conference of the Law of the Sea Institute, University of Rhode Island, 1968. Kingston, RI.

*   1970: *The law of the sea: national policy recommendations.* Proceedings of the 4th Annual Conference of the Law of the Sea Institute, University of Rhode Island, 1969. Kingston, RI.

*   1971: *The law of the sea: the United Nations and ocean management.* Proceedings of the 5th Annual Conference of the Law of the Sea Institute, University of Rhode Island, 1970. Kingston, RI.

*   1972: *The law of the sea: a new Geneva conference.* Proceedings of the 6th Annual Conference of the Law of the Sea Institute, University of Rhode Island, 1971. Kingston, RI.

*   1973: *The law of the sea: needs and interests of developing countries.* Proceedings of the 7th Annual Conference of the Law of the Sea Institute, University of Rhode Island, 1972. Kingston, RI.

*ANDRASSY, J. 1970: *International law and the resources of the sea.* New York: Columbia University Press.

*BOUCHEZ, L. J. 1964: *The regime of bays in international law.* Leyden: A. W. Sijthoff.

*BOWETT, D. W. 1970: *The law of international institutions.* 2nd edition. London: Stevens & Sons.

BRIERLEY, J. L. 1963: *The law of nations.* 6th edition, ed. Sir Humphrey Waldock. Oxford: Clarendon Press.

*BROWN, E. D. 1971: *The legal regime of hydrospace.* London: Stevens & Sons.

*BURKE, W. T. (editor) 1969: *Towards a better use of the ocean: contemporary legal problems in ocean development.* Stockholm: Almquist & Wiksell.

*BUTLER, W. E. 1971: *The Soviet Union and the law of the sea.* Baltimore: Johns Hopkins.

CHRISTY, F. T. JR 1973: Northwest Atlantic fisheries arrangements: a test of the species approach. *Ocean Development and Int'l Law J.* **1,** 65.

*COLOMBOS, C. J. 1967: *The international law of the sea.* 6th edition. London: Longman.

FENN, P. T. 1926: *The origin of the right of fisheries in territorial waters.* Cambridge, Mass.

*FRIEDMANN, W. 1971: *The future of the oceans.* New York: George Braziller.

*GARCIA AMADOR, F. V. 1963: *The exploitation and conservation of the resources of the sea: a study of contemporary international law.* 2nd edition. Leyden: A. W. Sijthoff.

1972: *Latin America and the law of the sea.* Kingston, RI: The Law of the Sea Institute, University of Rhode Island.

*GLASSNER, M. I. 1970: *Access to the sea for developing land-locked countries.* The Hague: Martinus Nijhoff.

*GOLD, E. 1972: *Oil pollution: a survey of worldwide legislation.* Arendal: Assuranceforeningen Gard.

GROTIUS, H. 1916: *On the freedom of the seas.* Ed. and transl. Hogoffin.

*GULLION, E. A. (editor) 1968: *Uses of the seas.* Englewood Cliffs, NJ: Prentice-Hall.

*HARGROVE, J. L. (editor) 1972: *Law, institutions and the global environment.* Dobbs Ferry, NY and Leyden: Oceana Publications, Inc. and A. W. Sijthoff.

*JOHNSTON, D. M. 1965: *The international law of fisheries: a framework for policy-oriented inquiries.* New Haven: Yale University Press.

1972–3: Marine pollution control: law, science and politics. *Int'l J.* **28,** 69.

JOHNSTON, D. M. and GOLD, E. 1973: *The economic zone in the law of the sea: survey, analysis and appraisal of current trends.* Kingston, RI: The Law of the Sea Institute, University of Rhode Island.

*KOERS, A. W. 1973: *International regulation of marine fisheries.* London: Fishing News (Books) Ltd.

*LAY, S. H., CHURCHILL, R., and NORDQUIST, M. (editors) 1973: *New directions in the law of the sea.* Vols 1 and 2; CHURCHILL, R., SIMMONS, K. R. and WELCH, J. (editors) Vol. 3, London and Dobbs Ferry, NY: The British Institute of International and Comparative Law and Oceana Publications, Inc.

LEGAULT, L. H. J. 1971: The freedom of the seas: a licence to pollute? *U. of Toronto L.J.* **21,** 211.

*LOGUE, J. J. (editor) 1972: *The fate of the oceans.* Villanova, Pa: Villanova University Press.

*MCDOUGAL, M. S. and BURKE, W. T. 1962: *The public order of the oceans: a contemporary international law of the sea.* New Haven: Yale University Press.

NANDA, V. 1967: The *Torrey Canyon* disaster: some legal aspects. *Denver L.J.* **44,** 400.

*ODA, S. (editor) 1972: *The international law of the ocean development: basic documents.* Leyden: A. W. Sijthoff.

OPPENHEIM, L. 1955: *International law.* Vol. 1, 8th edition. London.

*OUDENDIJK, J. K. 1970: *Status and extent of adjacent waters: a historical orientation.* Leyden: A. W. Sijthoff.

M

*SINGH, N. 1973: *International conventions of merchant shipping.* 2nd edition. London: Stevens & Sons.

*SLOUKA, Z. J. 1968: *International custom and the continental shelf: a study in the dynamics of customary rules of international law.* The Hague: Martinus Nijhoff.

SMITH, H. 1950: *The law and custom of the sea.* 2nd edition.

UNITED NATIONS 1970: *National legislation and treaties relating to the territorial sea, the contiguous zone, the continental shelf, the high seas and to fishing and conservation of the living resources of the sea* (UN Doc. ST/LEG/SER. B/15). New York.

References

ADAMS, J. K. and BUCHWALD, V. T. 1969: The generation of continental shelf waves. *J. Fluid Mech.* **35** (4), 815–26.

ALLISON, N. W. 1947: An investigation into the causes of the annual variation of mean sea level in the North Sea. *MNR Astron. Soc. Geophys.* **5,** 146–57.

ALVERSON, D. L., LANGHURST, A. R., and GULLARD, J. A. 1970: How much food in the sea? *Science* **168,** 503–5.

AMOS, A. F., GORDON, A. L., and SCHNEIDER, E. D. 1971: Water masses and circulation patterns in the region of the Blake–Bahama Outer Ridge. *Deep Sea Res.* **18** (2), 145–65.

ARONS, A. B. and STOMMEL, H. 1967: On the abyssal circulation of the World Ocean III. *Deep Sea Res.* **14,** 441–57.

ASSAF, G., GERARD, R., and GORDON, A. L. 1971: Some mechanism of oceanic mixing revealed in aerial photographs. *J. Geophys. Res.* **76** (27), 6550–72.

BAGNOLD, R. A. 1963: Beach and nearshore processes. In M. N. Hill (editor), *The sea.* Vol. **III.** New York: Wiley, 507–28.

BALLS, R. 1961: *Fish capture.* Buckland lecture for 1959. London: Edward Arnold.

BANG, N. D. 1971: The southern Benguela Current region in February 1966. Part II: Bathythermography and air–sea interactions. *Deep Sea Res.* **18** (2), 209–24.

BARBER, N. F. 1949: The behaviour of waves on tidal streams. *Proc. Roy. Soc.* A **198,** 81–93.
1954: Finding the direction of travel of sea waves. *Nature* **174,** 1048–50.

BARBER, N. F. and TUCKER, M. J. 1962: Wind waves. In M. N. Hill (editor), *The sea.* Vol. **I,** chap. 19. New York: Wiley, 664–99.

BARBER, N. F. and URSELL, F. 1948: The generation and propagation of ocean waves and swell. I: Wave periods and velocities. *Phil. Trans. Roy. Soc.* A **240,** 527–60.

BARDACH, J. 1968: *Harvest of the sea.* London: Allen and Unwin.

BARNES, F. A. 1952: The Trent Eagre. *Survey,* Univ. of Nottingham, **3,** 1–16.

BARNES, H. (editor) 1969: *Oceanog. and Mar. Biol. Ann. Rev.* **7,** London: Allen and Unwin.

BÉ, A. W. H., FORNS, J. M., and ROELS, O. A. 1971: Plankton abundance in the north Atlantic Ocean. In J. D. Costlow (editor), *Fertility of the sea.* Vol. **I.** New York: Gordon and Breach Sci. Pubs., 17–50.

BEERS, J. R. and STEWART, G. L. 1971: Micro-zooplankters in the plankton communities of the upper waters of the eastern tropical Pacific. *Deep Sea Res.* **18** (9), 861–83.

BELDERSON, R. H. and KENYON, N. H. 1969: Direct illustration of one-way sand transport by tidal currents. *J. Sedim. Petrol.* **39,** 1249–50.

BENSON, N. G. (editor) 1970: A century of fisheries in North America. *Sp. Pub.* **7.** Washington: Amer. Fish. Soc. 329 pp.

BEVERTON, R. J. H. and HOLT, S. J. 1957: On the dynamics of exploited fish populations. *Fish. Invest. London Ser.* **2,** 19.

BLACKBURN, M. 1965: Oceanography and ecology of tunas. *Oceanog. and Mar. Biol. Ann. Rev.* **3,** 299–322.

BLANDFORD, R. R. 1971: Boundary conditions in homogeneous ocean models. *Deep Sea Res.* **18** (7), 739–51.

BLAXTER, J. H. S. and HOLLIDAY, F. G. T. 1963: The behaviour and physiology of herring and other clupeids. In F. S. Russell (editor), *Adv. in Mar. Biol.* London: Academic Press, 261–393.

BODINE, B. R. 1969: Hurricane surge frequency estimated for the Gulf of Texas. *CERC Tech. Memo.* **26.** 32 pp.

BÖHNECKE, G. 1938: Temperatur, Saltgehalt und Dichte an der Oberfläche des Atlantische Exped. Meteor, 1925–27. *Wiss. Erg.* Bd. **5.** 62 pp.

BONEY, A. D. 1965: Aspects of the biology of the seaweeds of economic importance. In F. S. Russell (editor), *Adv. in Mar. Biol.* Vol. **3.** London: Academic Press, 105–253.

BONNEFILLE, R., CORMAULT, P., and VLEMBOIS, J. 1967: Progrès des méthodes de mesure de la houle naturelle au laboratoire national d'hydraulique. Chap. 9. *Proc. 10th Conf. Coastal Eng.* Tokyo, 115–26.

BOOTH, E. 1969: Seaweeds of commerce. In F. E. Firth (editor), *Encyclopedia of marine resources.* New York: Van Nostrand Reinhold, 626–30.

BOURNE, W. 1578: *The treasurer for travellers.* London.

BOWDEN, K. F. 1964: Turbulence. In H. Barnes (editor), *Oceanog. and Mar. Biol. Ann. Rev.* **2.** London: Allen and Unwin, 11–30.

BRATZ, J. F. (editor) 1968: *Ocean engineering.* New York: Wiley. 720 pp.

BREBNER, A. and KAMPHUIS, J. W. 1963: Model tests on relationships between deep-water wave characteristics and long-shore currents. *Queen's Univ. Civ. Eng. Res. Rep.* **3,** 1–25. Kingston, Ontario.

BRETSCHNEIDER, C. L. 1959: Wave variability and wave spectra for wind-generated gravity waves. *US Army BEB Tech. Memo.* **118.** Washington.

1965: The generation of waves by wind. State of the art. *Nat. Eng. Sci. Co. Off. Naval Res.* **SN** 134-6. 96 pp.

1966: On tides and longshore currents over the continental shelf due to winds blowing at an angle to the coast. *Nat. Eng. Sci. Co. Off. Naval Res.* **SN** 134-13.

(editor) 1969: *Topics in ocean engineering.* Houston: Gulf Publishing. 120 pp.

BROECKER, W. 1963: Radioisotopes and large scale oceanic mixing. In M. N. Hill (editor), *The sea.* Vol. **II,** chap. 4. New York: Wiley, 88–108.

BROWN, S. G. 1955: Fifty years of Antarctic whaling. *Naut. Mag.* **171,** 88–90.

1958: Whale marks recovered during the Antarctic whaling season 1957–58. *Norsk Hvalfangstid.* **10,** 503–7.

1959: Whale marks recovered in the Antarctic seasons 1955–56, 1958–59, and in South Africa 1958 and 1959. *Norsk Hvalfangstid.* **12,** 609–12.

1963: A review of Antarctic whaling. *Polar Record* **11,** 555–66.

BRYAN, K. and COX, D. 1967: A numerical investigation of the oceanic general circulation. *Tellus* **19,** 54–80.

BRYANT, A. 1944: *Years of victory, 1802–1812.* London: Collins.

BUTTON, D. K. 1971: Petroleum–biological effects in the marine environment. In D. W. Hood (editor), *Impingement of man on the oceans.* Chap. 14. New York: Wiley–Interscience, 421–9.

CALVERT, S. E. and PRICE, N. B. 1971: Upwelling and nutrient regeneration in the Benguela Current, October 1968. *Deep Sea Res.* **18,** 505–23.

CARPENTER, E. J. and SMITH, K. L. 1972: Plastics on the Sargasso Sea surface. *Science* **175** (4027), 1240–41.

CARRITT, D. E. 1971: Oceanic residence time and geobiochemical interactions. In D. W. Hood (editor), *Impingement of man on the oceans.* Chap. 6. New York: Wiley–Interscience, 191–9.

CARRUTHERS, J. N. 1954: Some inter-relationships of oceanography and fisheries. *Archiv für Met. Geophys. Bioklim.* **B 6,** 167–89.

1956: Fish, fisheries and environmental factors. *Oceanus* **4,** 14–20.

CARTER, L. J. 1968: Continental shelf: scramble for federal oil-lease revenues. *Science* **160,** 1431–2.

CARTWRIGHT, D. E. 1967: Modern studies of wind-generated waves. *Contemp. Phys.* **8,** 171–83.

1968: A unified analysis of tides and surges round north and east Britain. *Phil. Trans. Roy. Soc.* **A 263,** 1–55.

1969: Extraordinary tidal currents near St Kilda. *Nature* **223,** 928–32.

CARTWRIGHT, D. E. and TAYLOR, R. J. 1971: New computation of the tide-generating potential. *Geophys. J. Roy. Astron. Soc.* **13** (1), 45–74.

CARTWRIGHT, D. E. and WOODS, A. J. 1963: Measurements of upper and lower tidal currents at Banc de la Chapelle. *Deutsch. Hydrog. Zeit.* **16,** 64–76.

CHAMPION, H. and CORKAN, R. 1936: The bore on the Trent. *Proc. Roy. Soc.* **A 154,** 158.

CHARLESWORTH, J. K. 1957: *The Quaternary era.* Chap. 8. London: Edward Arnold, 277–308.

CHARNEY, J. G. and SPIEGEL, S. L. 1971: Structure of wind-driven equatorial currents in homogeneous oceans. *J. Phys. Oceanog.* **1** (3), 149–60.

CHARNOCK, H. 1958: A note on empirical wind-wave formulae. *Quart. J. Roy. Met. Soc.* **84,** 443–7.

1960: Ocean currents. *Sci. Prog.* **48,** 257–70.

CHESHER, R. H. 1969: Destruction of Pacific corals by the Sea Star *Acanthaster planci. Science* **165,** 280–82.

CHITALE, S. V. 1934: Bores in tidal rivers with a special reference to the Hoogly. *Irrig. and Power* **11,** 110–20.

CHRISTENSEN, N. 1971: Observations of the Cromwell Current near the Galapagos Islands. *Deep Sea Res.* **19** (1), 27–33.

CLARKE, M. R. 1955: A giant squid swallowed by a sperm whale. *Norsk Hvalganfstid.* **44,** 589–93.

1963: Economic importance of north Atlantic squids. *New Sci.* **17,** 568–70.

CLARKE, R. 1962: Research on marine resources in Chile, Ecuador and Peru. *Fishing News Int.* **1,** 44–50.

COASTAL ENGINEERING RESEARCH CENTER (CERC) 1966: Shore protection, planning and design. 3rd edition. *Tech. Rep.* **1**, 50–114.

COCHRANE, J. D. 1956: The frequency distribution of surface-water characteristics in the Pacific Ocean. *Deep Sea Res.* **4**, 45–53.

1958: The frequency distribution of water characteristics in the Pacific Ocean. *Deep Sea Res.* **5**, 111–27.

COLEBROOK, J. M. 1969: Variability in the plankton. Part II. In M. Sears (editor), *Progress in oceanography* **V.** Oxford: Pergamon, 115–25.

CORKAN, R. H. 1950: The levels in the North Sea associated with the storm disturbance of 8 January 1949. *Phil. Trans. Roy. Soc.* **A 242**, 483–525.

COSTLOW, J. D. (editor) 1971: *Fertility of the sea.* Vols. **I** and **II.** New York: Gordon and Breach Sci. Pubs.

COTÉ, L. J. *et al.* 1960: The directional spectrum of a wind-generated sea as determined from data obtained by SWOP. *Meteor Paper* **2** (6). New York. 88 pp.

COX, R. A. 1959: The chemistry of sea water. *New Scientist* **6**, 518–21.

COX, R. A. and SMITH, N. D. 1959: The specific heat of sea water. *Proc. Roy. Soc.* **A 252**, 51–62.

CRISP, D. T. 1962: The tonnages of whales taken by Antarctic pelagic operators during 20 seasons and an examination of the Blue Whale Unit. *Norsk Hvalfangstid.* **10**, 389–93.

CRONAN, D. S. *et al.* 1972: Iron-rich basal sediments from the eastern equatorial Pacific: Leg 16, deep sea drilling project. *Science* **175**, 61–3.

CUSHING, D. H. and BURD, A. C. 1957: On the herring of the southern North Sea. *Fish. Invest. London* Ser. **II**, 20.

DALTON, F. K. 1951: Fundy's prodigious tides and the Petitcodiac's tidal bore. *J. Roy. Astron. Soc. Canada* **45**, 225–31.

DARBYSHIRE, J. 1957a: A note on the comparison of proposed wave spectra formulae. *Deutsch. Hydrog. Zeit.* **10**, 184–90.

1957b: Attenuation of swell in the north Atlantic Ocean. *Quart. J. Roy. Met. Soc.* **83**, 351–9.

1959a: The spectra of coastal waves. *Deutsch. Hydrog. Zeit.* **12**, 153–67.

1959b: A further investigation of wind-generated waves. *Deutsch. Hydrog. Zeit.* **12**, 1–13.

1960: Microseisms and storms. *Adv. Sci.* **17**, 149–57.

1962: Microseisms. In M. N. Hill (editor), *The sea.* Vol. **I.** New York: Wiley, 700–19.

1963: The one-dimensional wave spectrum in the Atlantic Ocean and coastal waters. *Ocean Wave Spectra*, Rep. of Conf., Easton, Maryland, 27–31.

DARBYSHIRE, J. and DARBYSHIRE, M. 1956: Storm surges in the North Sea during the winter 1953–1954. *Proc. Roy. Soc.* **A 235**, 260–74.

DARBYSHIRE, M. 1958: Waves in the Irish Sea. *Dock and Harb. Auth.* **39**, 245–8.

DARBYSHIRE, M. and DRAPER, L. 1963: Forecasting wind-generated sea waves. *Eng.* **195**, 482–4.

DARLING, J. M. and DUMM, D. G. 1967: The wave record program at CERC. *CERC Misc. Pap.* **1–67**. 30 pp.

DA SILVA, P. DE CASTRO MOREIRA 1971: Upwelling and its biological effects in south Brazil. In J. D. Costlow (editor), *Fertility of the sea.* New York: Gordon and Breach Sci. Pubs.

DAVENPORT, J. M. 1971: Incentives for ocean mining: a case study of sand and gravel. *J. Mar. Tech. Soc.* **5** (4), 35–40.

DAVIES, J. L. 1959: Wave refraction and the evolution of shoreline curves. *Geog. Stud.* **5,** 1–14.

DEACON, G. E. R. 1959: The Antarctic Ocean. *Sci. Prog.* **47,** 637–60.

1960a: The southern cold temperate zone. *Proc. Roy. Soc.* **B 152,** 441–7.

1960b: The sea and its problems. In G. Sutton (editor), *The world around us.* Chap. 5. London: English Universities Press, 77–97.

1960c: International co-operation in marine science. *Sci. Prog.* **48,** 667–72.

1960d: The Indian Ocean Expedition. *Nature* **187,** 661–2.

DEACON, G. E. R., SVERDRUP, H. U., STOMMEL, H., and THORNETHWAITE, C. W. 1955: Discussion on the relations between meteorology and oceanography. *J. Mar. Res.* **14,** 499–515.

DEACON, M. 1971: *Scientists and the sea.* London: Academic Press.

DEFANT, A. 1961: *Physical oceanography.* Vol. **I.** Oxford: Pergamon.

DE LEONIBUS, P. S. 1963: Power spectra of surface wave heights from recordings made from a submerged hovering submarine. *Ocean Wave Spectra,* Rep. of Conf. Englewood Cliffs: Prentice-Hall, 243–50.

DICKSON, R. and LEE, A. 1969: Atmospheric and marine climatic fluctuations in the north Atlantic region. In M. Sears (editor), *Progress in oceanography* **V.** Oxford: Pergamon, 55–66.

DIETRICH, G. 1963: *General oceanography.* New York: Interscience Pubs.

DIETRICH, G. and KALLE, K. 1957: *Allgemeine Meereskunde.* Berlin: Borntraeger.

DILLON, O. W. 1970: The growing importance of commercial fish farming in the United States. *The Amer. Fish Farm.* **1,** 20.

DOODSON, A. T. 1956: Tides and storm surges in a long uniform gulf. *Proc. Roy. Soc.* **A 237,** 325–43.

DOODSON, A. T. and WARBURG, H. D. 1941: *Admiralty manual of tides.* London: HMSO.

DOODSON, A. T., ROSSITER, J. R., and CORKAN, R. H. 1954: Tidal charts based on coastal data, Irish Sea. *Proc. Roy. Soc. Edin.* **64,** 90–101.

DRAPER, L. 1964: 'Freak' ocean waves. *Oceanus* **10,** 13–15.

1967a: Instruments for measurement of wave height and direction in and around harbours. *Proc. Inst. Civ. Eng.* **37,** 213–19.

1967b: Wave activity at the sea bed around northwestern Europe. *Mar. Geol.* **5,** 133–40.

1967c: The analysis and presentation of wave data—a plea for uniformity. *Proc. 10th Conf. Coastal Eng.* Tokyo, 1966. Vol. **I,** 1–11.

DRAPER, L. and DOBSON, P. J. 1965: Rip currents on a Cornish beach. *Nature* **206** (4990), 1249.

DRAPER, L. and FRICKER, H. S. 1965: Waves off Land's End. *J. Inst. Navig.* **18,** 180–87.

DRAPER, L. and SQUIRE, E. M. 1967: Waves at ocean weather ship station 'India' (59°N, 19°W). *Trans. Roy. Inst. Nav. Arch.* **109,** 85–93.

DRAPER, L. and WHITAKER, M. A. B. 1965: Waves at ocean weather ship station 'Juliet' (52°30′N, 20°00′W). *Deutsch. Hydrog. Zeit.* **18,** 25–30.

DUGDALE, C. R. and GOERING, J. J. 1971: A model of nutrient-limited phytoplankton growth. In D. W. Hood (editor), *Impingement of man on the oceans.* New York: Wiley–Interscience, 589–600.

DUING, W. and JOHNSON, D. 1971: Southward flow under the Florida Current. *Science* **173,** 428–30.

EAGLESON, P. S. 1965: The theoretical study of longshore currents on a plane beach. *MIT Hydrodynamics Lab. Tech. Rep.* **82,** 1–31.

ECKART, C. 1953: The generation of wind waves over a water surface. *J. Appl. Phys.* **24,** 1485–94.

EDMUND, J. M., CHUNG, Y., and SCLATER, J. G. 1971: Pacific bottom water: penetration east around Hawaii. *J. Geophys. Res.* **76** (33), 8089–97.

EKMAN, V. W. 1905: On the influence of the earth's rotation on ocean currents. *Arkiv für Matem. Astr. and Fysik.* Stockholm **2,** 11. 53 pp.

EMERY, K. O. 1960: *The sea off southern California.* New York: Wiley.

EMILIANI, C. 1957: Temperature and age analysis of deep-sea cores. *Science* **125,** 383–7.

ESTES, J. E. and GOLOMB, B. 1970: Oil spills and method of measuring their extent on the sea surface. *Science* **169,** 676–8.

FAIRBURN, L. A. 1954: The semi-diurnal tides along the equator in the Indian Ocean. *Phil. Trans. Roy. Soc.* **A 247,** 191–212.

FARQUHARSON, W. J. 1954: Storm surges on the east coast of England. Proceedings of Conference on the North Sea Floods of 31 January–1 February 1953. *Inst. Civ. Eng.* **14.**

FELL, H. B. 1967: Cretaceous and tertiary surface currents of the oceans. *Oceanog. and Mar. Biol. Ann. Rev.* **5,** 317–41.

FERGUSON WOOD, E. J. 1971: Production and prediction. In J. D. Costlow (editor), *Fertility of the sea.* New York: Gordon and Breach Sci. Pubs., 617–22.

FILLOUX, J. 1970: Deep sea tides 1250 km off Baja California. *Science* **169,** 862–4.

FIRTH, F. E. (editor) 1969: *Encyclopedia of marine resources.* New York: Van Nostrand Reinhold.

FOYN, E. 1971: Municipal wastes. In D. W. Hood (editor), *Impingement of man on the oceans.* Chap. 16. New York: Wiley–Interscience, 445–59.

FRANCIS, J. R. D. 1959: Wind action on a water surface. *Proc. Inst. Civ. Eng.* **12,** 197–216.

FRASER, J. 1962: *Nature adrift. The story of marine plankton.* London: Foulis.

FUGILISTER, F. C. and WORTHINGTON, L. V. 1951: Some results of a multiple-ship survey of the Gulf Stream. *Tellus* **3,** 1–14.

FYE, P. M., MAXWELL, A. E., EMERY, K. O., and KATCHUM, B. H. 1968: Ocean science and marine resources. In E. A. Gullion (editor), *Uses of the sea.* Englewood Cliffs: Prentice-Hall, 17–68.

GALVIN, C. J. 1967: Longshore current velocity: a review of theory and data. *Rev. of Geophys.* **5,** 287–304.

GALVIN, C. J. and EAGLESON, P. S. 1965: Experimental study of longshore currents on a plane beach. *CERC Tech. Memo.* **10,** 80 pp.

GARVINE, R. W. 1971: A simple model of coastal upwelling dynamics. *J. Phys. Oceanog.* **1** (3), 169–79.

GEORGE, J. D. 1971: Can the seas survive? *Ecologist* **1** (9), 4–9.

GILL, A. E. 1971: The equatorial current in a homogeneous ocean. *Deep Sea Res.* **18** (4), 421–31.

GILL, A. E. and BRYAN, K. 1971: Effects of geometry on the circulation of a three-dimensional southern hemisphere ocean model. *Deep Sea Res.* **18** (7), 685–721.

GOLDBERG, E. D. 1954: Marine geochemistry. *J. Geol.* **62**, 249–65.

GORDON, A. L. and GERARD, R. D. 1970: North Pacific bottom potential temperature. *Geol. Soc. Amer. Mem.* **126**, 23–39.

GOULD, W. J. 1971: Observations of an event in some current measurements in the Bay of Biscay. *Deep Sea Res.* **18** (1), 35–49.

GRAHAM, M. 1948: *Rational fishing of cod in the North Sea.* Buckland lectures for 1939 London: Edward Arnold.

(editor) 1956: *Sea fisheries: their investigation in the United Kingdom.* London: Edward Arnold.

GREEN, F. H. W. 1951: Tidal phenomena with special reference to Southampton and Poole. *Dock and Harb. Auth.* **32**, 143–8.

GREVE, P. A. 1971: Chemical wastes in the sea: new forms of marine pollution. *Science* **173**, 1021–2.

GROEN, P. and GROVES, G. W. 1962: Surges. In M. N. Hill (editor), *The sea.* Vol. **I**, chap. 17. New York: Wiley, 611–46.

GUILLEN, O. 1971: The El Niño phenomena in 1965 and its relation with the productivity in coastal Peruvian waters. In J. D. Costlow (editor), *Fertility of the sea.* Vol. **I**. New York: Gordon and Breach Sci. Pubs., 187–96.

GUILLEN, O., DE MENDIOLA, B. R., and DE ROUDEN, R. I. 1971: Primary productivity and phytoplankton in the coastal Peruvian waters. In J. D. Costlow (editor), *Fertility of the sea.* Vol. **I**. New York: Gordon and Breach Sci. Pubs., 157–85.

GULLARD, J. A. and CARROZ, J. E. 1968: Management of fishery resources. *Adv. in Mar. Biol.* **6**, 1–73.

GULLION, E. A. (editor) 1968: *Uses of the seas.* The American Assembly, Columbia University. Englewood Cliffs: Prentice-Hall.

GUNTHER, E. R. 1936: A report on oceanographical investigations in the Peru coastal current. *Discovery Reports* **13**, 107–276.

HADLEY, H. L. 1964: Wave-induced bottom currents in the Celtic Sea. *Mar. Geol.* **2**, 164–167.

HAMON, B. V. 1965: The East Australian Current. *Deep Sea Res.* **12**, 899–921.

1967: Medium-scale temperature and salinity structure in the upper 1500 m in the Indian Ocean. *Deep Sea Res.* **14**, 169–81.

HARDEN JONES, F. R. 1968: *Fish migration.* London: Edward Arnold.

HARDY, A. 1956: *The open sea.* **I**: *The world of plankton.* London: Collins.

1959: *The open sea.* **II**: *Fish and fisheries.* London: Collins.

HARDY, J. R. 1964: The movement of beach material and wave action near Blakeney Point, Norfolk. *Trans. Inst. Brit. Geog.* **34**, 53–69.

1966: An ebb-flood channel system and coastal changes near Winterton, Norfolk. *East Mid. Geog.* **4**, 24–30.

HARRISON, W. 1968: Empirical equations for longshore current velocity. *J. Geophys. Res.* **73**, 6929–36.

HARRISON, W. and KRUMBEIN, W. C. 1964: Interaction of the beach–ocean–atmosphere system at Virginia Beach, Virginia. *CERC Tech. Memo.* **7**. 102 pp.

HARRISON, W., PORE, N. A., and TUCK, D. R. 1965: Predictor equations for beach processes and responses. *J. Geophys. Res.* **70**, 6013–19.

HARVEY, J. G. 1967: Drifter studies in the Irish Sea. *Liverpool Essays in Geography*. London: Longman, 137–56.

HASLE, G. R. 1959: A quantitative study of phytoplankton from the equatorial Pacific. *Deep Sea Res.* **6**, 38–59.

HATTORI, AKIHIJKO and WADA, EITARO 1971: Nitrite distribution and its regulating processes in the equatorial Pacific Ocean. *Deep Sea Res.* **18** (6), 557–68.

HEAPS, N. S. 1967: Storm surges. *Oceanog. and Mar. Biol. Ann. Rev.* **5**, 11–47.

    1969: A two-directional numerical sea model. *Phil. Trans. Roy. Soc.* **A 265**, 93–138.

HECHT, A. and WHITE, R. A. 1968: Temperature fluctuations in the upper layer of the ocean. *Deep Sea Res.* **15**, 339–54.

HEEZEN, B. C. 1957: Whales caught in deep-sea cables. *Deep Sea Res.* **4**, 105–15.

HEEZEN, B. C. and HOLLISTER, C. 1964: Deep-sea current evidence from abyssal sediment. *Mar. Geol.* **1**, 141–74.

HEEZEN, B. C. and HOLLISTER, C. D. 1971: *The face of the deep.* New York and London: Oxford University Press.

HELLAND-HANSEN, B. 1916: Nogen hydrografiske metoder. *Skand. Naturforsker möte,* Oslo.

HENKIN, L. 1968: Changing law for the changing seas. In E. A. Gullion (editor), *Uses of the seas.* Englewood Cliffs: Prentice-Hall, 69–97.

HEYERDAHL, T. 1950: The voyage of the raft Kon-tiki. *Geog. J.* **115**, 20–41.

HILL, M. N. (editor) 1963: *The sea.* Vol. **II.** New York: Wiley.

HINDE, B. J. and GAUNT, D. I. 1967: Microseisms. *Contemp. Phys.* **8**, 267–83.

HODGSON, W. C. 1957: *The herring and its fishery.* London: Routledge and Kegan Paul.

HOLLAND, W. R. 1967: On the wind-driven circulation in an ocean with bottom topography. *Tellus* **19**, 582–600.

HOLT, S. J. 1969: The food resources of the ocean. In *The ocean.* San Francisco: Freeman/Scientific American.

HOOD, D. W. (editor) 1971: *Impingement of man on the oceans.* New York: Wiley–Interscience.

HORIKAWA, K. and KUO, C. T. 1967: A study of wave transformation inside the surf zone. *Proc. 10th Conf. Coastal Eng.* Tokyo, 1966. Chap. 15, 217–33.

HORN, M. K. and ADAMS, J. A. S. 1966: Computer-derived geochemical balances and element abundances. *Geochem. Cosmochim. Acta* **30**, 279–97.

HUGHES, P. 1956: A determination of the relation between wind and sea-surface drift. *Quart. J. Roy. Met. Soc.* **82**, 494–502.

HULL, S. 1964: *Bountiful sea.* Englewood Cliffs: Prentice-Hall. 340 pp.

IDYLL, C. P. 1973: The Anchovy crisis. *Sci. Amer.* **228**, (6), 22–8.

IJIMA, T. and TANG, F. L. W. 1967: Numerical calculations of wind waves in shallow water. *Proc. 10th Conf. Coastal Eng.* Tokyo, 1966. Chap. 4, 38–49.

IKARD, F. N. 1967: Oily wastes at sea. *Science* **157**, 625.

INMAN, D. L. and BAGNOLD, R. A. 1963: Littoral processes. In M. N. Hill (editor), *The sea.* Vol. **III.** New York: Wiley, 529–53.

INMAN, D. L., MUNK, W. H., and BELAY, M. 1962: Spectra of low-frequency waves along the Argentine shelf. *Deep Sea Res.* **8**, 155–64.

INMAN, D. L., TAIT, R. J., and NORDSTROM, C. E. 1971: Mixing in the surf zone. *J. Geophys. Res.* **76** (15), 3493–3514.

INTERGOVERNMENTAL OCEANOG. COMM. UNESCO 1964: ICSU draft of a general scientific framework for world ocean study. Sci. Comm. Ocean. Res. 76 pp.

ISELIN, C. O'D. 1940: Preliminary report on long-term variation in the transport of the Gulf Stream system. *Pap. Phys. Oceanog. and Met.* **8**, 1. 40 pp.

ISHIGURO, S. 1962: An electronic analogue method for tides and storm surges, and some applications in the North Sea. *Proc. Symp. on Math. Hydrol. Math. of Phys. Oceanog.* **196**, 265–9.

IYER, H. M. 1959: Composition and propagation of storm microseisms. Ph.D. thesis, University of London.

JEFFREYS, H. 1925: On the formation of water waves by wind. *Proc. Roy. Soc.* **A 107**, 189–206.

JOHNS, B. 1965: Inertia currents. *Deep Sea Res.* **12**, 825–30.

JOHNSTONE, J., SCOTT, A., and CHADWICK, H. C. 1924: *The marine plankton.* London: Hodder and Stoughton.

JONES, P. W. G. 1971: The southern Benguela current region in February 1966. Part I, Chemical observations with particular reference to upwelling. *Deep Sea Res.* **18** (2), 193–208.

KING, C. A. M. 1972: *Beaches and coasts.* 2nd edition. London: Edward Arnold.

KINSMAN, B. 1965: *Wind waves, their generation and propagation on the ocean surface.* Englewood Cliffs: Prentice-Hall.

KNAUSS, J. A. 1960: Measurements of the Cromwell Current. *Deep Sea Res.* **6**, 265–86.
   1966: Further measurements and observations on the Cromwell Current. *J. Mar. Res.* **24**, 205–40.

KNAUSS, J. A. and TOFT, B. A. 1964: Equatorial undercurrent of the Indian Ocean. *Science* **143**, 354–6.

KORRINGA, P. 1971: Marine pollution and its biological consequences. In J. D. Costlow (editor), *Fertility of the sea.* Vol. **I.** New York: Gordon and Breach Sci. Pubs., 215–23.

KRAUSS, W. 1969: Typical features of internal wave spectra. In M. Sears (editor), *Progress in oceanography* **V.** Oxford: Pergamon.

KRUEGER, R. B. 1971: International and national regulations of pollution from offshore oil production. In D. W. Hood (editor), *Impingement of man on the oceans.* Chap. 24. New York: Wiley–Interscience, 603–34.

KUENEN, P. H. 1955: *Realms of water.* London: Cleaver-Hume.

KVINGE, T. 1969: Moored oceanographic buoys. Reliability, representability and application. In M. Sears (editor), *Progress in oceanography* **V.** Oxford: Pergamon.

LACOMBE, H. 1970: Physical oceanography of the eastern boundary current of the Atlantic Ocean. In F. M. Delaney (editor), *Geology of the eastern Atlantic continental margin.* IGS 70/13. London: HMSO.

LAFOND, E. C. 1950: Sand movement near the beach in relation to tides and waves. *Scripps Inst. Oceanog. Pub.* **107.**

LAFOND, E. C. and LAFOND, K. G. 1971: Oceanography and its relation to marine organic production. In J. D. Costlow (editor), *Fertility of the sea.* Vol. **I.** New York: Gordon and Breach Sci. Pubs., 241–65.

LAUGHTON, A. S. 1959a: Photography of the ocean floor. *Endeavour* **18**, 175–85.
   1959b: The sea floor. *Sci. Prog.* **47**, 230–48.

LAWFORD, A. L. and VELEY, V. F. C. 1956: Change in the relationship between wind and surface water movement at higher wind speeds. *Trans. Amer. Geophys. Un.* **37**, 691–3.

LAWRENCE, L. G. 1967: *Electronics in oceanography.* Indianapolis: Sams. 288 pp.

LEE, A. and ELLETT, D. 1967: On the water masses of the northwest Atlantic Ocean. *Deep Sea Res.* **14**, 183–90.

LEE, A. J. 1958: Marine life in relation to the physical–chemical environment. In J. A. Steers (editor), *Physical geography.* 4th edition. Cambridge University Press, 420–55.

LEE, W. H. K. and COX, C. S. 1966: Time variation of ocean temperature and its relation to internal waves and oceanic heat flow measurements. *J. Geophys. Res.* **71**, 2101–12.

LIGHTHILL, M. J. 1962: Physical interpretations of the mathematical theory of wave generation by wind. *J. Fluid Mech.* **14**, 385–98.

LOFTAS, T. 1969: *The last resource.* London: Hamilton. 256 pp.

LONGUET-HIGGINS, M. S. 1953: Mass transport in water waves. *Phil. Trans. Roy. Soc.* **A 245**, 535–81.

1962: The direction spectra of ocean waves, and processes of wave generation. *Proc. Roy. Soc.* **A 265**, 286–315.

1965a: Some dynamical aspects of ocean currents. *Quart. J. Roy. Met. Soc.* **91**, 425–51.

1965b: The response of a stratified ocean to stationary or moving wind systems. *Deep Sea Res.* **12**, 923–73.

LONGUET-HIGGINS, M. S., CARTWRIGHT, D. E., and SMITH, N. D. 1963: Observations of the directional spectrum of sea waves using the motions of a floating buoy. *Ocean Wave Spectra,* Rep. of Conf., Englewood Cliffs: Prentice-Hall, 111–31.

LONGUET-HIGGINS, M. S. and STEWART, R. W. 1962: Radiation stress and mass transport in gravity waves, with applications to 'surf beats'. *J. Fluid Mech.* **13**, 481–504.

LYNN, R. J. and REID, J. L. 1968: Characteristics and circulation of deep abyssal waters. *Deep Sea Res.* **15**, 577–98.

MCCONNAUGHEY, B. H. 1970: *Introduction to marine biology.* St Louis: Mosby.

MACDONALD, G. J. F. 1968: An American strategy for the oceans. In E. A. Gullion (editor), *Uses of the seas.* Englewood Cliffs: Prentice-Hall, 163–94.

MACINTYRE, F. 1970: Why the sea is salt. In J. R. Moore (editor), *Oceanography.* San Francisco: Freeman, Sci. Amer. Readings. 110–21.

MCKEE, A. 1967: *Farming the sea.* London: Souvenir Press. 314 pp.

MCKENZIE, F. 1958: Rip-current systems. *J. Geol.* **66**, 103–13.

MACKINTOSH, N. A. 1946: The natural history of whalebone whales. *Biol. Rev.* **21**, 60–74.

MACMILLAN, D. H. 1949: Tidal features of Southampton Water. *Dock and Harb. Auth.* **29**, 259–64.

1966: *Tides.* London: C.R. Books.

MANN, C. R. 1967: The termination of the Gulf Stream and the beginning of the North Atlantic Current. *Deep Sea Res.* **14**, 337–59.

MARMER, H. A. 1943: Tidal investigations on the west coast of South America. *Geog. Rev.* **33**, 299–303.

MARTIN, L. W. and BULL, H. 1968: The strategic consequences of Britain's revised naval role. In E. A. Gullion (editor), *Uses of the seas.* Englewood Cliffs: Prentice-Hall, 113–37.

MARX, W. 1967: *The frail ocean.* New York: Ballantine Books. 274 pp.

MASSACHUSETTS INSTITUTE OF TECHNOLOGY 1970: *Man's impact on the global environment.* Cambridge, Mass.: MIT Press.

MAUCHLINE, J. 1969: The biology of euphausiids. *Adv. in Mar. Biol.* **7,** 1–454.

MEHAUTÉ, B. J. and KOH, R. C. Y. 1966: On the breaking of waves arriving at an angle to the shore. *Nat. Eng. Sci. Co. Off. Nav. Res.* **SN,** 134-10. 80 pp.

MENZEL, D. W. and RYTHER, J. H. 1960: The annual cycle of primary production in the Sargasso Sea off Bermuda. *Deep Sea Res.* **6,** 351–67.

MERLINI, M. 1971: Heavy-metal contamination. In D. W. Hood (editor), *Impingement of man on the oceans.* New York: Wiley–Interscience, 461–86.

MERO, J. L. 1965: *Mineral resources of the sea.* Amsterdam: Elsevier. 312 pp.

MEYERS, J. J., HOLM, C. H., and MCALLISTER, R. F. 1969: *Handbook of ocean and underwater engineering.* New York: McGraw-Hill.

MILES, J. W. 1965: A note on the interaction between surface waves and wind profiles. *J. Fluid Mech.* **22,** 823–7.

1967: On the generation of surface waves by shear flows. Part V. *J. Fluid Mech.* **30,** 163–75.

MIYAKE, YASUO 1971: Radioactive models. In D. W. Hood (editor), *Impingement of man on the oceans.* New York: Wiley–Interscience, 565–88.

MOBAREK, I. E. and WIEGEL, R. L. 1967: Diffraction of wind-generated water waves. *Proc. 10th Conf. Coastal Eng.* Tokyo, 1966. Vol. **I,** chap. 13, 185–206.

MONTGOMERY, R. B. 1958: Water characteristics of the Atlantic Ocean and of the world ocean. *Deep Sea Res.* **5,** 134–48.

MOORE, J. R. 1971: *Oceanography—Readings from Scientific American,* with introductions by J. R. Moore. San Francisco: Freeman. 417 pp.

MORGAN, G. W. 1956: On the wind-driven ocean circulation. *Tellus* **8,** 301–20.

MORRIS, I. 1970: Restraints on the big fish-in. *New Sci.* **48,** 373–5.

MOSBY, H. 1936: Verdunstung und Strahlung auf dem Meere. *Ann. d. Hydrog. u. Mar. Meteorol.* **64,** 281–6.

MOSKOWITZ, L. 1964: Estimates of the power spectrums for fully developed sea for wind speeds of 20 to 40 knots. *J. Geophys. Res.* **69,** 5161–79.

MOSS, J. E. 1971: Petroleum—the problem. In D. W. Hood (editor), *Impingement of man on the oceans.* New York: Wiley–Interscience, 381–419.

MOWROOZI, A. A., EWING, M., NAFE, J. E., and FLIEGEL, M. 1968: Deep ocean current and its correlation with the ocean tide off the coast of north California. *J. Geophys. Res.* **73,** 1921–32.

MULLIM, M. M. 1969: Production of zooplankton in the oceans: present status and problems. *Oceanog. and Mar. Biol. Ann. Rev.* **7,** 292–314.

MUNK, W. H. 1949: Surf beats. *Trans. Amer. Geophys. Un.* **30,** 849–54.

1950: On the wind-driven ocean circulation. *J. Meteor.* **7,** 79–93.

1962: Long ocean waves. In M. N. Hill (editor), *The sea.* Vol. **I,** chap. 18. New York: Wiley, 647–63.

1969: Deep sea tides. In M. Sears (editor), *Progress in oceanography* **V.** Oxford: Pergamon, 67–70.

MUNK, W. H. and CARTWRIGHT, D. E. 1966: Tidal spectroscopy and prediction. *Phil. Trans. Roy. Soc.* **A 259,** 533–81.

MUNK, W. H., MILLER, G. R., SNODGRASS, F. E., and BARBER, N. F. 1963: Directional recording of swell from distant storms. *Phil. Trans. Roy. Soc.* **A 255,** 505–84.

MUNK, W. H. and REVELLE, R. 1952: Sea level and the rotation of the earth. *Amer. J. Sci.* **250,** 829–33.

MURRAY, J. and RENARD, A. F. 1891: *The scientific results of the voyage of HMS Challenger.* Deep Sea Deposits, London: HMSO, 1–525.

NELSON-SMITH, A. 1970: The problem of oil pollution of the sea. *Adv. in Mar. Biol.* **8,** 215–306.

NEUMANN, G. 1953: An ocean wave spectra and a new method of forecasting wind-generated sea. *BEB Tech. Memo.* **43.**

— 1960: Evidence for an equatorial undercurrent in the Atlantic. *Deep Sea Res.* **6,** 328–34.

— 1968: *Ocean currents.* Amsterdam: Elsevier. 352 pp.

NEUMANN, G. and PIERSON, W. J. 1957: A detailed comparison of theoretical wave spectra and wave forecasting methods. *Deutsch. Hydrog. Zeit.* **10** (3), 73–92; **10** (4), 134–46.

— 1966: *Principles of physical oceanography.* Englewood Cliffs: Prentice-Hall.

NEUMANN, G. and WILLIAMS, R. E. 1965: Observations of the equatorial undercurrent in the Atlantic at 15°w, equalant I. *J. Geophys. Res.* **70,** 297–304.

NORDBURG, W. 1965: Geophysical observations from Nimbus I. *Science* **150,** 559–71.

NOWLIN, W. D. 1967: A steady, wind-driven, frictionless model of two moving layers in a rectangular ocean basin. *Deep Sea Res.* **14,** 89–110.

NOWROZZI, A. A., EWING, M., NAFE, J. E., and FLIEGEL, M. 1968: Deep ocean current and its correlation with the ocean tide off the coast of North Carolina. *J. Geophys. Res.* **73,** 1921–32.

PARKER, F. L. and BERGER, W. H. 1971: Faunal and solution patterns of planktonic foraminifera in surface sediments of the south Pacific. *Deep Sea Res.* **18** (1), 73–107.

PARRISH, B. B. and SAVILLE, A. 1967: Changes in the fisheries of the North Sea and the Atlanto-Scandian herring stocks and their causes. *Oceanog. and Mar. Biol. Ann. Rev.* **5,** 409–47.

PATTERSON, C. 1971: Lead. In D. W. Hood (editor), *Impingement of man on the oceans.* New York: Wiley–Interscience, 245–58.

PEKERIS, C. L. and ACCAD, M. 1969: Solution of Laplace's equations for the M_2 tide in the world oceans. *Phil. Trans. Roy. Soc.* **A 265,** 413–36.

PETERSEN, C. G. J. 1918: The sea bottom and its production of fish food. *Rep. Dan. Biol. Sta.* **23,** 29–32.

PETTERSSON, O. 1926: Currents and fish migration in the transition area. *J. Cons. perm. int. Mer* **1,** 322–6.

PHILLIPS, A. W. 1969: Sea-bed water movements in Morecambe Bay. *Dock and Harb. Auth.* **49,** 379–82.

— 1970: The use of the Woodhead sea bed drifters. *Brit. Geom. Res. Gr. Tech. Bull.* **4.** 29 pp.

PHILLIPS, O. M. 1957: On the generation of waves by turbulent wind. *J. Fluid Mech.* **2,** 417–45.

— 1966: *The dynamics of the upper ocean.* Cambridge University Press.

PIERSON, W. J., NEUMANN, G., and JAMES, R. W. 1955: Practical methods of observing and forecasting ocean waves by means of wave spectra and statistics. *Hydrog. Off. Pub.* **603.** US Navy. Washington, DC.

PILPEL, N. 1972: Oil pollution of the sea. *Ecologist* **2** (3), 4–7.

POCHAPSKY, T. E. 1963: Measurements of small-scale oceanic motion with neutrally buoyant floats. *Tellus* **4,** 352–62.

1966: Measurement of deep water movements with instrumented neutrally buoyant floats. *J. Geophys. Res.* **71,** 2491–2504.

POLLAK, M. J. 1958: Frequency distribution of potential temperature and salinities in the Indian Ocean. *Deep Sea Res.* **5,** 128–33.

POSTMA, H. 1971: Distribution of nutrients in the sea and the oceanic nutrient cycle. In J. D. Costlow (editor), *Fertility of the sea.* Vol. **II.** New York: Gordon and Breach Sci. Pubs., 337–49.

POUNDER, E. R. 1965: *Physics of ice.* Chaps. 2–4. Oxford: Pergamon. 151 pp.

POWERS, W. H., DRAPER, L., and BRIGGS, P. M. 1969: Waves at Camp Pendleton, California. *Proc. 11th Conf. Coastal Eng.* London, 1968, chap. 1.

PRIVETT, D. W. and FRANCIS, J. R. D. 1959: The movement of sailing ships as a climatological tool. *Mariner's Mirror* **45,** 292–300.

PROUDMAN, J. 1944: The tides of the Atlantic Ocean. *MNR Astron. Soc.* **104,** 244–56.

1953: *Dynamical oceanography.* London: Methuen.

PUTNAM, J. A., MUNK, W. H., and TRAYLOR, M. A. 1949: The prediction of longshore currents. *Trans. Amer. Geophys. Un.* **30** (3), 337–45.

RAE, K. M. 1957: Continuous plankton records: a relationship between wind, plankton distribution and haddock brood strength. *Bull. Mar. Ecol.* **5,** 247–69.

RAMPTON, V. 1970: The brown seaweed industries in the British Isles. In R. H. Osborne, F. A. Barnes, and J. C. Doornkamp (editors), *Geographical essays in honour of K. C. Edwards.* Department of Geography, University of Nottingham, 233–241.

REID, J. L. and LYNN, R. J. 1971: On the influence of the Norwegian–Greenland and Weddell seas upon the bottom waters of the Indian and Pacific Oceans. *Deep Sea Res.* **18** (11), 1063–88.

REID, J. L. and NOWLIN, W. D. 1971: Transport of water through Drake Passage. *Deep Sea Res.* **18** (1), 51–64.

RICE, T. R. and WOLFE, D. A. 1971: Radioactivity—chemical and biological aspects. In D. W. Hood (editor), *Impingement of man on the oceans.* New York: Wiley–Interscience, 325–79.

RICHARDSON, P. L. and KNAUSS, J. A. 1971: Gulf Stream and western boundary undercurrent observations at Cape Hatteras. *Deep Sea Res.* **18** (11), 1039–1109.

RISEBOROUGH, R. W. 1971: Chlorinated hydrocarbons. In D. W. Hood (editor), *Impingement of man on the oceans.* New York: Wiley–Interscience, 259–86.

ROBINSON, A. H. W. 1960: Ebb-flood systems in sandy bays and estuaries. *Geog.* **45,** 183–99.

1961a: The hydrography of Start Bay and its relationship to beach changes at Hallsands. *Geog. J.* **127,** 63–77.

1961b: The Pacific tsunami of May 22nd, 1960. *Geog.* **46,** 18–24.

1968: The use of sea bed drifters in coastal studies with particular reference to the Humber. *Zeit. für Geomorph.* **NF 7,** 1–23.

ROBINSON, A. R. 1965: A three-dimensional model of inertial currents in a variable density ocean. *J. Fluid Mech.* **21,** 211.

ROELS, O. A., GERARD, R. D., and BÉ, A. H. W. 1971: Fertilizing the sea by pumping nutrient-rich deep water to the surface. In J. D. Costlow (editor), *Fertility of the sea.* Vol. **II.** New York: Gordon and Breach Sci. Pubs., 401–15.

ROSSBY, C. G. 1951: A comparison of current patterns in the atmosphere and in the ocean basins. Assn of Met. Brussels, *UGGI* **9.**

ROSSITER, J. R. 1954: The North Sea storm surge of 31 January and 1 February 1953. *Phil. Trans. Roy. Soc.* **A 246,** 371–99.

1959: Methods of forecasting storm surges on the east and south coast of Great Britain. *Quart. J. Roy. Met. Soc.* **85,** 262–77.

ROTSCHI, H. and LEMASSON, L. 1967: Oceanography of the Coral and Tasman Seas. *Oceanog. and Mar. Biol. Ann. Rev.* **5,** 49–97.

ROWE, G. T. 1971a: Observations on bottom currents and epibenthic populations in Hatteras submarine canyon. *Deep Sea Res.* **18** (6), 569–81.

1971b: Benthic biomass and surface productivity. In J. D. Costlow (editor), *Fertility of the sea.* Vol. **II.** New York: Gordon and Breach Sci. Pubs., 441–54.

ROWE, G. T. and MENZIES, R. J. 1968: Deep bottom currents off the coast of North Carolina. *Deep Sea Res.* **15,** 711–19.

RUBIOFF, I. 1968: Sea-level canal across central America. *Science* **161,** 857–61.

RUDDIMAN, W. F. 1968: Historical stability of the Gulf Stream meander belt: foraminiferal evidence. *Deep Sea Res.* **15,** 137–48.

RUSSELL, F. S. 1926–7: The vertical distribution of macroplankton. IV, The apparent importance of light intensity as a controlling factor in the behaviour of certain species in the Plymouth area. *J. Mar. Biol. Assn* **14,** 415–40; 1928, **15,** 81–99.

RUSSELL, R. C. H. and MACMILLAN, D. H. 1952: *Waves and tides.* London: Hutchinson.

RUSSELL, R. C. H. and OSORIO, J. D. C. 1958: An experimental investigation of drift profiles in a closed channel. *Proc. 6th Conf. Coastal Eng.* Miami, 1957, 171–93.

RYTHER, J. H. 1969: Photosynthesis and fish production in the sea. *Science* **166,** 72–6.

1972: Photosynthesis and fish production in the sea. In R. L. Smith (editor), *The ecology of man: an ecosystem approach.* New York: Harper and Row.

SALSMAN, G. G., TOLBERT, W. H., and VILLARS, R. G. 1966: Sand-ridge migration in St Andrews Bay, Florida. *Mar. Geol.* **4,** 11–20.

SASAKI, TADAYOSHI, WATANABE, SEIICHI, and OSHIBA, GOHACHIRO 1967: Current measurements on the bottom in the deep water in the west Pacific. *Deep Sea Res.* **14,** 159–167.

SCHAEFER, M. B. 1968: Methods of estimating effects of fishing on fish populations. *Trans. Amer. Fish. Soc.* **97,** 231–41.

SCHMIDT, J. 1932: *Danish eel investigations during 25 years, 1905–1930.* Carlsberg Foundation. 16 pp.

SCHMITT, W. R. and ISAACS, J. D. 1971: Enhancement of marine protein production. In J. D. Costlow (editor), *Fertility of the sea.* Vol. **II.** New York: Gordon and Breach Sci. Pubs., 455–62.

SCHMITZ, W. J. and RICHARDSON, W. S. 1968: On the transport of the Florida Current. *Deep. Sea Res.* **15,** 679–94.

SCHOTT, G. 1935: *Geographie des Indischen und Stillen Ozeans.* Hamburg: Boysen.

SCHWARTZ, M. L. 1967: Littoral zone tidal cycle sedimentation. *J. Sedim. Petrol.* **37,** 677–83.

SEARS, M. 1954: Notes on the Peruvian coastal current. I, An introduction to the ecology of Pisco Bay. *Deep Sea Res.* **1,** 141–69.

(editor) 1969: *Progress in oceanography* **V.** Oxford: Pergamon.

SHEARER, J. M., MACNAB, R. F., PELLETIER, B. R., and SMITH, T. B. 1971: Submarine pingos in the Beaufort Sea. *Science* **174** (4011), 816–18.

SHEMDIN, O. H. and HSU, E. Y. 1967: Direct measurements of aerodynamic pressure above a single progressive gravity wave. *J. Fluid Mech.* **30,** 403–16.

SHEPARD, F. P. 1959: *The earth beneath the sea.* Baltimore: Johns Hopkins.

SHEPARD, F. P. and INMAN, D. L. 1950: Nearshore circulation related to bottom topography and wave refraction. *Trans. Amer. Geophys. Un.* **31,** 196–212.

SHULMAN, M. D. 1968: The Soviet turn to the sea. In E. A. Gullion (editor), *Uses of the sea.* Englewood Cliffs: Prentice-Hall, 138–62.

SKOLNIKOFF, E. B. 1968: National and international organization for the seas. In E. A. Gullion (editor), *Uses of the sea.* Englewood Cliffs: Prentice-Hall, 98–112.

SMALL, G. L. 1971: *The blue whale.* New York: Columbia University Press. 248 pp.

SMITH, R. L. (editor) 1972: *The ecology of man: an ecosystem approach.* New York: Harper and Row.

SMITH, R. L., MOOERS, C. N. K., and ENFIELD, D. B. 1971: Mesoscale studies of the physical oceanography in two coastal upwelling regions: Oregon and Peru. In J. D. Costlow (editor), *Fertility of the sea.* Vol. **II.** New York: Gordon and Breach Sci. Pubs., 513–35.

SNODGRASS, F. E. 1964: Precision digital tide gauge. *Science* **146,** 198–208.

SNODGRASS, F. E., GROVES, G. W., HASSELMAN, K. F., MILLER, G. R., MUNK, W. H., and POWERS, W. H. 1966: Propagation of ocean swell across the Pacific. *Phil. Trans. Roy. Soc.* **A 259,** 431–97.

SNYDER, R. L. and COX, C. S. 1966: A field study of the wind generation of ocean waves. *J. Mar. Res.* **24,** 141–78.

SONU, C. J., MCCLOY, J. M., and MCARTHUR, D. S. 1967: Longshore currents and nearshore topography. *Proc. 10th Conf. Coastal Eng.* Tokyo, 1966. Chap. 32, 524–49.

STALCUP, M. C. and PARKE, C. E. 1965: Drogue measurements of shallow currents on the equator in the west Atlantic Ocean. *Deep Sea Res.* **12,** 535–6.

STEFANSSON, U., ATKINSON, L. P., and BUMPUS, D. F. 1971: Hydrographic properties and circulation of the North Carolina shelf and slope waters. *Deep Sea Res.* **18** (4), 383–420.

STERNECK, R. V. 1921: Die Gezeiten der Ozeane. *Sitz. Ber. Akad. Wiss. Wien.* **129,** 130, 131–50, 363–71.

STEWART, R. W. 1969: The atmosphere and the ocean. In, *The ocean.* San Francisco: Freeman/Scientific American.

STOMMEL, H. 1957: A survey of ocean current theory. *Deep Sea Res.* **4,** 149–84.

1958a: *The Gulf Stream.* 2nd edition, 1965. Berkeley: University of California Press.

1958b: Abyssal circulation. *Deep Sea Res.* **5,** 80–82.

STOMMEL, H. and ARONS, A. B. 1960: On the abyssal circulation of the world oceanic basins. *Deep Sea Res.* **6,** 217–33.

STOMMEL, H. and ROOTH, C. 1968: On the interaction of gravitational and dynamic forcing in simple circulation models. *Deep Sea Res.* **15,** 165–70.

STRIDE, A. H. 1959: A pattern of sediment transport for sea floors around south Britain. *Dock and Harb. Auth.* **467,** 145–7.

SVERDRUP, H. U., JOHNSON, M. W., and FLEMING, R. H. 1946: *The oceans*. Englewood Cliffs: Prentice-Hall.

SVERDRUP, H. U. and MUNK, W. H. 1947: Wind, sea and swell—theory of relationships in forecasting. *Hydrog. Off. Pub.* **601**. US Navy. Washington, DC.

SWALLOW, J. C. 1957: Some further deep current measurements using neutrally buoyant floats. *Deep Sea Res.* **4**, 93–104.

1962: Ocean circulation. *Proc. Roy. Soc.* **A 265**, 326–8.

1964: Equatorial undercurrent in the west Indian Ocean. *Nature* **204**, 436–7.

SWALLOW, J. C. and BRUCE, J. G. 1966: Current measurements off the Somali coast during the southwest monsoon of 1964. *Deep Sea Res.* **13**, 861–88.

SWALLOW, J. C. and HAMON, B. V. 1960: Some measurements of deep currents in the eastern north Atlantic. *Deep Sea Res.* **6**, 155–68.

SWALLOW, J. C. and WORTHINGTON, L. V. 1957a: Measurements of deep currents in the western north Atlantic. *Nature* **179**, 1183–4.

1957b: Measurements of deep currents in the eastern Atlantic. *Nature* **179**, 1183–4.

TADAYOSHI, SASAKI, WATANABE, SEIISHI, and OSHIBA, GOHACHIRO 1965: New current meters for great depths. *Deep Sea Res.* **12**, 815–24.

TAFT, B. 1971: Ocean circulation in monsoon areas. In J. D. Costlow (editor), *Fertility of the sea*. Vol. **II**. New York: Gordon and Breach Sci. Pubs., 565–79.

THORSON, G. 1971: *Life in the sea*. London: Weidenfeld and Nicolson. 256 pp.

TRESSLER, D. K. and LEMON, J. M. 1951: *Marine products of commerce*. New York: Reinhold. 782 pp.

TRICKER, R. A. R. 1964: *Bores, breakers, waves and wakes*. London: Mills and Boon.

TUCKER, D. W. 1959: A new solution to the Atlantic eel problem. *Nature* **183**, 495–501.

TUCKER, M. J. 1950: Surf beats: sea waves of 1 to 5 minutes period. *Proc. Roy. Soc.* **A 202**, 565–73.

1956: A ship-borne wave recorder. *Trans. Inst. Nav. Arch. Lond.* **98**, 236–50.

1963: Recent measurements and analysis techniques developed at the National Institute of Oceanography. *Ocean Wave Spectra*, Rep. of Conf. Englewood Cliffs: Prentice-Hall, 219–26.

UNOKI, S. and IOZAKI, I. 1967: A possibility of generation of surf beats. *Proc. 10th Conf. Coastal Eng.* Tokyo, 1966. Chap. 14, 207–16.

WALDEN, H. 1963: Comparison of one-dimensional wave spectra recorded in the German Bight with various 'theoretical spectra'. *Ocean Wave Spectra*, Rep. of Conf. Englewood Cliffs: Prentice-Hall, 67–80.

WARREN, B., STOMMEL, H., and SWALLOW, J. C. 1966: Water masses and pattern of flow in the Somali Basin during the southwest monsoon. *Deep Sea Res.* **13**, 825–60.

WARREN, B. A. 1963: Topographic influences on the path of the Gulf Stream. *Tellus* **15**, 167–83.

WEBSTER, F. 1965: Measurements of eddy fluxes of momentum in the surface layers of the Gulf Stream. *Tellus* **11**, 309–18.

1969: On the representativeness of direct deep-sea current measurements. In M. Sears (editor), *Progress in oceanography* **V**. Oxford: Pergamon, 3–16.

1971: On the intensity of horizontal ocean currents. *Deep Sea Res.* **18** (9), 885–93.

WENK, E. 1969: The physical resources of the oceans. *Sci. Amer.* **221**, 166–76.

WIEGEL, R. L. and KIMBERLEY, H. L. 1950: Southern swell observed at Oceanside, California. *Trans. Amer. Geophys. Un.* **31,** 717–22.

WILLIAMS, L. C. 1969: CERC wave gauges. *CERC Tech. Memo.* **30.** 117 pp.

WILLIAMS, W. W. 1947: The determination of the gradient of enemy-held beaches. *Geog. J.* **109,** 76–93.

WILSON, B. W. 1960: The prediction of hurricane storm-tides in New York Bay. *BEB Tech. Memo.* **120.** 107 pp.

WILSON, B. W. and TØRUM, A. 1968: The tsunami of the Alaskan earthquake, 1964: engineering evaluation. *CERC Tech. Memo.* **25.** 401 pp.

WIMPENNY, R. S. 1953: *The plaice.* Buckland lecture for 1949. London: Edward Arnold.

WOOSTER, W. S. 1969: The ocean and man. *Sci. Amer.* **221,** 218–314.

WORTHINGTON, L. V. 1954a: A preliminary note on the time scale in the north Atlantic circulation. *Deep Sea Res.* **1,** 244–51.

  1954b: Three detailed cross-sections of the Gulf Stream. *Tellus* **6,** 116–**23.**

WORTHINGTON, L. V. and VOLKMANN, G. H. 1965: The volume transport of the Norwegian Sea overflow water in the north Atlantic. *Deep Sea Res.* **12,** 667–76.

WÜST, G. 1935: Schichtung und Zirkulation des Atlantischen Ozeans. Die Stratosphere. *Wiss. Ergebn. Deut. Atlant. Exped. 'Meteor' 1925–27.* Vol. **6,** Teil. 1 and 2.

ZIMMERMAN, H. B. 1971: Bottom currents on the New England continental rise. *J. Geophys. Res.* **76,** 5865–76.

Index